Introduction to
Quantitative
Chemical Analysis

INTRODUCTION TO
QUANTITATIVE

Ernest Grunwald

Brandeis University

Louis J. Kirschenbaum

University of Rhode Island

CHEMICAL
ANALYSIS

Prentice-Hall, Inc., Englewood Cliffs, New Jersey

INTRODUCTION TO
QUANTITATIVE CHEMICAL ANALYSIS
Ernest Grunwald
Louis J. Kirshenbaum

© 1972 by Prentice-Hall, Inc.
Englewood Cliffs, N.J.

10 9 8 7 6 5 4 3 2 1

ISBN: 0-13-493825-9

Library of Congress Catalog Card Number 75-38308

Printed in the United States of America

PRENTICE-HALL INTERNATIONAL, INC., *London*
PRENTICE-HALL OF AUSTRALIA PTY. LTD., *Sydney*
PRENTICE-HALL OF CANADA, LTD., *Toronto*
PRENTICE-HALL OF INDIA PRIVATE LIMITED, *New Delhi*
PRENTICE-HALL OF JAPAN, INC., *Tokyo*

Contents

xi   **PREFACE**

xiii   **GLOSSARY OF SYMBOLS**

Part One

1   **CONCEPTS AND METHODS OF CHEMICAL ANALYSIS**

1

3   **METHODS OF QUANTITATIVE ANALYSIS**

4   Chemical Methods and Physical Methods

9   Stoichiometry

12   Practical Classification of Chemical Methods

12   Gravimetric Analysis

16   Volumetric Titration

16   Composition and Reaction Capacity

20   Problems

v

2

TITRATION AND THE EQUILIBRIUM CONSTANT   24

Formulating the Equilibrium Constant   25
References   31
Relationships among Equilibrium Constants   32
Equilibrium Constant and Titration Error   33
Buffers   39
Pseudo-Equilibrium Constants   41
Problems   42

3

TITRATION OF COMPLEX SUBSTRATES AND MIXTURES   44

Complexity on the Molecular Level   45
Complexity on the Stoichiometric Level   51
Substrates with Several Reactive Sites   56
Chemical Equation for Selective Titration   59
Problems   64

4

VOLUMETRIC ANALYSIS IN AQUEOUS SOLUTION   66

Acid-Base Titration in Aqueous Solution   67
Precipitation of Insoluble Silver Salts   75
Oxidation-Reduction   80
Complex-Ion Formation with Ethylenediamine Tetraacetate Ion   90
References   96
Problems   96

5

EFFECT OF THE SOLVENT ON THE CHEMISTRY OF IONS   101

Charged-Sphere Model of an Ion and the Dielectric Constant   102
Some Common Nonaqueous Solvents   103
Formation of Ion Pairs   103
Chemical Consequences of Ion-Pair Formation   109
Effect of Dielectric Constant on Chemical Equilibrium   112
Chemical Models of Ions and Ion Pairs in Solution   115
Hydrogen Bonding   116
Coordination   120
Use of Nonaqueous Solvents in Analytical Chemistry   121
Problems   122

6

125 THE SOLVENT AS AN ACID OR BASE

125 The Brønsted Definitions
132 Acidity and Basicity of Solutions
135 Acid-Base Properties of Common Solvents
137 Acid-Base Titrations in Nonaqueous Solvents
146 References
146 Problems

7

148 TITRATION CURVES

149 A Few Examples
154 Calculations of Molar Concentrations from Formal Concentrations
156 Setting up the Algebraic Equations
159 Solving the Algebraic Equations
164 Calculation of Titration Curves
169 Problems

8

172 INNATE TITRATION ERROR

173 Dependence of Error on Slope of Titration Curve
176 Solution by Means of the Differential Calculus
178 General Treatment of the Equivalence Point
182 Feasibility and Design of Titrations
183 Mixtures of Substrates
187 When the Effective Titrant is a Solid
189 Problems

9

191 DETECTING THE EQUIVALENCE POINT WITH COLOR INDICATORS

192 The Color of Solutions
195 Colored Titrant or Substrate as Self-Indicator
196 Stepwise Reaction Sequence with Color Change
198 Color Indicators at High Dilution
200 Acid-Base Indicators
214 Problems

10

INSTRUMENTAL ANALYSIS OF LIQUID SOLUTIONS 216

Chemists and "Black Boxes" 217
Bulk Solution Properties 217
Solute Concentrations from Bulk
  Solution Properties 219
Stoichiometry from Bulk Solution Properties 226
Equilibrium Constants from Spectral
  Absorption Data 229
Electrochemical Cells 236
Solute Concentrations from EMF Measurements
  on Electrochemical Cells 241
Potentiometric Titration 249
References 255
Problems 256

Part Two

LABORATORY EXPERIMENTS AND PROCEDURES 261

11

S.O.P. (STANDARD OPERATING PROCEDURE) 263

Laboratory Apparatus 264
Preparation and Dissolution of the Sample 274
A Final Note Before You Proceed 275

12

GRAVIMETRIC ANALYSIS 276

Decomposition of a Solid 277
Precipitation of an Ion from Solution 278
Precipitation of Barium Sulfate 280

13

ACID-BASE TITRATIONS 284

Choice and Test of Indicator 285
Standard Solutions of Acid and Base 287
Some Applications of Aqueous Acid-Base Analysis 290
Nonaqueous Titrations with Methanol as Solvent 294

14

297   PRECIPITATION TITRATIONS WITH SILVER NITRATE: VOLUMETRIC CHLORIDE ANALYSIS
298   Mohr's Method
299   Volhard's Method
300   Fajan's Method

15

302   COMPLEXOMETRIC TITRATIONS
303   Standard Reagents
304   Direct Titration with EDTA
306   Substitution Titration of Calcium(II) with EDTA
307   Determination of Cobalt(II) by Back-Titration
308   Determination of Water Hardness
310   Other EDTA Titrations

16

313   OXIDATION-REDUCTION TITRATIONS
314   Standard Potassium Permanganate Solution
316   Permanganate Titrations in Acid Media
317   The Iodine-Iodide Couple: Standard Solutions
320   Titrations Involving the Iodine-Iodide Couple

17

323   CHEMICAL KINETICS
324   Reaction to be Studied
325   Analytical Method
326   Reagents and Control Experiments
327   Kinetic Experiments

18

330   POTENTIOMETRIC TITRATIONS
331   Operation of the pH Meter
333   Aqueous Acid-Base Titrations
334   Nonaqueous Acid-Base Titrations
336   Potentiometric Redox Titrations
337   Potentiometric Determination of Mixed Halides
338   Concluding Remarks

19

SOME ADDITIONAL INSTRUMENTAL METHODS   339

Spectrophotometric Analysis   339
Conductometric Titrations   343
Amperometric Titrations   346

Part Three
APPENDIX   351

INDEX   364

Preface

In this book we attempt a logical simplification of the traditional introduction to quantitative chemical analysis so that the entire subject can be covered in a one-semester college course on the freshman or sophomore level. In developing the concepts, we abandon the traditional textbook which is organized according to chemical reaction (acid-base reaction, precipitation, redox, and so on) and deal simply with the reaction of a substrate with a titrant. To provide a perspective of the entire field, we illustrate each step in the logical development with examples drawn from a variety of analytical methods.

On the whole, our approach remains within the traditional conceptual framework, but some generalization of concepts has been achieved. The equilibrium constant becomes, quite naturally, the equilibrium constant for titration, whose value is obtained from traditional constants such as K_A, K_{SP}, or standard electrode potentials. The feasibility of titration is discussed in terms of the innate titration error, which in turn is related to the equilibrium constant for titration. In dealing with complex sub-strates and mixtures, we make frequent use of a simple type of principal-species diagram. In the discussion of selective and stepwise titration, we simplify the logical analysis by finding the effective titrant.

The book is organized into two parts: Part I describes concepts and methods, and Part II describes laboratory experiments and procedures. In Part I, Chapters 1 through 4 contain the essential minimum for an abbreviated course of instruction. Chapters 1 through 3 present principles (illustrated by examples), and Chapter 4 describes volumetric analysis in aqueous solution. This "basic core" is followed by chapters on nonaqueous solutions, theory of titration, color indicators, and instrumental analysis. The organization of Part I is such that the instructor can be extremely flexible in setting up the laboratory schedule.

Part II consists of a chapter on general laboratory technique, followed by chapters that describe specific experiments in gravimetric, volumetric and instrumental analysis. These experiments are largely traditional, re-quiring only apparatus that most colleges are likely to have, and are not intended to be comprehensive. We expect that the instructor will introduce additional experiments, to suit his educational objectives and the facilities available to him.

The book reflects our strongly-held opinion that the students' first exposure to quantitative analysis should emphasize chemical methods rather than physical methods. Students taking this course usually have the necessary background to understand chemical reactions at a fairly sophisticated level. However, they know little or no physics and are not really prepared to appreciate accurate physical measuring instruments, which tend to be delicate and expensive. In our own teaching we steer a middle course, using such physical instruments as pH-meters and color-imeters, the cost of which is relatively modest, and which are relatively student-proof.

We are indebted to a number of colleagues and to the editors of *Analytical Chemistry* and of the *Journal of Chemical Education* for per-mission to reproduce numerical data and graphs; specific acknowledgements have been made in the text. We are grateful to Harold F. Walton for constructive criticism of the manuscript, and to Stephen E. Gould for checking the problems and making sure that they are workable. Above all, we thank our wives for encouragement and help with the preparation of the manuscript.

E. G.

L. J. K.

Glossary of Symbols

α	Degree of conversion of substrate to reaction product
a	Characteristic distance of approach of an ion pair
A	Absorbance
$\mathring{A}$	Angstrom unit (10^{-8} cm, 0.1 nm, 0.1 mμ)
c_i	Formal concentration (formula weights per liter) of the ith component
c_S	($= n_S/V$) formal concentration of substrate component
c_T	($= n_T/V$) formal concentration of titrant component
Δ	Change in whatever follows the delta symbol owing to the conversion of the reactants to the reaction products
ϵ	Molar extinction coefficient
ε	Dielectric constant
$\mathbf{E}$	Electromotive force
$\mathbf{E}^\circ$	Standard electrode potential
$\mathbf{E}^\circ_{ox}$ ($\mathbf{E}^\circ_{red}$)	Standard half-reaction potential for oxidation (reduction)
$\mathbf{E}^*$	"Cell constant" for the electrical measuring cell
F	Formal; formality (formula weights per liter)
$\mathbf{F}$	Faraday's electrochemical constant (96,494 coulombs per chemical equivalent)
(g)	Gas

I	Light intensity transmitted by a solution
i_d	Diffusion current (in Chapter 19)
I_0	Light intensity transmitted by the solvent; also, light intensity entering a solution
K	Equilibrium constant; equilibrium constant for titration
K_A	Acid dissociation constant
K_{A1}, K_{A2}	Stepwise acid dissociation constants
K_{assoc}	Association constant
K_B	Base dissociation constant
K_d	Dissociation constant
K_i	Ionization constant, as explained in (6.11) and (6.13)
K_{IP}	Ion-product constant of the solvent
K_{SP}	Solubility-product constant
K_W	Ion product of water
K_ψ	Pseudo-equilibrium constant, as explained in (2.25)
l	Length (of light path)
(l)	Liquid
M	Molar; molarity (moles per liter)
m.d.	Mean deviation
n	Number of electrons in half-reaction; number of Faradays of charge; number of formula weights; refractive index
n_i, n_S, n_T	Number of formula weights of the ith component, of the substrate component, of the titrant component
N	Normal; normality (equivalents per liter)
p	$(-\log_{10})$
pH	$-\log [H_3O^+]$; $-\log$ [lyonium ion]
P	Pressure
P.	Procedure (in Chapters 12–19)
$\mathcal{P}$	Any bulk solution property
$\mathcal{P}_0$	Same as $\mathcal{P}$, measured for the pure solvent
P, Q	Reaction products
[P], [Q]	Molar concentration of reaction products
q	Electrical charge
ρ	Density
r	Titrant/substrate ratio [Eq. (7.47)]
R	Gas constant per mole; electrical resistance (in Chapter 19)
(s)	Solid
S	Substrate
[S]	Molar concentration of substrate
t	One mole of substrate reacts with t moles of titrant (in Chapter 8); time (in Chapter 17)
T	Temperature (°K)
T	Titrant
[T]	Molar concentration of titrant
V	Volume (of solution)
V_e	Volume of titrant at the equivalence point in a titration
x	Unknown variable
x	Molar concentration of unreacted titrant (in Chapters 7 to 10)
y	Unknown variable
z	Unknown variable
z	Charge number (An integer)

PART ONE

Concepts and Methods
of
Chemical Analysis

Methods of
Quantitative Analysis

According to many historians, modern chemistry began when Antoine Lavoisier discovered the essentials of quantitative analysis and thereby disproved the phlogiston theory. Today's chemists may or may not agree, but no one questions the importance of analytical chemistry in modern chemistry. Natural matter is usually a mixture of substances, millions of which are known. It is the business of analytical chemistry to determine the composition of matter and to identify and characterize the component substances. And if our knowledge of matter is to be quantitative, the analysis must be quantitative.

In quantitative analysis there are certain rules of common sense that it is prudent to observe. First, get some idea of the composition of the sample! This does not mean that you must carry out a complete qualitative analysis—indeed, in most cases that would be impractical. But you *should* learn enough about the sample to be able to choose a valid analytical method for the substance of interest, and to control the experimental conditions. Start by asking some questions. If the sample is a synthetic mixture, what are the original components? What are the likely impurities? What reactions may have taken place? If the sample comes from a natural source, what is its history? Then carry out qualitative tests if necessary to confirm or supplement that information.

In choosing a method for quantitative analysis, choose the most specific feasible method for the given substance, so that the identity of the substance can be confirmed. Suppose you are presented with a strange liquid, told it is a solution of HCl in water, and asked to find the HCl concentration. A possible approach would be to measure the density of the liquid. For known solutions of HCl in water, the density is known accurately as a function of both concentration and temperature, so that there should be no problem about finding the HCl concentration of the unknown liquid when its density at a given temperature is known. Unfortunately, that approach can succeed only if the unknown liquid is indeed a pure solution of HCl in water; and unfortunately again, the density measurement itself provides no proof of that. On the other hand, if the unknown liquid is analyzed by titration for acid or for chloride ion, then the titration not only provides the desired quantitative analysis but proves at the same time that the solute *reacts* like HCl. Such further proof of identity is always desirable because it reduces the chance for error. Experienced chemists, who know how much time can be wasted in research if an analysis is wrong, usually insist on some redundancy of information.

CHEMICAL METHODS AND PHYSICAL METHODS

Methods of quantitative analysis can be classified into chemical and physical methods, depending on whether the method employs stoichiometry in a chemical reaction or a suitable physical property of the sample. Each class has certain advantages and disadvantages.

The chemical methods are always specific, although the degree of that specificity varies. For example, in the analysis of HCl in water by acid-base titration, a method which can distinguish between strong acid and weak acid is evidently more revealing than one which cannot. The equipment used in the chemical methods tends to be simple, inexpensive, and easy to maintain. The scope and reaction conditions are, in most cases, well understood, and an accuracy of better than 0.5 percent is readily attainable. On the other hand, the chemical methods are ineffective for substances that are chemically inert, and they do not differentiate easily between substances with similar chemical properties, for instance acetic acid and benzoic acid.

Physical methods become specific when there are at least as many measurements of independent properties as there are substances in the sample. In the case of HCl in water, the density at a given temperature measures the HCl concentration only if the qualitative composition—HCl in water—is taken for granted. The validity of that assumption can be tested by measuring a second property, such as the refractive index, whose value is also known to vary in a characteristic fashion with the concentration. If the HCl concentration inferred from the density agrees with that inferred from the refractive index, then both inferences are probably correct.

The best physical methods are those in which an entire set of physical properties is measured conveniently with a single instrument. Consider, for example, the transmission of light by a liquid solution. Simple observation will tell whether the liquid, as seen in white light, is clear or turbid, whether it is colorless ("water-white") or colored. If the liquid is both clear and colorless, then the visible radiation entering it is transmitted with little or no loss. However, if the liquid is turbid, then some of the radiation is scattered in all directions by suspended particles such as bacteria, dust, or finely divided precipitates; and if it is colored, then part of the visible spectrum is absorbed.

Of special interest are those cases in which the addition of a solute to a clear, colorless solvent results in a clear (nonturbid) but *colored* solution. The absorption of light is then due to the solute, and quantitative measurement of the light absorption as a function of wavelength can establish both the identity and the concentration of the solute. The measuring instrument is called a *spectrophotometer*.

Spectrophotometry

The schematic diagram of a spectrophotometer is shown in Fig. 1.1. A beam of white light from a source of constant intensity is dispersed into a continuous spectrum by passing through a prism, and a narrow spectral band, centered at a known wavelength, is selected for measurement by means of a small slit. The selected radiation is allowed to pass through the measuring liquid to a detector, where the transmitted intensity is measured. One usually measures two liquids, a control and an unknown, contained in carefully matched test tubes or optical cells. The control is usually the clear solvent. The transmitted intensity I_0 of the control is taken as 100 percent transmission. The transmitted intensity I of the unknown is then compared with I_0 to yield either the percent transmission, $100(I/I_0)$, or the *absorbance* or *optical density*, A, defined in Eq. (1.1).

$$A = \log\left(\frac{I_0}{I}\right) \qquad (1.1)$$

In practice, the absorbance is more convenient because it is precisely proportional to the length l of the lightpath in the solution and also, in most cases, to the concentration c of the light-absorbing solute. This relation-

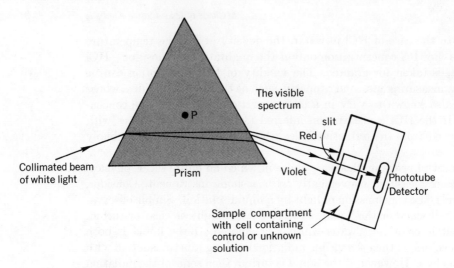

Fig. 1.1. Schematic diagram of a spectrophotometer, as described in the text. The wavelength of the light passing through the slit is varied by rotating the prism about P.

ship is expressed by Beer's law, Eq. (1.2). The proportionality constant ϵ is called the *extinction coefficient* or *absorbancy index*.

$$A = \epsilon l c \qquad (1.2)$$

The relationship between absorption of radiation and wavelength is called the *absorption spectrum*. For example, Fig. 1.2 shows the absorbance

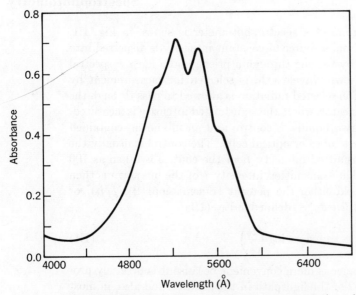

Fig. 1.2. Absorption spectrum of an aqueous solution of potassium permanganate (3.08×10^{-4} formula weights per liter). The length of the light-path in the solution is 1 cm. (Courtesy of James Guzinski.)

A as a function of wavelength for an aqueous solution of potassium permanganate whose color is a medium purple. The solute concentration is 3.08×10^{-4} formula weights per liter, and the light-path in the solution is 1 cm. The absorption spectrum shows several distinctive features: two maxima, at 5250 Å and at 5460 Å; a minimum at 5360 Å; and shoulders at 5050 Å and at 5650 Å. These features are characteristic of permanganate ion, and their appearance in the absorption spectrum is a much stronger proof that permanganate ion is present than is the purple color of the solution. A purple color might have been produced by some other solute, such as bromocresol purple (an acid-base indicator) in alkaline solution, but a distinctive and shapely absorption spectrum is characteristic of just one substance. Thus the absorption spectra of purple solutions of potassium permanganate and of alkaline bromocresol purple are compared in Fig. 1.3 and are indeed quite different.

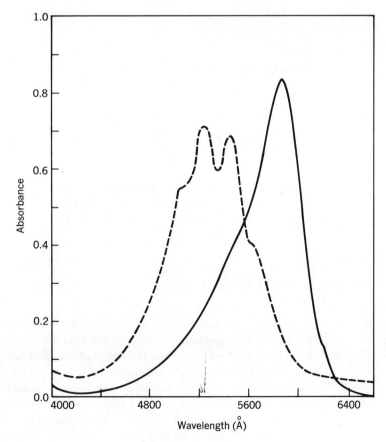

Fig. 1.3. Absorption spectra of aqueous solutions. Solid curve, bromocresol purple; dashed curve, potassium permanganate. The concentrations have been adjusted so that the absorbance maxima are similar. (Courtesy of James Guzinski.)

While the qualitative features of the absorption spectrum—the maxima, minima, and shoulders—identify the solute, the magnitude of the absorbance measures the concentration. For instance, Fig. 1.4 shows absorbance

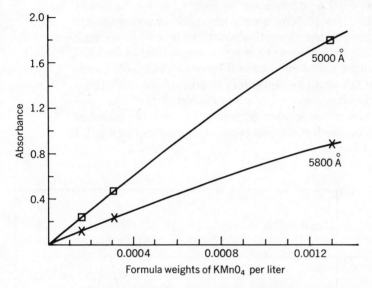

Fig. 1.4. Plot of absorbance versus concentration for potassium permanganate in aqueous solution at two wavelengths. The length of the light-path in the solutions is 1 cm. (Courtesy of James Guzinski.)

as a function of concentration of potassium permanganate in aqueous solution, measured in 1-cm optical cells at two different wavelengths. The relationships are nearly proportional, as suggested by Eq. (1.2).

Critique of physical methods

With recent advances in technology, the labor involved in measuring an absorption spectrum has been greatly reduced. Spectrophotometers have been devised that scan the spectral range automatically and record the measured absorbance on calibrated graph paper. Such automatic recording instruments are available not only for absorption spectra in the ultraviolet, visible, and infrared regions of the spectrum, but for a wide variety of characteristic physical properties, and they have added immeasurably to the efficiency and effectiveness of the science. Chemists have learned to relate physical properties to molecular structure, just as they had earlier learned to relate chemical properties to molecular structure. Because no serious inconsistencies have arisen, it is fair to say that the molecular-structural theory of chemistry has been tested anew by means of physical properties, and been reaffirmed. Chemists now attach the same importance to evidence based on physical properties that they have always attached to evidence based on chemical properties.

Physical methods of analysis are indispensable when the sample is chemically inert, or when it is too precious to be destroyed in a chemical reaction. Accuracy of 1 percent or better is attainable in many quantitative applications. In practice there are some inconveniences, owing to the high cost and complexity of good measuring instruments. Few laboratories can afford to supply each chemist with a spectrophotometer for his personal use, in the same way that they supply him with burets and weighing bottles. It is more usual for expensive instruments to be shared by the entire staff. To avoid costly breakdowns, the instruments are likely to be set up in a special instrument laboratory under full-time supervision, and to be operated only by experienced professionals.

In modern practice, chemical and physical methods of quantitative analysis are of coordinate importance. However, in the teaching of quantitative analysis it is reasonable to emphasize the chemical methods first, to give you time to learn more physics and acquire more laboratory experience before attempting any serious work with complex physical measuring instruments. This does not mean, however, that instrumental methods will be ignored—quite the contrary. Instrumental methods will be described that are invaluable for detecting the equivalence point in chemical reactions, and which require only relatively simple measuring instruments.

STOICHIOMETRY

Most chemical methods of quantitative analysis are based on stoichiometry. The substance whose amount is to be determined, the *substrate*, is allowed to undergo a known reaction, and the amount of a product, or the equivalent amount of a second reactant, is determined. Given the balanced chemical equation for the reaction, the amount of substrate can then be deduced.

Many analytical reactions are simple enough so that their balanced equations can be derived by inspection from the molecular formulas of the reactants and products. In more complicated cases, especially when an oxidation-reduction (redox) process is involved, it is easier and more foolproof to use a more logical approach. All chemical equations in this book will be written so that they involve *one formula weight of substrate*.

Algebraic method for balancing chemical equations

Consider, for example, the reaction of iodate ion (IO_3^-) with iodide ion to produce *triiodide* ion (I_3^-) in aqueous acid. We begin by writing a tentative chemical equation, using one formula weight of substrate (IO_3^-) and using letters (a, b, c, d) to denote the unknown coefficients.

$$1\ IO_3^- + a\ I^- + b\ H_3O^+ \longrightarrow c\ I_3^- + d\ H_2O \qquad (1.3)$$

We then solve for a, b, c, d by insisting on conservation of atoms and of electrical charge, as follows.

To conserve:	*We must write:*	
I atoms	$1 + a = 3c$	(1.4)
O atoms	$3 + b = d$	(1.5)
H atoms	$3b = 2d$	(1.6)
Electrical charge	$-1 - a + b = -c$	(1.7)

The equations that result are always sufficient to evaluate the unknown coefficients. In the present case, there are four unknowns and four equations.

Equations (1.4) through (1.7) are solved by standard methods of algebra.

From (1.6): $d = 1.5b$
Then, from (1.5): $3 = 1.5b - b = 0.5b$
Therefore: $b = 6; d = 9$

From (1.7): $a - c = b - 1 = 5$
From (1.4): $a - 3c = -1$
Therefore: $2c = 6; c = 3; a = 8$

On substituting these values in (1.3), we obtain the balanced equation (1.8).

$$1\ IO_3^- + 8\ I^- + 6\ H_3O^+ = 3\ I_3^- + 9\ H_2O \qquad (1.8)$$

This algebraic method avoids the use of valences, oxidation states, or half-reactions and has the advantage that you can check each step of algebra at your leisure. It also enables you to find omissions in the initial tentative chemical equation. If a reactant or product has been left out, conservation of atoms as well as of electrical charge will be impossible, and at least one of the conservation equations will be obviously incomplete.

Because all chemical equations in this book are written so that they involve one formula weight of substrate, the coefficients obtained for the other chemical species need not be integers. They may be ratios of integers. If a coefficient turns out to be a negative number, then the corresponding species has been written on the wrong side of the chemical equation. In that case, transpose the species to the other side of the equation and enter the same coefficient with a positive sign.

Half-reaction method for balancing redox equations

The half-reaction method is especially useful when we need to know the relationship between formula weight and equivalent weight or normality in oxidation-reduction reactions. You are probably already familiar with it. However, to lay the foundation for later discussion, we wish to review it briefly.

As in the algebraic method, we are concerned with conservation of atoms and of electrical charge, but we divide the reaction into separate processes of oxidation and reduction. For example, consider the oxidation of tin from stannous ion (Sn^{2+}) to stannic ion (Sn^{4+}) by dichromate ion ($Cr_2O_7^{2-}$) in acid solution; the $Cr_2O_7^{2-}$ is reduced to Cr^{3+}. The oxidation of Sn^{2+} requires just two electrons and is represented by Eq. (1.9), where e^- stands for an electron.

$$Sn^{2+} \longrightarrow Sn^{4+} + 2\,e^- \tag{1.9}$$

Note that (1.9) is a balanced equation, because both atoms and electrical charges are balanced.

For the reduction of $Cr_2O_7^{2-}$, we first balance the Cr atoms, writing

$$Cr_2O_7^{2-} \longrightarrow 2\,Cr^{3+} \quad \text{(not yet balanced)}$$

In order to balance the oxygen atoms, we place seven H_2O molecules on the right-hand side. This leaves a deficiency of 14 hydrogen atoms on the left, which is taken care of with H_3O^+ ions and 14 more H_2O molecules on the right.

$$14\,H_3O^+ + Cr_2O_7^{2-} \longrightarrow 2\,Cr^{3+} + 21\,H_2O$$
$$\text{(atoms balanced; charges not yet balanced)}$$

Now we have balanced all atoms but have a net charge of $+14 - 2 = +12$ on the left, and $2 \times (+3) = +6$ on the right. The discrepancy is remedied by adding electrons to give the balanced half-reaction (1.10).

$$14\,H_3O^+ + Cr_2O_7^{2-} + 6\,e^- \longrightarrow 2\,Cr^{3+} + 21\,H_2O \tag{1.10}$$

Equations (1.9) and (1.10) are independent of each other: Eq. (1.9) represents the oxidation of Sn^{2+} to Sn^{4+} not only by $Cr_2O_7^{2-}$, but by other oxidizing agents as well. Similarly, Eq. (1.10) represents the reduction of $Cr_2O_7^{2-}$ in acid by a variety of reducing agents.

To complete our equation, we need to add the proper multiples of (1.9) and (1.10) so that no electrons remain. If Sn^{2+} is the substrate, we wish to write the balanced equation so that just one formula weight of tin reacts.

We therefore multiply (1.10) by $\frac{1}{3}$ before adding the two equations. The balanced equation for the entire redox reaction is then (1.11).

$$\text{Sn}^{2+} + \tfrac{14}{3}\,\text{H}_3\text{O}^+ + \tfrac{1}{3}\,\text{Cr}_2\text{O}_7{}^{2-} \longrightarrow \text{Sn}^{4+} + \tfrac{2}{3}\,\text{Cr}^{3+} + 7\,\text{H}_2\text{O} \quad (1.11)$$

Although the half-reactions (1.9) and (1.10) in the preceding example were balanced by inspection, equations for half-reactions can also be balanced by the method of algebra if the electron is treated as a pure charge. Thus to derive (1.10), we begin as follows:

$$1\,\text{Cr}_2\text{O}_7{}^{2-} + a\,\text{H}_3\text{O}^+ + b\,\text{e}^- \longrightarrow c\,\text{Cr}^{3+} + d\,\text{H}_2\text{O} \quad (1.12)$$

To solve for the four unknown coefficients a, b, c, and d, we write three equations to express the conservation of Cr, H and O atoms, and a fourth equation, (1.13), to express conservation of electrical charge. Equation (1.13) includes the charge due to b electrons.

To conserve	*We must write*
Cr atoms	$2 = c$
H atoms	$3a = 2d$
O atoms	$7 + a = d$
Electrical charge	$-2 + a - b = (+3) \times c \quad (1.13)$

On solving these equations, we obtain: $a = 14$; $b = 6$; $c = 2$; $d = 21$. These values are, of course, identical to the corresponding coefficients in (1.10).

PRACTICAL CLASSIFICATION OF CHEMICAL METHODS

Chemical methods in which one measures a product are technically quite different from those in which one measures an added reactant. If the product is a solid, one isolates it and weighs it. If it is a gas, one collects it and measures its weight or volume at a known temperature and pressure. In each case the measurement is preceded by some kind of quantitative separation. Direct measurement of the product, without isolation from the reaction mixture, is normally attempted only if the product brings on some distinctive new property, such as color or electrical conductivity.

By contrast, the equivalent amount of a second reactant can be measured by direct titration and does not require any quantitative separation.

GRAVIMETRIC ANALYSIS

Methods of analysis in which a product is isolated and weighed, called *gravimetric methods*, are of great historical importance because they provided the facts upon which Dalton's atomic theory is based. Indeed, they still are the standard operations in the determination of chemical atomic weights. However, they tend to be laborious and their practical importance has been declining for those analyses where an alternative method is available.

Decomposition of solids at high temperatures

Operationally simplest are those methods in which a solid compound decomposes on heating to give a new solid that is weighed, and a gas that escapes into the atmosphere. For instance, calcium carbonate and magnesium carbonate decompose at elevated temperatures into the respective oxides and CO_2 gas. Therefore, in the analysis of limestone rocks, one of the standard tests is to heat the sample to somewhat above 900°C. The gas escaping under those conditions is almost entirely CO_2 and the loss in weight provides a good measure of the carbonate content of the rock. In practice, an oven-dried sample of the crushed rock is weighed into a tared platinum crucible (platinum remains chemically inert at high temperatures and has a melting point of 1770°C), and the crucible containing the sample is heated in an electric furnace whose temperature is maintained at about 950°C. After a few minutes at that temperature, the crucible, which now contains the decomposition product, is removed, allowed to cool in a desiccator, and weighed. To show that decomposition is complete, the cycle of heating, cooling, and weighing may be repeated.

For solids that decompose, the decomposition temperature is one of the characteristic properties. Consider, for example, the decomposition of calcium carbonate according to (1.14).

$$CaCO_3(s) \; \rightleftharpoons \; CaO(s) + CO_2(g) \qquad (1.14)$$

If the solid that decomposes and the solid product form two distinct phases, the equilibrium pressure of CO_2 in the gas phase is analogous to, say, the vapor pressure of a pure liquid. At each temperature there is only one pressure at which $CO_2(g)$ is in equilibrium with both $CaCO_3(s)$ and $CaO(s)$. That pressure is called the *decomposition pressure*. If the actual pressure of CO_2 in the gas phase is less than the decomposition pressure, then $CaCO_3(s)$ will decompose with the evolution of CO_2 gas. If the gas is pumped off, or if the gas phase is very large, then the decomposition will go to completion because the pressure of CO_2 cannot reach the decomposition pressure. At room temperature the decomposition pressure of calcium carbonate is 10^{-23} atm, incredibly small, and of course much smaller than the normal CO_2 content of air. But above 800°C the decomposition pressure exceeds 0.2 atm, and at 900°C it reaches 1 atm. If heat is supplied slowly at atmospheric pressure, carbon dioxide will "boil out" when the temperature reaches 900°C, and the temperature will rise further only after the endothermic decomposition is complete.

Representative data for the decomposition of a few inorganic solids are listed in Table 1.1. Because of the wide range in decomposition tem-

Table 1.1   DECOMPOSITION OF SOME INORGANIC SOLIDS

Reaction	Decomposition pressure (atm)	Temperature (°C)
$2\,NaHCO_3(s) \longrightarrow Na_2CO_3(s) + H_2O(g) + CO_2(g)$	1	270
$2\,CsHCO_3(s) \longrightarrow Cs_2CO_3(s) + H_2O(g) + CO_2(g)$	1	175
$Li_2CO_3(s) \longrightarrow Li_2O(s) + CO_2(g)$	0.05	910
$Na_2CO_3(s) \longrightarrow Na_2O(s) + CO_2(g)$	0.05	1180
$Cs_2CO_3(s) \longrightarrow Cs_2O(s) + CO_2(g)$	0.05	990
$MgCO_3(s) \longrightarrow MgO(s) + CO_2(g)$	1	550
$CaCO_3(s) \longrightarrow CaO(s) + CO_2(g)$	1	900
$2\,BaCO_3(s) \longrightarrow BaCO_3 \cdot BaO(l) + CO_2(g)$	1	*ca.* 1350
$Ag_2CO_3(s) \longrightarrow Ag_2O(s) + CO_2(g)$	1	220
$Ag_2O(s) \longrightarrow Ag(s) + \frac{1}{2}O_2(g)$	Dec.	300
$HgO(s) \longrightarrow Hg(l) + \frac{1}{2}O_2(g)$	Dec.	100
$BaO_2(s) \longrightarrow BaO(s) + \frac{1}{2}O_2(g)$	1	840
$LiCl \cdot H_2O(s) \longrightarrow LiCl(s) + H_2O(g)$	0.13	100
$CaC_2O_4(s) \longrightarrow CaCO_3(s) + Co(g)$	Dec. (1-2 hrs.)	475

peratures, the analysis of solid mixtures by selective decomposition is evidently quite feasible. For example, a mixture of $NaHCO_3$, Na_2CO_3, $MgCO_3$, and $CaCO_3$ can be analyzed by raising the temperature to somewhat above the decomposition point, first of $NaHCO_3(270°C)$, then of $MgCO_3(550°C)$, and finally of $CaCO_3(900°C)$. However, in order for this procedure to work, the sample must be a physical mixture of discrete solids rather than a solid solution.

Absorption of gaseous products

Of course, gravimetric analysis is not limited to decomposition reactions: Any chemical reaction that gives a quantitative yield of a product that can be isolated in a pure state and weighed is suitable. Thus, a gaseous product can be swept out of solution with a stream of nitrogen, absorbed by a specific absorbant, and weighed. For instance, a mixture of carbon dioxide gas and water vapor may be allowed to pass, first through a weighed drying tube containing anhydrous magnesium perchlorate to absorb the water, then through a weighed tube containing Ascarite (asbestos fibers impregnated with sodium hydroxide) to absorb the CO_2.

Quantitative separation of precipitates

Many of the separation and precipitation procedures that are used for qualitative analysis can also be used for quantitative analysis. For instance, to determine sulfate ion, one method is to dissolve a known weight of the sample in water, acidify with nitric acid, and add excess barium nitrate. The precipitate of barium sulfate is then separated by filtration, washed with distilled water, dried, and weighed. If necessary, the filtrate and washings can be combined and analyzed further.

The determination of sulfate from the weight of barium sulfate produced involves the same principles of stoichiometry as does the determination of, say, calcium carbonate from the weight of CO_2 that is expelled on decomposition. However, while the two methods do not differ in principle, they differ greatly in ease of application. In practice, analysis by decomposition involves a few weighings and some fairly straightforward handling of a crucible. On the other hand, analysis by precipitation runs the gamut of chemical operations: The precipitate must be formed quantitatively and in a high state of purity. It must be transferred completely onto a filter and washed free of impurities. Finally, it must be dried under conditions where decomposition and oxidation are negligible, and then weighed.

Skill is required at every step. Perhaps the most difficult part of the analysis is the collection, washing and drying of the precipitate, because *all* of it must be collected and *none* of it may be lost. The precipitate is usually collected on ashless filter paper in a cone-shaped funnel, or on a mechanically strong filter mat forming the bottom of a filter crucible that is mounted on a funnel. The quantitative transfer of the precipitate onto the filter with the aid of wash bottle and rubber scraper requires good coordination of hand and eye, and mental concentration. If the precipitate has been collected on ashless paper, the paper cone containing the precipitate must be transferred to a tared crucible and heated gradually to about 800°C. In that way the precipitate becomes thoroughly dry and the paper burns off. If the precipitate is collected in a tared filter crucible, filtration can be accelerated by the use of gentle suction and the precipitate can be dried under milder conditions, most commonly in an oven at 110°C or in a vacuum desiccator.

Because of the labor involved, the gravimetric analysis of insoluble precipitates is becoming more and more a method of last resort. Nevertheless, there are good reasons for including at least one such analysis in the laboratory work of this course. After all, the crystallization, isolation, and drying of solids are among the basic laboratory operations of chemistry, and it will pay you to learn to do them quantitatively.

VOLUMETRIC TITRATION

Among the chemical methods of quantitative analysis, volumetric titration stands out for versatility, accuracy, and convenience. In a titration one measures the amount of a substrate by allowing that substrate to react with a known reactant, the *titrant*, and finding the amount of titrant that is consumed. The equivalence point for the reaction is rendered visible by means of a *color indicator*—a substance that produces a color change at or near the equivalence point—or by means of a measuring instrument. With calibrated apparatus, and if the signal for the equivalence point is sharp, a precision of better than 0.2 percent is readily attained. Highly purified primary chemical standards are available from Bureaus of Standards around the globe to test the accuracy of various titration methods.

The reactions that can be used for titration span the whole range of chemistry. They include acid-base, oxidation-reduction, addition, substitution and displacement, precipitation, complex-ion formation, and ion-exchange reactions. We shall survey the more common methods of titration in Chapter 4.

COMPOSITION AND REACTION CAPACITY

In stating the chemical composition of a homogeneous sample, we must distinguish between the synthetic, the analytical, and the molecular composition. The *synthetic* composition specifies the relative amounts and identity of the component substances from which the sample was or might have been prepared. The *analytical* composition expresses the results of chemical analysis. The *molecular* composition expresses our conclusions concerning the identity and relative amounts of the molecular species that are actually present.

Formal and molar concentration

In volumetric analysis one usually works with reactants in liquid solution, and it is convenient to express the different kinds of composition in terms of *concentration per liter of solution*. To specify the synthetic composition, we characterize each solute component by an appropriate chemical formula and express the concentration in formula weights per liter. To specify the microscopic composition, we express the concentration of each molecular species in moles per liter.

The concentration in formula weights per liter is also called the *formal concentration* or *formality*. The concentration in moles per liter is called the *molar concentration* or *molarity*. For example, a solution prepared by dissolving 0.01 formula weight of NaCl in enough water to make one liter of solution is said to be 0.01 formal (0.01 F) in NaCl. Because the sodium chloride in 0.01 F aqueous NaCl is almost completely dissociated to sodium ions and chloride ions, such a solution is in effect 0.01 molar (0.01 M) in Na^+ and 0.01 M in Cl^-; the concentration of undissociated NaCl is very much less than 0.01 M.

When there are two or more solute components, we specify the formality of each, as in "0.002 F $FeCl_3$ and 0.05 F HCl in water." The molecular composition of such a solution happens to be very complex. The two components form a mixture of at least ten different solute species, including H_3O^+, Cl^-, $Fe(OH_2)_6^{3+}$, and a variety of hydrated ferric chloride coordination complexes and ion pairs.

We shall use the following notation to distinguish between molar and formal concentrations:

We shall use brackets around the chemical formula of a molecular species to denote the molar concentration of that species, as in $[Na^+]$ = 0.01 M, or $[H_3O^+] = 1 \times 10^{-5}$ M.

We shall use the symbol c to denote the formal concentration, with a subscript to denote the solute component, as in $c_{NaCl} = 0.01$ F, or c_{FeCl_3} = 0.002 F.

Reaction capacity and the chemical equivalent

The methods of chemical analysis are rarely so specific that they identify the substrate completely. For example, titration with silver nitrate can be made specific for ionic chloride, but it gives no clear-cut information about the cation of the chloride salt.

To report the analytical composition of a solution, we therefore need a scientific unit that expresses reaction capacity, yet implies only a generic description of the reactive substrates. Such a unit is the *chemical equivalent*. As originally defined, one chemical equivalent represents a reaction capacity equal to, or identifiable with, that of one gram atom of hydrogen in a similar reaction. For ionic reactions, it is convenient to restate that definition in terms of the faraday of charge: One chemical equivalent represents a reaction capacity causing the transfer of, or combination of, charged particles bearing one faraday of charge.

Values of the chemical equivalent for various classes of ionic reactions are given in Table 1.2. Two principles emerge from that table.

1. *For a pure substance, the number of equivalents per formula weight depends on the reaction that the substance undergoes.* For instance, potassium permanganate has five equivalents per formula weight in an oxidation-reduction reaction in which MnO_4^- is reduced to Mn^{2+}, but only one equiv-

Table 1.2 VALUE OF THE CHEMICAL EQUIVALENT IN SOME IONIC REACTIONS

Reaction	Examples
1. One equivalent in Brønsted acid-base reactions. Acid: transfers one faraday (Avogadro's number) of protons to base Base: accepts one faraday of protons from acid	1. $H_3O^+ + NH_3 \longrightarrow H_2O + NH_4^+$ H_3O^+: one equivalent per mole NH_3: one equivalent per mole
2. One equivalent in oxidation-reduction reactions. Oxidizing agent: accepts one faraday (Avogadro's number) of electrons from reducing agent Reducing agent: donates one faraday of electrons to oxidizing agent	2. Must write half-reactions: $MnO_4^- + 8 H_3O^+ + 5 e^- \longrightarrow Mn^{2+} + 12 H_2O$ MnO_4^-: five equivalents per mole $H_2AsO_3^- + 3 H_2O \longrightarrow H_2AsO_4^- + 2 H_3O^+ + 2 e^-$ $H_2AsO_3^-$: two equivalents per mole
3. One equivalent in the combination of cations with anions to form uncharged molecules or an insoluble precipitate. Cation: combines with anions bearing one faraday of charge. Anion: combines with cations bearing one faraday of charge.	3. $Pb^{2+}(aq) + 2 Ac^-(aq) \longrightarrow PbAc_2(aq)$ $Pb^{2+}(aq) + SO_4^{2-}(aq) \longrightarrow PbSO_4(s)$ $Pb^{2+}(aq)$: two equivalents per mole $Ac^-(aq)$: one equivalent per mole $SO_4^{2-}(aq)$: two equivalents per mole
4. One equivalent in Lewis acid-base reactions. Acid: coordinates with one mole (Avogadro's number) of basic (donor) sites Base: coordinates with one mole (Avogadro's number) of acidic (acceptor) sites	4. $Ag^+(aq) + 2 NH_3(aq) \longrightarrow Ag(NH_3)_2^+(aq)$ $Ag^+(aq)$: two equivalents per mole $NH_3(aq)$: one equivalent per mole

alent per formula weight in a reaction in which the K^+ ion (or the MnO_4^- ion) combines with a suitable counter-ion to produce an insoluble precipitate.

2. *Substances always react in a one-to-one ratio of equivalents.* This relationship, which follows directly from the definition of the chemical equivalent, is fundamental to all chemical analysis. When applied to volumetric titration, it leads to Eq. (1.15).

At the equivalence point in a titration:

$$\text{Equivalents of substrate} = \text{equivalents of titrant} \qquad (1.15)$$

The definition of the chemical equivalent leads to another concept, the *equivalent weight,* which is the number of grams in one equivalent of the substance, Eq. (1.16).

$$\text{Equivalent weight} = \frac{\text{weight (in grams)}}{\text{number of equivalents}} \qquad (1.16)$$

The equivalent weight of a reactant is related to its formula weight according to (1.17).

$$\text{Equivalent weight} = \frac{\text{formula weight}}{\text{number of equivalents per formula weight}} \quad (1.17)$$

For example, the equivalent weight of $KMnO_4$, in a reaction in which MnO_4^- is reduced to Mn^{2+}, is one-fifth of the formula weight of $KMnO_4$, or $158.04/5 = 31.61$ g per equivalent.

Normality

The analytical composition of liquid solutions is expressed conveniently in terms of the *normal concentration* or *normality*, which is equal to the number of equivalents per liter. Let V denote the volume of the solution in liters; the normality N is then given by (1.18).

$$N = \text{number of equivalents per liter}$$
$$= \frac{(\text{number of equivalents in volume } V)}{V} \quad (1.18)$$

On applying this definition to the reactants in a volumetric titration, we obtain Eqs. (1.19) and (1.20).

$$\text{Equivalents of substrate} = N_{\text{substrate}} \cdot V_{\text{substrate}} \quad (1.19)$$
$$\text{Equivalents of titrant} = N_{\text{titrant}} \cdot V_{\text{titrant}} \quad (1.20)$$

At the equivalence point, the equivalents of the reactants are equal. On introducing (1.19) and (1.20) in (1.15), we obtain (1.21).

At the equivalence point in a titration:

$$N_{\text{substrate}} \cdot V_{\text{substrate}} = N_{\text{titrant}} \cdot V_{\text{titrant}} \quad (1.21)$$

Thus, when the normality of the titrant is known, the normality of the substrate can be easily evaluated.

Does titration measure formal or molar concentrations?

For many purposes, the equivalent weight or the normality are entirely adequate for expressing the analytical composition of homogeneous substances. However, when the identity of the reactive substrate is known, it may be desirable to convert these units into formal or molar composition units. We now wish to show that *titration measures the formal composition.*

For definiteness, consider the titration of acetic acid with sodium hydroxide in aqueous solution. It is customary to represent that titration simply by Eq. (1.22).

$$\text{HAc(aq)} + \text{OH}^-\text{(aq)} \longrightarrow \text{Ac}^-\text{(aq)} + \text{HOH} \quad (1.22)$$

However, it is clear from the electrical conductivity, pH, and colligative properties of acetic acid in water that there is some acid dissociation, as well as some association of acetic acid to a dimer [Eqs. (1.23) and (1.24)].

$$HOH + HAc(aq) \; \rightleftharpoons \; H_3O^+(aq) + Ac^-(aq) \qquad (1.23)$$

$$2\,HAc(aq) \; \rightleftharpoons \; (HAc)_2(aq) \qquad (1.24)$$

If we aim at 0.1 percent accuracy, these reactions are not negligible. For example, 0.1000 F acetic acid in water is an equilibrium mixture consisting of 98.0 percent un-ionized HAc, 1.3 percent $H_3O^+ + Ac^-$, and 0.7 percent $(HAc)_2$.

Now let us ask: "What is the difference between measuring the molar and the formal concentration of acetic acid?" If we were measuring the molar concentration, we would be titrating the un-ionized HAc molecules only; the molecules of $H_3O^+ + Ac^-$ and of $(HAc)_2$ would not react. If we were measuring the formal concentration, then *all* acidic species produced from, and in equilibrium with, un-ionized HAc would be neutralized, so that each formula weight of acetic acid used in making up the solution will consume one equivalent of hydroxide ion. When the question is put in this way, the answer becomes obvious: The un-ionized HAc will remain in equilibrium with $H_3O^+ + Ac^-$ and with $(HAc)^2$ throughout the titration. (*Remember Le Chatelier's principle!*) The half-time for equilibration in (1.23) and (1.24) is less than a millisecond—much faster than any addition of hydroxide in an ordinary titration—and the various acidic species therefore disappear together. In short, we do not titrate the un-ionized HAc alone; we titrate the entire equilibrium mixture resulting from acetic acid in water, and thus measure the formal concentration.

In general, titration will measure the formal concentration of a substrate whenever the molecular species derived from the substrate remain in equilibrium throughout the titration. Of course, to deduce the formal concentration from the normality, we must also know the number of equivalents per formula weight.

PROBLEMS

1. On page 41 it is stated that physical methods of analysis become specific when there are at least as many measurements of independent properties as there are substances in the sample. What is the mathematical basis for this statement? Why is the measurement of an entire absorption spectrum so informative?

2. The following absorbance data (A_{5250}) were collected for potassium permanganate in a 1-cm cell at 5250 Å:

Concentration	A_{5250}
9.94×10^{-6}	0.022
5.03×10^{-5}	0.109
8.00×10^{-5}	0.175
9.90×10^{-5}	0.223
4.05×10^{-4}	0.890

Plot the data (see Fig. 1.4) to convince yourself that Beer's law (Eq. 1.2) is obeyed for this range of concentrations. Calculate the extinction coefficient (ϵ) for MnO_4^- at 5250 Å. Predict the absorbance (at 5250 Å) of a solution that is $6.25 \times 10^{-5}\ F$ in MnO_4^-. Estimate the concentration of a permanganate solution for which A_{5250} is 0.617.

3. Balance the following oxidation-reduction equations, supplying H_3O^+ (or OH^-) and H_2O where needed. Try to gain proficiency in both the algebraic and half-reaction methods. Assume that the substrate is the first reactant given.

(a) $Cr_2O_7^{2-} + Fe^{+2} \longrightarrow Cr^{3+} + Fe^{3+}$
(b) $H_2SO_3 + I_2 \longrightarrow SO_4^{2-} + I^-$
(c) $NO_3^- + Cl^- \longrightarrow NOCl + Cl_2$
(d) $C_2O_4^{2-} + MnO_4^- \longrightarrow CO_2 + Mn^{2+}$
(e) $NO_3^- + Zn \longrightarrow NH_4^+ + Zn^{2+}$
(f) $H_3AsO_3 + ICl \longrightarrow HAsO_4^{2-} + I^- + Cl^-$
(g) $NO_3^- + I_2 \longrightarrow NO + IO_3^-$

4. *Gravimetric Factors.* In a chemical reaction the reactants disappear, and the products appear, in definite ratios, which are identical to the ratios of the respective equivalent weights. In gravimetric analysis such ratios of equivalent weights are called *gravimetric factors.* For instance, a common method for the gravimetric analysis of chloride involves the reaction: $Cl^- + Ag^+(\text{in excess}) \rightarrow AgCl(s)$. This reaction is quantitative. The precipitate of silver chloride is separated, dried, and weighed. The weight of chloride in the precipitate is then obtained by applying the appropriate gravimetric factor, $f_{Cl/AgCl}$.

$$\frac{\text{Wt. of Cl}}{\text{Wt. of AgCl}} = f_{Cl/AgCl} = \frac{\text{at. wt. of Cl}}{\text{mol. wt. of AgCl}}$$

For the general case, one writes

$$\frac{\text{Wt. of substance sought}}{\text{Wt. of substance weighed}} = f$$

where

$$f = \frac{\text{equiv. wt. of substance sought}}{\text{equiv. wt. of substance weighed}}$$

On the basis of the above, find the gravimetric factors for the following:

Substance sought	Substance weighed
Chloride ion	$AgCl^a$
Sulfate ion	$BaSO_4$
Ferric ion	$Fe_2O_3^b$
O_2 gas	Ag_2O

a Solution: $35.45/143.32 = 0.2473$.
b Don't forget that there are two iron atoms in Fe_2O_3.

5. A chemist wishes to analyze a solid sample for chloride ion. He weighs out 0.4887 g of the solid, dissolves it in nitric acid, and precipitates the chloride with excess $AgNO_3$. The dried AgCl precipitate weighs 0.1050 g. What was the percent of chloride in the original sample?

6. A common natural source of iron is magnetite, which is mainly Fe_3O_4. One method of analysis of this ore involves fusion with potassium bisulfate ($KHSO_4$), oxidation of all the iron to Fe^{3+}, and precipitation with NH_4OH. The precipitate is a gelatinous mass containing hydrous ferric oxide, $Fe_2O_3 \cdot xH_2O$ or $Fe(OH)_3 \cdot yH_2O$. Ignition (heating to a very high temperature) drives off water, leaving Fe_2O_3. Calculate the percent of iron in an impure sample of magnetite if 1.489 g of ore yields 1.410 g of Fe_2O_3. What is the percent of Fe_3O_4 in this sample?

7. When one wishes to analyze a mixture of two salts containing a common ion, it is possible to determine the composition by analyzing for the common ion and using the proper weight relationships to set up *two* algebraic equations. Consider a sample which is a mixture of NH_4Cl and KCl. If 0.6400 g of this mixture gives 1.350 g of AgCl precipitate, what is the percent of each salt in the sample? Why does this determination rely on the fact that there are no impurities in the sample?

8. When a sample is to be analyzed for two constituents but also contains inert material (impurities), then *two* independent measurements are required. Why? For a mixture of two halide ions, a volumetric *and* a gravimetric analysis together give the information needed for analysis. A 1.000 g sample containing both chloride and bromide salts yields 0.9160 g of precipitate consisting of AgCl and AgBr. The total number of equivalents of chloride and bromide is then determined by titration: A second 1.000 g sample requires 55.00 ml of 0.1000 N silver nitrate solution. Calculate the percentages of chloride and bromide in the sample.

9. Baking soda consists largely of sodium bicarbonate, $NaHCO_3$, which decomposes at temperatures above 270°C.

$$2\ NaHCO_3(s) \rightleftharpoons Na_2CO_3(s) + H_2O(g) + CO_2(g)$$

It is found that a 5.010 g sample of a particular brand of baking soda (which has first been dried at 110°C) yields 3.003 g of Na_2CO_3 (washing soda) after heating with a Bunsen burner. What is the percent of $NaHCO_3$ in this sample? If you have tasted both baking soda and washing soda, you should now know two reasons why you should never add more of the former than is called for in a recipe.

10. Explain the difference between formality F and molarity M. Under what conditions are the molarity and formality of a solute the same? Can you think of an actual example?

11. Calculate the equivalent weight of the following compounds or ions.

    (a) $HClO_4$ to be used in acid-base titrations

    (b) H_2SO_4 to be used in acid-base titrations

    (c) H_2SO_4 to be used to precipitate barium as $BaSO_4$

    (d) Sn^{2+} used as a reducing agent where tin is oxidized to Sn^{4+}

    (e) $Cr_2O_7^{2-}$ used as an oxidizing agent where Cr^{3+} is produced

12. What weight of $KMnO_4$ would be required to make 500 ml of 0.1000 N solution if the permanganate is to be used as an oxidizing agent under conditions where it is reduced to Mn^{2+}?

13. *Dilution of Liquid Solutions.* Dilution with pure solvent does not change the *absolute amount* of a solute, but changes only its concentration. Thus, the number of equivalents (or grams, or formula weights) of the solute before and after dilution must be equal. Let V and V' denote the volume of the solution before and after dilution, N and N' the corresponding normality, and F and F' the corresponding formal concentration. Then, by reasoning analogous to that in Eqs. (1.15) through (1.21), we obtain $NV = N'V'$ and $FV = F'V'$.

(a) 10.0 ml of 0.1 F H_2SO_4 is diluted to a final volume of 135 ml. What is the final formal concentration?

(b) 0.4 F tin(II) is to be used as a reducing agent, according to (1.9). What volume of 0.4 F tin(II) is required to prepare (by dilution with pure water) 1.0 l of 0.2 N Sn^{2+}?

14. 10.00 ml of a NaOH solution requires 9.48 ml of 0.1000 F HCl solution for neutralization. What is the normality of the NaOH solution? What volume of 0.0800 F H_2SO_4 solution will be required to neutralize 10.00 ml of this NaOH solution?

15. A 0.2000 g sample of a (pure) solid acid was dissolved in water and titrated with 0.1500 N sodium hydroxide. What is the equivalent weight of the acid if 42.50 ml of titrant are required for the titration? What can you say about its formula weight?

16. What weight of ferrous iron can be oxidized to Fe^{3+} by 10.00 ml of 0.1000 F $K_2Cr_2O_7$ solution? Dichromate is reduced to Cr^{3+}.

chapter 2

Titration and
the Equilibrium Constant

A titration, carried out properly, is more than a method of chemical analysis: It is a chemical reaction going forward under perfect control. The control involves both the speed and the position of equilibrium. Conditions must exist under which the reaction between substrate and titrant is conveniently fast, sufficiently quantitative and free from complicating side reactions, and under which its progress can be monitored. In nearly all practical titrations, the reaction is fast enough so that after each addition of titrant, equilibrium according to the new composition is reached in a few seconds or less. The progress of the titration can therefore be monitored at chemical equilibrium.

$$\text{Substrate} + \text{Titrant} \; \rightleftharpoons \; \text{Product(s)} \tag{2.1}$$

The advantages of titration at chemical equilibrium are so compelling that titration is rarely attempted otherwise. If necessary, the original sample is subjected to elaborate chemical procedures to convert it into substances that *can* be titrated at equilibrium.

In this chapter we shall begin a scientific discussion of titration, based on the principles of chemical equilibrium. Since a reaction between substrate and titrant can be quantitative only if the equilibrium constant for the process shown in (2.1) is sufficiently large, a good starting point for our discussion will be the equilibrium constant—how to formulate it, how to find its numerical value, and how to use that information to control, and predict the possible accuracy of, an actual titration.

FORMULATING THE EQUILIBRIUM CONSTANT

The first step in formulating the equilibrium constant is to write the chemical equation for the titration. The equilibrium constant is then formulated in precisely the same way as for any reaction, with molar concentrations of the products appearing in the numerator and molar concentrations of the reactants appearing in the denominator.* We shall always write the chemical equation so that it involves *one mole of substrate*.

In writing the chemical equation, we try to show those molecular species that best represent the substrate, titrant, and reaction products under the conditions of the actual titration. For example, acetic acid in water consists mostly of un-ionized HAc molecules. The titration of acetic acid with sodium hydroxide is therefore represented by Eq. (2.2) and the equilibrium constant by Eq. (2.3).

$$\text{HAc(aq)} + \text{OH}^-\text{(aq)} \longrightarrow \text{Ac}^-\text{(aq)} + \text{HOH} \tag{2.2}$$

$$K = \frac{[\text{Ac}^-]}{[\text{HAc}][\text{OH}^-]} = 1.8 \times 10^9 (\text{M}^{-1}) \text{ at } 25°\text{C} \tag{2.3}$$

To give another familiar example, the titration of a water-soluble bromide salt with aqueous silver nitrate to give an insoluble precipitate of silver bromide is represented by Eq. (2.4) and the equilibrium constant by Eq. (2.5).

$$\text{Br}^-\text{(aq)} + \text{Ag}^+\text{(aq)} \longrightarrow \text{AgBr(s)} \tag{2.4}$$

$$K = \frac{1}{[\text{Br}^-][\text{Ag}^+]} = 1.3 \times 10^{12} (\text{M}^{-2}) \text{ at } 25°\text{C} \tag{2.5}$$

Equations (2.3) and (2.5) also illustrate the usual practice not to include

---

* In rigorous thermodynamic calculations one uses *activities* rather than molar concentrations. In dilute solution, the difference between activities and molar concentrations is not great, and we shall neglect it. However, we shall be careful to take into account all equilibria among the solutes.

solvent molecules and solid precipitates in the mass-action expressions for K.

In the presence of excess potassium iodide, iodine in aqueous solution is converted largely to nonvolatile, water-soluble triiodide ion, I_3^-. The titration of iodine with sodium thiosulfate in the presence of excess potassium iodide is therefore represented by (2.6) and the equilibrium constant by (2.7).

$$I_3^- + 2\,S_2O_3^{2-} \longrightarrow 3\,I^- + S_4O_6^{2-} \tag{2.6}$$

$$K = \frac{[S_4O_6^{2-}][I^-]^3}{[I_3^-][S_2O_3^{2-}]^2} = 2.8 \times 10^{15}\ (M)\ \text{at } 25°C \tag{2.7}$$

Equation (2.7) illustrates that the algebraic expressions for K are sometimes fairly complex.

Data for equilibrium constants

In appraising the worth of a specific titration, it is not enough to formulate the equilibrium constant. We must know its actual value. Numerical data are available for many reactions, particularly in aqueous solution, but unfortunately the data are not often tabulated in a form that is convenient for our purpose. During the last century, as the study of chemical equilibrium developed into a quantitative science, certain traditions grew up about the format in which equilibrium constants are reported. As a result, there are standard equilibrium expressions, employing standard symbols, that have been in use in the chemical literature for many decades, and which it would cause utter confusion to tamper with. The most pertinent of these expressions are summarized in Table 2.1 (page 28). Equilibrium constants for titration are derived by appropriate algebraic manipulation of these expressions.

For acids and bases in water and water-like solvents, the traditional constants are the acid dissociation constant (K_A) and the base dissociation constant (K_B). It is now common practice to use the Brønsted definitions.* Accordingly, Table 2.1 shows examples in which the acid (proton donor) or base (proton acceptor) happens to be an ion, or in which the solvent is other than water.

According to the Brønsted definitions for conjugate acid-base pairs, K_A is related to K_B through the ion product (K_{IP}) of the solvent, as shown in (2.8).

$$K_A \cdot K_B \text{ (for conjugate acid-base pair)} = K_{IP} \tag{2.8}$$

The derivation of (2.8) is illustrated for the NH_4^+-NH_3 pair in water.

---

* A detailed description will be given in Chapter 6.

$$NH_4^+ + H_2O \rightleftharpoons NH_3 + H_3O^+; \quad K_A = \frac{[H_3O^+][NH_3]}{[NH_4^+]}$$

$$NH_3 + H_2O \rightleftharpoons NH_4^+ + OH^-; \quad K_B = \frac{[NH_4^+][OH^-]}{[NH_3]}$$

$$K_A \cdot K_B = \frac{[H_3O^+][NH_3]}{[NH_4^+]} \cdot \frac{[NH_4^+][OH^-]}{[NH_3]}$$

$$= [H_3O^+][OH^-] = K_W$$

The ion product of water is usually denoted by the symbol K_W. The term *ion product* is used synonymously with two other terms, *autoprotolysis constant* and *self-ionization constant*.

Precipitation and solubility equilibria are usually discussed in terms of the solubility-product constant (K_{SP}). Association, complex-ion formation, and dissociation are similarly discussed in terms of the association constant and dissociation constant. The association constant for complex ion formation is also called *stability constant* and *formation constant*.

Oxidation-reduction equilibria are usually treated in terms of standard electrode potentials for half-reactions. This practice developed because much of our information comes from emf measurements on electrochemical cells. Fortunately, you need not be an expert on electrochemical cells in order to follow the remainder of this section. However, if the subject is totally unfamiliar, you should look it up in your general chemistry textbook before going on.

Two kinds of standard electrode potentials are in common use: *oxidation potentials*, E_{ox}°, for half-reactions in which the substrate is oxidized, and *reduction potentials*, E_{red}° for half-reactions in which the substrate is reduced.* Some examples of standard oxidation potentials follow.

Some Standard Oxidation Potentials

(Reduced form $\rightleftharpoons$ oxidized form + n electrons)

$Li(s) \rightleftharpoons Li^+(aq) + e^-;$	$E_{ox}^\circ = +3.045$ V
$2 H_2O + H_2(g) \rightleftharpoons 2 H_3O^+(aq) + 2 e^-;$	$E_{ox}^\circ = 0.000$ V
$2 I^-(aq) \rightleftharpoons I_2(s) + 2 e^-;$	$E_{ox}^\circ = -0.5355$ V

If the half-reaction is reversed so that the reduced form appears on the right, the sign of E_{ox}° must also be reversed, and we obtain standard reduction potentials.

---

* We use the following symbols for standard electrode potentials:

E_{ox}° for oxidation half-reactions
E_{red}° for reduction half-reactions
E° for half-reactions in general or for full reactions

Table 2.1   SOME TRADITIONAL EQUILIBRIUM EXPRESSIONS

Reaction	Example	Equilibrium expression
1. Acid dissociation:		Acid dissociation constant:
in water, uncharged acid	$HCN + H_2O \rightleftharpoons H_3O^+ + CN^-$	$K_A = \dfrac{[H_3O^+][CN^-]}{[HCN]}$
in water, charge of acid $+1$	$NH_4^+ + H_2O \rightleftharpoons H_3O^+ + NH_3$	$K_A = \dfrac{[H_3O^+][NH_3]}{[NH_4^+]}$
in methanol, uncharged acid	$HAc + MeOH \rightleftharpoons MeOH_2^+ + Ac^-$	$K_A = \dfrac{[MeOH_2^+][Ac^-]}{[HAc]}$
in water, dibasic acid	$H_2S + H_2O \rightleftharpoons HS^- + H_3O^+$	$K_{A1} = \dfrac{[H_3O^+][HS^-]}{[H_2S]}$
	$HS^- + H_2O \rightleftharpoons S^{2-} + H_3O^+$	$K_{A2} = \dfrac{[H_3O^+][S^{2-}]}{[HS^-]}$
2. Base dissociation:		Base dissociation constant:
in water, uncharged base	$NH_3 + H_2O \rightleftharpoons NH_4^+ + OH^-$	$K_B = \dfrac{[OH^-][NH_4^+]}{[NH_3]}$
in water, charge of base -1	$Ac^- + H_2O \rightleftharpoons HAc + OH^-$	$K_B = \dfrac{[OH^-][HAc]}{[Ac^-]}$
in liquid acetic acid, uncharged base	$NH_3 + HAc \rightleftharpoons NH_4^+ + Ac^-$	$K_B = \dfrac{[Ac^-][NH_4^+]}{[NH_3]}$
3. Self-ionization of solvent:		Ion product of solvent:
in water	$H_2O + H_2O \rightleftharpoons H_3O^+ + OH^-$	$K_W = [H_3O^+][OH^-]$
in liquid ammonia	$NH_3 + NH_3 \rightleftharpoons NH_4^+ + NH_2^-$	$K_{IP} = [NH_4^+][NH_2^-]$

Reaction	Example	Equilibrium expression
4. Solubility of salts:		Solubility-product constant:
univalent ions	$AgBr(s) \rightleftharpoons Ag^+ + Br^-$	$K_{SP} = [Ag^+][Br^-]$
salts of n-m charge type	$PbI_2(s) \rightleftharpoons Pb^{2+} + 2\,I^-$	$K_{SP} = [Pb^{2+}][I^-]^2$
5. Association:		Association constant:
molecular	$2\,HAc \rightleftharpoons (HAc)_2$	$K_{assoc} = \dfrac{[(HAc)_2]}{[HAc]^2}$
ionic	$K^+ + Ac^+ \rightleftharpoons K^+Ac^-$	$K_{assoc} = \dfrac{[K^+Ac^-]}{[K^+][Ac^-]}$
6. Complex-ion formation	$Ag^+ + 2\,NH_3 \rightleftharpoons Ag(NH_3)_2{}^+$	Association constant for complex formation: $K_{assoc} = \dfrac{[Ag(NH_3)_2{}^+]}{[Ag^+][NH_3]^2}$
7. Dissociation	$NaSO_4{}^- \rightleftharpoons Na^+ + SO_4{}^{2-}$	Dissociation constant: $K_d = \dfrac{[Na^+][SO_4{}^{2-}]}{[NaSO_4{}^-]}$
8. Oxidation reduction:		Standard potential, $\mathbf{E}°$;
half-reactions	$I_3^- + 2\,e^- \rightleftharpoons 3\,I^-; \; \mathbf{E}_{red}° = 0.536$ V $S_4O_6{}^{2-} + 2\,e^- \rightleftharpoons 2\,S_2O_3{}^{2-}; \; \mathbf{E}_{red}° = 0.08$ V	Number of faradays, n; $RT \ln K = n\mathbf{E}\mathbf{F}$
full reaction	$I_3^- + 2\,S_2O_3{}^{2-} \rightleftharpoons 3\,I^- + S_4O_6{}^{2-};$ $\mathbf{E}° = 0.536 - 0.08 = 0.456$ V	

Some Standard Reduction Potentials

(Oxidized form + n electrons $\rightleftharpoons$ reduced form)

$$Li^+(aq) + e^- \rightleftharpoons Li(s); \qquad\qquad E^\circ_{red} = -3.045 \text{ V}$$

$$2 H_3O^+(aq) + 2 e^- \rightleftharpoons H_2(g) + 2 H_2O; \qquad E^\circ_{red} = 0.000 \text{ V}$$

$$I_2(s) + 2 e^- \rightleftharpoons 2 I^-(aq); \qquad\qquad E^\circ_{red} = +0.5355 \text{ V}$$

Standard oxidation and standard reduction potentials evidently convey the same information, since one is merely the negative of the other. In American textbooks (including this one) it is customary to tabulate only the standard reduction potentials. The standard potential for the reduction of hydrogen ion (H_3O^+) to hydrogen gas in aqueous solution is zero, by definition.*

The standard electrode potential, whether of oxidation or reduction, is an intrinsic property and is therefore independent of the size of the system. Thus, multiplying the equation for a half-reaction by an arbitrary scaling factor z has no effect on the value of E°. An example follows.

$$2 I^-(aq) \rightleftharpoons I_2(s) + 2 e^-; \qquad E^\circ_{ox} = -0.5355 \text{ V}$$

$$I^-(aq) \rightleftharpoons \tfrac{1}{2} I_2(s) + e^-; \qquad E^\circ = E^\circ_{ox} = -0.5355 \text{ V}$$

$$z I^-(aq) \rightleftharpoons \frac{z}{2} I_2(s) + z e^-; \qquad E^\circ = E^\circ_{ox} = -0.5355 \text{ V}$$

The equilibrium constant for an oxidation-reduction titration is calculated from standard potentials by the following procedure:

Write the half-reaction for one mole of substrate and note the number n of electrons.

Write the half-reaction for an equivalent amount of titrant.

Find the standard emf, E°, for the full reaction by appropriate addition or subtraction of E° values for the half-reactions.

Calculate $\log K$ from (2.9), where R is the gas constant per mole, T the absolute temperature, and F the Faraday constant.

$$\log K = \frac{n F E^\circ}{2.303 \, RT} \tag{2.9}$$

On expressing E° in volts and solving (2.9) specifically for 25°C, one obtains (2.10).

---

* Unfortunately, two opposite conventions are in use in the chemical literature for stating the sign of electrode potentials. According to the *American convention*, which is used in this book, the standard reduction potential will be most negative for the best reducing agent and most positive for the best oxidizing agent. According to the *European convention*, these signs are reversed. Thus, for the half-reaction, $Li^+(aq) + e^- \rightleftharpoons Li(s)$, $E^\circ_{red} = -3.045$ V according to the American convention, and $E^\circ_{red} = +3.045$ V according to the European convention. The use of two opposite conventions causes much unnecessary confusion. Never use electrode potentials without first ascertaining the convention according to which E° is defined.

$$\log K = \frac{n\mathbf{E}^\circ(V)}{0.059} \qquad \text{at } 25°C \qquad (2.10)$$

The procedure is illustrated for the reaction $I^{3-} + 2\ S_2O_3^{2-} \rightleftharpoons 3\ I^- + S_4O_6^{2-}$ at 25°C.

$I_3^- + 2\,e^- \rightleftharpoons 3\,I^-;$	$\mathbf{E}^\circ_{red} = 0.536\ V;$	$n = 2$
$2\,S_2O_3^{2-} \rightleftharpoons S_4O_6^{2-} + 2\,e^-;$	$\mathbf{E}^\circ_{ox} = -0.08\ V;$	

Add: $I_3^- + 2\,S_2O_3^{2-} \rightleftharpoons 3\,I^- + S_4O_6^{2-};$ $\mathbf{E}^\circ = 0.456\ V;$ $n = 2$

Note that the first half-reaction is written in the form of a standard reduction potential, while the second is written in the form of a standard oxidation potential. As a result, the two half-reactions are *added*. An alternate procedure is to write both half-reactions in the form of a standard reduction potential. In that case, the half-reactions are *subtracted*, as shown in Table 2.1. On substituting in Eq. (2.10), we obtain

$$\log K = \frac{2 \times 0.456}{0.059} = 15.46; \qquad K = 2.8 \times 10^{15}$$

REFERENCES

Some commonly-used data for equilibrium constants and standard reduction potentials are compiled in the Appendix. Further tables of data can be found in laboratory handbooks, International Critical Tables, reviews, and reference books, of which a small reference list is given below.

G. Kortüm, W. Vogel, and K. Andrussow, *Dissociation Constants of Organic Acids in Aqueous Solution* (before 1963: *Electrochemical Data*), Butterworth & Co., Ltd., London, 1963.

D. D. Perrin, *Dissociation Constants of Organic Bases in Aqueous Solution*, Butterworth & Co., Ltd., London, 1965.

L. G. Sillén and A. E. Martell, *Stability Constants of Metal-Ion Complexes*, The Chemical Society, London, 1964, Special Publication No. 17. Supplement, 1971, Special Publication No. 25.

W. M. Latimer, *Oxidation Potentials*, 2nd Ed., Prentice-Hall, Inc., Englewood Cliffs, N.J., 1952. (Uses the American convention.)

A. J. deBethune and N. A. Swendeman Loud, *Standard Aqueous Electrode Potentials*, C. A. Hampel, Skokie, Ill., 1964. (Uses the European convention.)

A. Seidell, *Solubilities of Inorganic and Metal-Organic Compounds*, 4th Ed., D. Van Nostrand, Co., Inc., Princeton, N.J., 1958.

The entire literature of chemistry, that is, the results of all published chemical research, is summarized in Chemical Abstracts. It will be worth your while to become familiar with this important reference tool.

RELATIONSHIPS AMONG EQUILIBRIUM CONSTANTS

Equilibrium constants for titration are deduced from equilibrium expressions such as those in Table 2.1 by straightforward algebra. Sometimes the relationship is obvious simply by inspection. For example, for the titration of bromide ion with silver ion [Eq. (2.4)],

$$K = \frac{1}{[Br^-][Ag^+]} = \frac{1}{K_{SP}} \qquad \text{for silver bromide}$$

In other cases we compare the equilibrium expression for the titration with equilibrium expressions whose value is already known. For example, for the titration of formic acid (HFo) with ammonia in aqueous solution, K is given by (2.11).

$$HFo + NH_3 \rightleftharpoons Fo^- + NH_4^+$$

$$K = \frac{[Fo^-][NH_4^+]}{[HFo][NH_3]} \tag{2.11}$$

Since the value of K_A for formic acid is known, we generate an expression containing K_A for formic acid by multiplying the right-hand side of (2.11) by $[H_3O^+]/[H_3O^+]$ and rearranging:

$$K = \frac{[H_3O^+][Fo^-]}{[HFo]} \times \frac{[NH_4^+]}{[H_3O^+][NH_3]} = \frac{K_A(\text{for HFo})}{K_A(\text{for NH}_4^+)}$$

The result is found by inspection to be simply the ratio of known K_A values. On substituting actual numbers from Table A.1, we find that $K = 3.2 \times 10^5$.

For another example, consider the reaction of the silver-ammonia complex with hydrogen ion in aqueous solution, for which K is given in (2.12).

$$Ag(NH_3)_2^+ + 2\,H_3O^+ \rightleftharpoons Ag^+ + 2\,NH_4^+$$

$$K = \frac{[Ag^+][NH_4^+]^2}{[Ag(NH_3)_2^+][H_3O^+]^2} \tag{2.12}$$

To generate an expression containing the formation constant of $Ag(NH_3)_2^+$, multiply the right-hand side by $[NH_3]^2/[NH_3]^2$ and rearrange:

$$K = \frac{[NH_3]^2[Ag^+]}{[Ag(NH_3)_2^+]} \times \frac{[NH_4^+]^2}{[NH_3]^2[H_3O^+]^2}$$

$$= \frac{1}{K_{\text{form}} (\text{for Ag(NH}_3)_2^+)} \times \frac{1}{K_A^2 (\text{for NH}_4^+)}$$

On substituting actual numbers from Tables A.1 and A.4, we find that $K = 1.5 \times 10^{11}$.

Finally, to illustrate the use of equilibrium constants for stepwise acid dissociation, consider the reaction of H_3PO_4 with HPO_4^{2-}, for which K is given in (2.13).

$$H_3PO_4 + HPO_4^{2-} \rightleftharpoons 2 H_2PO_4^-$$

$$K = \frac{[H_2PO_4^-]^2}{[H_3PO_4][HPO_4^{2-}]} \tag{2.13}$$

To generate an expression containing acid dissociation constants, multiply the right-hand side of (2.13) by $[H_3O^+]/[H_3O^+]$ and rearrange:

$$K = \frac{[H_3O^+][H_2PO_4^-]}{[H_3PO_4]} \times \frac{[H_2PO_4^-]}{[H_3O^+][HPO_4^{2-}]} = \frac{K_{A1}}{K_{A2}}$$

We thus find that $K = K_{A1}/K_{A2}$, the ratio of the first to the second acid dissociation constant. On substituting actual numbers from Table A.1, we obtain $K = 1.2 \times 10^5$.

EQUILIBRIUM CONSTANT AND TITRATION ERROR

Accurate titration is possible only if the reaction between substrate and titrant is sufficiently quantitative, and this condition implies that the equilibrium constant for the titration must be large. In the preceding section we described some reactions that are suitable for titration [Eqs. (2.2) through (2.7)]. You probably noticed that in each case the experimental value listed for K is a large number. We now wish to give some "engineering formulas" that relate the titration error to the magnitude of K.

Kinds of experimental error

We begin with a brief discussion of experimental error in general. There are three kinds of experimental error: *determinate* or systematic error, *indeterminate* or random error, and *fatuous* or stupid error.

Determinate error is any error that can be eliminated by careful experimentation—by calibration of measuring instruments, by purification of reagents, and by checking that each step in the procedure is accurately under control.

Indeterminate error represents the random statistical fluctuations that occur whenever measurements are repeated, owing to the limited precision of measuring systems. This kind of error is well-known in physical measurements, where the precision is limited by the finite sensitivity of the measuring instruments. In chemical analysis, the precision is limited also by any failure of the chemical reaction to be perfectly quantitative at the equivalence point.

Fatuous error is due to the human element in the laboratory. When you become tired at the end of a hard day, you are prone to make inadvertent mistakes. For instance, you might record the result as 69.218 even though you measured 62.918; or you might omit that critical 2 ml of nitric acid without which the method will not work. Fatuous error creeps in whenever you relax your concentration on the experiment—when you are sleepy, tired, or in a hurry; when you daydream; or whenever your mind is preoccupied elsewhere.

Repetition of experiments. Average and mean deviation

Repetition of experiments is desirable to improve the accuracy of the final result, to demonstrate the reproducibility of the experimental method, and to guard against fatuous error.

A correctly computed *average*, or medial value based on several measurements, is always more accurate than any single measurement, in the sense that the average has a higher probability of being near the true value. This correctly computed average should be a *weighted* average, in which each result is given a statistical weight that reflects its inherent reliability.

In this chapter we shall consider only the simplest kind of average, with which you are already familiar, in which each result is given the same statistical weight.

When the statistical weights of the results are equal,
the statistical average is simply the algebraic mean.

The algebraic mean is the correct average when experiments are repeated precisely, with the same method and equipment, because then each result should be just as reliable as any other. Equation (2.14) shows the familiar formula for computing the algebraic mean $\bar{x}$. $x_1, x_2, \ldots, x_n$ denote the results of n replicate measurements.

$$\bar{x} = \frac{x_1 + x_2 + \ldots + x_n}{n} = \frac{\sum\limits_{i=1}^{n} x_i}{n} \tag{2.14}$$

Repetition of experiments is important also to establish the reproducibility, or precision, of the method. When the average is given by the algebraic mean $\bar{x}$, then a convenient measure of precision is the *mean magnitude of the deviations from the mean*, or *mean deviation* for short. The mean deviation (m.d.) is defined in (2.15). An actual example to illustrate the calculation of $\bar{x}$ and m.d. is given in Table 2.2.

$$\text{m.d.} = \frac{\sum\limits_{i=1}^{n} |x_i - \bar{x}|}{n} \tag{2.15}$$

In reporting the results of replicate experiments, it is convenient to report $\bar{x} \pm$ m.d. For example, the data in Table 2.2 would be reported as 7.21 ± 0.02.

MEAN AND MEAN DEVIATION   **Table 2.2**

Experiment number	Result x_i	Deviation from $\bar{x}$ $\|x_i - 7.21\|$
1	7.21	0.00
2	7.18	0.03
3	7.24	0.03
4	7.20	0.01
Mean	7.20_{75}	0.01_{75}

Significant figures of mean and m.d.:

$$\bar{x} = 7.21 \qquad\qquad \text{m.d.} = 0.02$$

The mean deviation provides an *absolute* measure of the experimental precision. However, we often need a *relative* measure of the experimental precision, especially when we wish to estimate the probable percentage error of the final result. A convenient measure of relative precision is the mean percentage deviation (% m.d.), that is, the mean deviation expressed as percent of $\bar{x}$. The % m.d. is defined in (2.16).

$$\% \text{ m.d.} = \frac{100 \text{ m.d.}}{\bar{x}} = \frac{100 \sum\limits_{i=1}^{n} |x_i - \bar{x}|}{\sum\limits_{i=1}^{n} x_i} \qquad (2.16)$$

Thus, for the data in Table 2.2, the % m.d. is $(100 \times 0.02)/7.21$, or 0.3 percent.

Repetition of experiments is desirable also because it guards against fatuous error. It is quite unlikely that the same inadvertent mistake will be made over and over again. Therefore, if replicate measurements are in acceptable agreement, you may be confident that fatuous error is absent. On the other hand, if one measurement in a series is far out of line, and if your calculations and procedure have no obvious flaws, then fatuous error must be suspected. For beginners we recommend the following procedure: If the suspicious result is really "wild," discard it; if it deviates from the mean of the others by less than four times their mean deviation, retain it.

Experienced scientists develop work habits that minimize the likelihood of fatuous error, and unexplained wild deviations should be extremely rare. An experienced chemist who repeatedly finds it necessary to discard suspicious results is not much of a scientist.

Significant figures. In combining data to compute a sum, difference, or other function, the answer can never be known more accurately than is the *least* reliable of the measurements involved. Thus, if you measure the length of a rectangle to the nearest inch and the width to within one-sixteenth inch, you may calculate the perimeter only to the nearest inch.

In the decimal notation used in analytical chemistry, we say that a particular value is reliable to a certain number of *significant figures*. We count significant figures by starting at the first nonzero numeral on the left and proceeding to the right. For instance, 0.1000, 2.101, 2470, all have *four* significant figures while 0.100, 0.210, 0.0100, 247 have *three*. In this notation, the last figure we write is uncertain, owing to experimental error, but all other figures are known with certainty. For example, we say that an analytical balance weighs to four decimal places (four significant figures for samples between 0.1000 and 0.9999 grams) even though the fourth place is only *estimated* from a scale and contains a possible error of ± 1 unit.

When working with large numbers, we may have to use exponential notation to specify which zeros are significant. For instance, the number "two thousand" in "two thousand blind mice" may be a round number accurate to the nearest thousand, in which case we write 2×10^3; or it may be accurate to ± 1 or ± 2, in which case we write 2.000×10^3. If the number of blind mice is known to be exactly two thousand, then exponential notation is not used, but the number is written 2,000. In cases of doubt, it is usually clear from the context whether the number is a round number subject to error, or whether it is an exact counting number.

When adding or subtracting experimental values that are expressed to the correct number of significant figures, the answer should contain the same number of *decimal places* as the least reliable piece of data. In multiplication and division, the answer should contain as many *significant figures* as does the piece of data with the fewest significant figures. These rules are best explained by means of a few examples.

Addition. $1.491 + 0.2$.

> Round off to the same number of decimal places
> and add: $1.5 + 0.2 = 1.7$

Subtraction. $122.6 - 0.081$.

> Round off to the same number of decimal places
> and subtract: $122.6 - 0.1 = 122.5 = 1.225 \times 10^2$

Multiplication. $1.05 \times 6.1 \times 10^3 = 6.4 \times 10^3$.

> Multiply $(1.05 \times 6.1 = 6.405)$ and round off to
> two significant figures.

Division. $153 \div 1.48 = 103 = 1.03 \times 10^2$.

> Divide $(153 \div 1.48 = 103.3 \ldots)$, round off to three significant
> figures, and (optional) express in exponential notation.

Innate titration error

All methods for finding the equivalence point in a titration are tantamount to watching the concentration of unreacted titrant. If the equilibrium constant for titration is large, the concentration of unreacted titrant will remain quite small until the equivalence point is reached and then increase sharply as titrant becomes in excess. An endpoint indicator will sense this sudden increase and give a sharp visual signal.

On the other hand, if the equilibrium constant is not quite so large, the concentration of unreacted titrant will not stay quite so small as the equivalence point is reached, and it will increase less abruptly as titrant becomes in excess. As a result, the signal given by an endpoint indicator will be less sharp, and the endpoint will be less well defined. The resulting error will be called *innate titration error*.

For example, in the titration of acetic acid with sodium hydroxide [Eq. (2.2)], the innate titration error is less than 0.1 percent. As a result, in the usual method of titration with phenolphthalein endpoint indicator, the color change from colorless to red-purple is sharp and the equivalence point can be located with an uncertainty of one drop of titrant or less.

On the other hand, if acetic acid is titrated with the weaker base, ammonia [Eq. (2.17)], the innate titration error becomes quite significant.

$$HAc(aq) + NH_3(aq) \rightleftharpoons Ac^-(aq) + NH_4^+(aq) \qquad (2.17)$$

$$K = \frac{[Ac^-][NH_4^+]}{[HAc][NH_3]} = 3.1 \times 10^4 \qquad at\ 25°C \qquad (2.18)$$

A suitable endpoint indicator is bromothymol blue. However, the color change from green to blue, which marks the equivalence point, is indistinct, several drops of titrant being required, and the equivalence point is correspondingly uncertain. Although the precision can be improved by use of a more sensitive instrumental method of endpoint detection, the titration of acetic acid with ammonia can never have the same elegance of easy precision as that of acetic acid with sodium hydroxide.

The mathematical analysis of innate titration error is within the scope of this book and will be taken up in Chapter 8. However, at this point it is useful to indicate some of the results.

1. The innate titration error is proportional to the fraction of the original substrate that remains unreacted at the equivalence point. The constant of proportionality will vary with the sensitivity of the endpoint indicator. However, since we define the innate titration error to be independent of the method of endpoint detection, all calculations will be made with a fixed constant of proportionality. [See Eq. (8.5).] This constant is chosen so that realistic results will be obtained for typical, common methods of endpoint detection.

2. The mathematical expression for the innate titration error depends on the stoichiometry—that is, on the number of reactant and product molecules that are solutes—in the balanced chemical equation for the titration. In most cases, the innate titration error depends on the substrate concentration and is affected adversely by dilution.

3. In the following, S denotes the substrate, T the titrant, P and Q are

soluble titration products, and K is the equilibrium constant; c_S denotes the formal concentration of substrate and is given by,

$$c_S = \frac{n_S}{V} \qquad (2.19)$$

where

n_S = number of formula weights of substrate to be titrated
V = volume of solution at the equivalence point

At the beginning of the titration, before any titrant has been added, the concentrations of P and Q are assumed to be zero.

(a) If the chemical equation takes the form

$$S + T \rightleftharpoons P + Q$$

so that

$$K = \frac{[P][Q]}{[S][T]}$$

then the innate titration error is estimated to be

$$\text{less than } 1\% \quad \text{if } K > 3 \times 10^4$$
$$\text{less than } 0.25\% \quad \text{if } K > 5 \times 10^5 \qquad (2.19a)$$

(b) If the chemical equation takes the forms

$$S + T \rightleftharpoons P$$
$$S + T \rightleftharpoons P + \text{solvent}$$

so that

$$K = \frac{[P]}{[S][T]}$$

then the innate titration error is estimated to be

$$\text{less than } 1\% \quad \text{if } c_S \cdot K > 3 \times 10^4$$
$$\text{less than } 0.25\% \quad \text{if } c_S \cdot K > 5 \times 10^5 \qquad (2.19b)$$

(c) If the chemical equation takes the forms

$$S + T \rightleftharpoons \text{solvent}$$
$$S + T \rightleftharpoons \text{insoluble product}$$

so that

$$K = \frac{1}{[S][T]}$$

then the innate titration error is estimated to be

$$\text{less than } 1\% \quad \text{if } c_S^2 \cdot K > 3 \times 10^4$$
$$\text{less than } 0.25\% \quad \text{if } c_S^2 \cdot K > 5 \times 10^5 \qquad (2.19c)$$

It should be understood that the numbers given above ($>3 \times 10^4$ for 1 percent and $>5 \times 10^5$ for 0.25 percent) are *typical values only*. In actual cases there will be specific variations depending on the sensitivity of the endpoint indicator. However, if the estimate of the innate error is less than 0.25 percent, then the titration can be recommended with confidence; and if it is greater than 1 percent, the titration is likely to be difficult.

Actual examples given earlier in this chapter are:

For case 3(a), Eqs. (2.17) and (2.18).
For case 3(b), Eqs. (2.2) and (2.3).
For case 3(c), Eqs. (2.4) and (2.5).

It is left as an exercise to the reader to examine the innate error in these titrations on the basis of the formulas given. Note that in titrations (2.2) and (2.4), the error will vary with c_S.

BUFFERS

As any experienced chemist knows, it is extremely difficult to make up accurate solutions if the solute concentration must be less than about 10^{-5} molar. After all, one milliliter of a 10^{-5} molar solution contains only 10 *nanomole* (10×10^{-9} mole) of solute. At that concentration it does not take much reaction with impurities in the solvent or dust in the flask or with reactants in the atmosphere to destroy the solute and cause major error. Indeed, even when chemical purity is rigorously maintained, there are problems because containers have surfaces, and no surface is truly inert. For instance, if the molecular weight of the solute is around 100, one cm^2 of smooth surface can remove as much as one nanomole of solute just by physical adsorption, without there being any formal chemical reaction.

To protect a low solute concentration from change, one uses a chemical buffer. In a typical buffer, the solute at low concentration is brought into equilibrium with other solutes whose concentration is much higher.

For example, to buffer 1.0×10^{-5} M H_3O^+ in water, one might use acetic acid and acetate ion and establish the equilibrium

$$HAc + H_2O \; \rightleftharpoons \; H_3O^+ + Ac^-$$

The hydrogen ion concentration is then related to the concentration of acetic acid and acetate ion according to Eq. (2.20), which is simply a rearranged version of the mathematical expression for K_A.

$$[H_3O^+] = K_A \frac{[HAc]}{[Ac^-]} \tag{2.20}$$

In water at 25°C, $K_A = 1.8 \times 10^{-5}$. Thus, for $[H_3O^+]$ to be 1.0×10^{-5} M, $[HAc]/[Ac^-]$ must be 1/1.8. Note that $[H_3O^+]$ depends on the *ratio* of $[HAc]$ to $[Ac^-]$.

To show how the buffer works, let us use 0.10 M HAc and 0.18 M Ac$^-$ to obtain 1.0×10^{-5} M H_3O^+ at equilibrium. One milliliter of the buffered solution therefore contains 10 nanomole (0.01 micromole) of H_3O^+, 100 micromole of HAc, and 180 micromole of Ac$^-$. Now suppose that a speck of dust containing 0.01 micromole of alkali falls into that solution. If the solution were not buffered, the alkali would be just equivalent to the H_3O^+ present, and the hydrogen ion would be neutralized. But because the solution is buffered, we may assume that the alkali reacts with the overwhelmingly larger amount of HAc. The ratio [HAc]/[Ac$^-$] therefore changes from 100/180 to 99.99/180.01, or by 0.016 percent. In view of Eq. (2.20), [H_3O^+] changes proportionately. Thus, instead of being wiped out, the hydrogen ion changes only by 0.016 percent.

The most important use of buffers is to maintain low concentrations of specific solutes in chemical reactions. In order for a reaction to be sufficiently complete at equilibrium, it may be necessary to maintain a product at a very low concentration. For example, a standard method for determining the concentration of a triiodide solution is by titration with arsenious acid whose anhydride, arsenious oxide (As_2O_3), is available in high purity as a primary oxidimetric standard. The desired reaction takes place in aqueous solution and is shown in Eq. (2.21).

$$H_3AsO_3 + I_3^- + 4\,H_2O \;\rightleftharpoons\; H_2AsO_4^- + 3\,I^- + 3\,H_3O^+ \quad (2.21)$$

However, the equilibrium constant is unfavorably small, 1.0×10^{-3}, and calculation shows that quantitative results can be obtained only if the hydrogen ion concentration is kept below $10^{-4}\,M$. In practice, one titrates in the presence of a buffer consisting of *ca.* 0.2 M bicarbonate and dissolved CO_2. As reaction proceeds, the liberated H_3O^+ reacts with bicarbonate, and CO_2 bubbles out of the saturated solution.

$$H_3O^+ + HCO_3^- \;\longrightarrow\; CO_2(g) + 2\,H_2O$$

Thus the hydrogen ion concentration is kept at a low value.

To see why the pH remains buffered, we recall that CO_2 in aqueous solution exists in equilibrium with carbonic acid and this acts as a typical weak dibasic acid.

$$CO_2 + H_2O \;\rightleftharpoons\; H_2CO_3$$

The ratio of H_2CO_3 to CO_2 at equilibrium is small (about $\frac{1}{300}$ at 25°C), and we shall therefore write [CO_2] to denote the total molar concentration of CO_2 plus H_2CO_3. The hydrogen ion concentration in a solution containing both CO_2 and bicarbonate is then given by (2.22), where K_{A1} denotes the first acid dissociation constant.

$$[H_3O^+] = \frac{[CO_2]}{[HCO_3^-]} \times K_{A1} \text{ (for aqueous } CO_2) \quad (2.22)$$

The solubility of CO_2 in water at 1 atm is 0.034 M, and $K_{A1} = 4.6 \times 10^{-7}$. The bicarbonate concentration decreases during the titration, say from 0.2 M at the beginning of the titration to 0.1 M at the end. The hydrogen ion concentration therefore varies between an initial value of

$$\frac{0.034}{0.2} \times 4.6 \times 10^{-7} = 0.8 \times 10^{-8}\,M$$

and a final value of

$$\frac{0.034}{0.1} \times 4.6 \times 10^{-7} = 1.6 \times 10^{-7} \ M$$

It is convenient to express hydrogen ion concentrations in units of pH, defined for aqueous solutions in (2.23).

$$pH = -\log \text{[solvated hydrogen ion]}$$
$$= -\log [H_3O^+] \qquad \text{for aqueous solutions} \qquad (2.23)$$

The symbol p in pH is a mathematical operator and stands for $-\log$. It enjoys wide use in chemistry. For instance, $pK = -\log K$; $pCl = -\log [Cl^-]$.

Any small concentration can be buffered if the given solute can be brought into equilibrium with relatively large amounts of other substances, either in the same solution or in a separate phase. For example, reaction (2.24) can be used to buffer the chloride concentration:

$$AgCl(s) \ \rightleftharpoons \ Ag^+(aq) + Cl^-(aq)$$

$$[Cl^-] = \frac{K_{SP}}{[Ag^+]}; \qquad K_{SP} = 1.0 \times 10^{-10} \qquad (2.24)$$

Thus, an $0.1 \ F$ solution of silver nitrate in contact with solid silver chloride can be regarded as a buffer that maintains the chloride concentration at $10^{-9} \ M$.

In choosing a reaction for buffering, one must choose a reaction with the right equilibrium constant. For instance, pH buffers normally consist of a weak acid (HA) and its conjugate base (A^-), so that $[H_3O^+] = K_A[HA]/[A^-]$. However, there are practical limits on the concentrations of HA and A^-: Neither concentration should be less than about $0.001 \ M$, lest the buffer capacity become too small; and neither concentration should be greater than about $0.2 \ M$, lest the solution become too concentrated. The ratio of these limits is 0.2/0.001, or 200. Hence $[HA]/[A^-]$ can vary between 200 and 1/200, and $[H_3O^+]$ can vary between $200K_A$ and $K_A/200$. Ideally, for best buffering, $[HA]/[A^-]$ should be near unity, so that the best buffer is an acid-base pair for which K_A is approximately equal to the desired $[H_3O^+]$. In other words, to maintain a pH of 6.0, pK_A of the buffer should be near 6.0. Of course, there are other considerations, too. The buffer must be chemically compatible with other substances in the solution.

PSEUDO-EQUILIBRIUM CONSTANTS

In the traditional method of formulating equilibrium constants, substances whose mass-action effect is *necessarily* constant are not written explicitly. Thus, Eq. (2.2) does not include the water concentration, because water is the solvent and the solutions are dilute. Equation (2.4) does not include the concentration of silver bromide, because silver bromide is present as a pure solid.

In some problems it is desirable also to take out of the mass-action expression those substances whose concentration is kept constant deliberately by means of a buffer. We shall call the constant value of the remaining mass-action expression a *pseudo-equilibrium constant* (or *conditional* equilibrium constant) and denote it by K_ψ. For instance, if in Eq. (2.21) the concentration of H_3O^+ is maintained constant at 1.6×10^{-7} M by means of a buffer, then it is desirable to introduce K_ψ as in (2.25)

$$K_\psi = \frac{[H_2AsO_4^-][I^-]^3}{[H_3AsO_3][I_3^-]} = \frac{K}{[H_3O^+]^3} = \frac{1.0 \times 10^{-3}}{(1.6 \times 10^{-7})^3} = 2.4 \times 10^{17} \quad (2.25)$$

Thus, although K itself is small, the large value obtained for K_ψ reflects correctly the large tendency for the reaction to proceed at that pH.

PROBLEMS

1. For each of the following (aqueous) titration reactions:
  (i) Write a balanced chemical equation, showing one mole of substrate.
  (ii) Formulate an expression for the equilibrium constant.
  (iii) Evaluate the equilibrium constant; use numerical data listed in the Appendix.
  (a) Titration of hydrochloric acid (the substrate) with sodium hydroxide
  (b) Titration of benzoic acid with ammonia (ammonium hydroxide)
  (c) Oxidation of ferrous ion with ceric ion (Ce^{4+})
  (d) Precipitation of sulfate ion with Ba^{2+}
  (e) Oxidation of bromide ion by permanganate in acid (Assume that the products are Br_2 and Mn^{2+}.)
Are all of these titrations likely to be successful? Consider the innate titration errors.

2. Derive an expression for the equilibrium constant for $3\ H_3O^+ + PO_4^{3-} \rightleftharpoons H_3PO_4 + 3\ H_2O$ in terms of the three acid dissociation constants of phosphoric acid.

3. What would be the effect on the value of the equilibrium constant if the balanced chemical equation were written in terms of *two moles* of substrate? What would happen if the equation contained one mole of *titrant?* How would these changes (from our normal practice of having one mole of substrate in the balanced equation) affect the $E°$ value for a redox titration?

4. You have just received your grade for an experiment in which you titrated HCl with sodium hydroxide. It is an "F." Your reported results were 0.1100 N, 0.1103 N, and 0.1102 N for three trials. What do you suppose might have gone wrong? What would you do before repeating the experiment?

5. Four replicate determinations of the normality of an HCl solution give the following results: 0.1005 N, 0.1019 N, 0.1020 N, 0.1008 N. Calculate the mean and the mean percentage deviation (% m.d.). Would you report the mean of all four values to your instructor? If not, which of the four values would you reject?

6. Your laboratory partner only did one titration of an unknown, reported the same number three times, and got an "A." Do you think he will get an "A" next time? Why?

7. An early chemist named VanHelmont did the following experiment to "prove" that water could be turned into solid material. He planted a five pound willow tree in 200 pounds of dry earth. After fifteen years during which the tree was fed only water, the willow's weight had increased to 169 pounds while the earth lost only two ounces. Thus, according to VanHelmont, water had been converted into 164 pounds of tree. Criticize this experiment in terms of its intent and approach to the problem, its analytical validity, and the interpretation of results. What type of errors do you think might be involved? Design an experiment which would convince you that water could (or could not) be turned into a solid (other than ice).

8. Perform the following calculations expressing only *significant* figures in your answer. Assume that the last figure in each number has an uncertainty of ±1.

(a) $\begin{array}{r} 2.13 \\ +\ 53.2 \\ +115.317 \\ \hline \end{array}$
(b) $\begin{array}{r} 0.0001 \\ +1.0 \\ \hline \end{array}$
(c) $\begin{array}{r} 0.13 \\ -0.001 \\ \hline \end{array}$

(d) 3.14×2.0

(e) $\dfrac{21.6 \times 1.01 \times 10^4}{2.315}$

9. Sometimes one would like to have a more precise estimation of the uncertainty of a determination than is afforded by just knowing which figures are significant. It then becomes necessary to keep a close account of the actual uncertainty after each step of the calculation, up to the final answer. Suppose that a rectangular solid whose "true" dimensions are height $(H) = 24.3$ cm, length $(L) = 19.3$ cm, width $(W) = 10.2$ cm is measured incorrectly as having $H = 24.0$ cm, $L = 19.9$ cm, $W = 10.0$ cm. Calculate the error, both absolute and relative, if these values are used to determine: (a) the sum of all twelve edges of the rectangular solid, and (b) its volume.

10. Analytical precision of ±0.25 percent, or better than three parts per thousand, is obtainable in a titration under proper conditions. Estimate the absolute and relative error in each step of a titration: pipetting of substrate into a beaker, preparation of titrant solution from a solid primary standard, and titration from a buret. You may assume that 50.00 ml of substrate is being titrated with approximately the same amount of titrant and that the endpoint is very sharply defined. Which operations would you expect to be most subject to error? Why? You may want to examine the equipment in your laboratory desk before deciding on your answer.

11. (a) How would you prepare a benzoic acid-sodium benzoate buffer of $pH = 5$?
(b) How would you prepare an ammonia-ammonium chloride buffer of $pH = 10$?

12. Describe how you would prepare an acetate buffer of $pH = 5.45$ starting with two solutions, one containing 0.1 F NaOH and the other containing 0.05 F HAc.

13. Consult Table A.5 of the Appendix and, by adding half-reactions, find a redox reaction that you would expect to proceed from right to left at low pH and from left to right at high pH. Calculate the pseudo-equilibrium constant at pH 1.00 and pH 8.00 for this reaction.

chapter 3

Titration of
Complex Substrates
and Mixtures

Complexity exists in chemistry both on the stoichiometric and on the molecular level. Stoichiometric complexity implies an unwanted wealth of *components*. For instance, a mixture of sodium nitrate and sodium chloride can be analyzed simply by titration for chloride, while a mixture of sodium and potassium nitrates and chlorides requires a much more involved analytical procedure. Molecular complexity implies an unwanted wealth of *molecular species*. For instance, acetic acid in water has been represented as an equilibrium mixture of at least four species: un-ionized HAc, H_3O^+, Ac^-, and acetic acid dimer $(HAc)_2$.

Liquid solutions are often complex on the molecular level. The number of molecular species is nearly always greater than the number of components, and many liquid solutions give evidence for the presence of so many species that their molecular composition is only vaguely understood. Fortunately for quantitative chemistry, molecular complexity need not imply stoichiometric complexity. For instance, cupric sulfate in aqueous solution is an equilibrium mixture consisting of at least three copper(II) species: hydrated $Cu(OH_2)_4{}^{2+}$ ions, and two distinct species of hydrated $CuSO_4$ molecules or ion pairs. Yet in the presence of potassium iodide in dilute sulfuric acid, the entire copper(II) component is reduced to cuprous iodide.

$$2\,CuSO_4(aq) + 4\,KI(aq) \xrightarrow[\text{dilute H}_2\text{SO}_4]{\text{excess KI}}$$

$$Cu_2I_2(s) + I_2(aq) + 2\,K_2SO_4(aq) \quad (3.1)$$

The clean stoichiometry of this reaction is indicated by Eq. (3.1). The reaction is often used to determine copper(II), since the liberated iodine can be titrated quantitatively with thiosulfate.

The chemist's most reliable tool for stripping away complexity and displaying the essence of a problem is the chemical equation. Philosophically speaking, no chemical reaction is ever so simple that it can be represented in all its aspects by a single chemical equation: We write different equations to represent different aspects. Consider again our friend, the reaction of acetic acid with sodium hydroxide in aqueous solution. If we wish to represent the formal reactants and the stoichiometry, we write the *stoichiometric* or *formal* equation [Eq. (3.2)].

$$HAc(aq) + NaOH(aq) \rightleftharpoons NaAc(aq) + HOH \quad (3.2)$$

If we wish to show the principal molecular species formed from the reactants and products in aqueous solution, we write the *molecular* equation, as follows.

$$HAc + OH^- \rightleftharpoons Ac^- + HOH \quad (3.3)$$

Still other molecular equations may be needed to depict the reaction mechanism or to describe the presence of other molecular species that are in equilibrium with the principal reactants. Thus, if we wish to account for the pH of the reaction mixture, we write Eq. (3.4).

$$HAc + HOH \rightleftharpoons H_3O^+ + Ac^- \quad (3.4)$$

Since each chemical equation depicts only one aspect of the whole truth, the chemical equation is always, in some degree, an approximation.

In this chapter we wish to discuss chemical complexity as it arises in titration, and how to cope with it through the formulation of suitable chemical equations.

COMPLEXITY ON THE MOLECULAR LEVEL

Reactions that are complex on the molecular level but proceed with simple stoichiometry are often represented by stoichiometric or formal equations. Examples of such representations are Eqs. (3.1) and (3.2). The stoichiometric equation is a balanced equation showing the conventional chemical formula and (in parentheses) the physical state for each reactant and product. The conventional formula is the chemical formula by which the substance is commonly known. For nonelectrolytes it is usually the formula of the vapor under ideal-gas conditions; for electrolytes it is usually the simplest formula. The stoichiometric equation is sufficient for all problems involving stoichiometry, including the calculation of formal concentration from normality as measured by titration. However, if principles of chemical equilibrium are to be applied, the stoichiometric equation is inadequate. We then need molecular equations.

Principal molecular species

Whenever a solute component reacts with simple stoichiometry, then *all* the molecular species that comprise the component are converted to product. To give a complete description of the reaction on the molecular level, we should therefore write a number of simultaneous molecular equations, one for each reacting species. However, for many practical purposes such a description is unnecessarily complex, and we make the following simplifying approximations.

1. We represent the reaction by a single molecular equation in which each reactant and product is represented by its *principal molecular species* under the conditions of the reaction. For solutes that are complex on the molecular level, the principal species is the *most probable* molecular species, that is, the molecular species that accounts for the largest fraction of the solute component.

2. We express the equilibrium constant for the reaction in terms of principal molecular species. In numerical calculations that require molar concentrations, we proceed as if each solute existed entirely in the form of its principal species.

We repeat, for emphasis, that the conclusions reached on this basis can be of approximate validity only. However, the approximation is often entirely adequate, especially in considerations involving the innate error of titration. And the conceptual and mathematical simplification that results will more than make up for any loss in scientific accuracy.

In many cases, the principal molecular species is strongly prevalent and accounts for nearly all of the solute component. In those cases, the approximation of letting the principal species represent the entire solute component is excellent. A familiar example is Eq. (3.3). To give a less familiar example, in the electro-gravimetric determination of nickel, metallic nickel is plated onto a platinum cathode from an aqueous ammonia-ammonium sulfate buffer in which the principal nickel species is $Ni(NH_3)_6^{2+}$. The electrolytic reduction is therefore represented by the molecular equation (3.5).

$$Ni(NH_3)_6^{2+} + 2\,e^- \;\rightleftharpoons\; Ni(s) + 6\,NH_3 \qquad (3.5)$$

However, even though only one equation is written, it is understood that $Ni(NH_3)_6^{2+}$ is merely the principal species of an equilibrium mixture. Other species that may be present are partially hydrated or sulfated nickel-ammonia complexes, such as $Ni(NH_3)_5OH_2^{2+}$ or $Ni(NH_3)_5SO_4$, and ion pairs with sulfate ion, such as $Ni(NH_3)_6^{2+}SO_4^{2-}$.

The approximation of letting the principal species represent the entire solute component becomes less good as that species becomes less prevalent, and it becomes impractical if the principal species is ill-defined or unknown. In that case, one is obliged to write an equation that is at least partially formal or stoichiometric. The problem arises most often in the case of metal ions in aqueous solution. For example, the titration of iron(II) with ceric ammonium sulfate in aqueous acid is usually represented either by (3.6) or by (3.7).

$$Fe(II) + Ce(IV) \rightleftharpoons Fe(III) + Ce(III) \tag{3.6}$$

$$Fe^{2+} + Ce^{4+} \rightleftharpoons Fe^{3+} + Ce^{3+} \tag{3.7}$$

In this case, the microscopic equilibria involving the metal ions are very complex, and the principal species vary with the nature of the mineral acid and with the concentrations of the solute components. The format shown in (3.6), with each oxidation state denoted by a Roman numeral following the formula of the element, conforms to the nomenclature adopted by the International Union of Pure and Applied Chemistry. The format shown in (3.7), with the oxidation state shown as a quasi-charge, simplifies the balancing of oxidation-reduction equations and is often used in textbooks. The reader is expected to assume that the principal species of a divalent or higher-valent metal ion is not likely to be the bare ion, so that the bare ion that appears in the chemical equation must denote the *formal* solute component.

In writing chemical equations it is often advisable to be inconsistent, representing some reactants as principal molecular species and others as formal solutes, in order to give a maximum amount of information about the molecular species. For example, the precipitation of indium(III) by potassium ferrocyanide from dilute aqueous acid may be represented by (3.8), where In^{3+} denotes the formal solute and $H_2Fe(CN)_6^{2-}$ denotes the principal molecular species.

$$4\,In^{3+} + 3\,H_2Fe(CN)_6^{2-} + 6\,H_2O \rightleftharpoons In_4[Fe(CN)_6]_3(s) + 6\,H_3O^+$$

$$\tag{3.8}$$

Change of principal species

Complexity on the molecular level is often desirable: By changing the conditions under which the solute exists we may be able to change the principal species. A change of principal species is often accompanied by a marked, perhaps even dramatic, effect on chemical equilibrium or reaction rate.

For example, the principal species of silver(I) in water, the hydrated ion $Ag(OH_2)_2^+$, is converted upon addition of ammonia to $Ag(NH_3)_2^+$, with the well-known result that water-insoluble silver salts such as AgCl become quite soluble. We represent these facts by writing different molecular equations.

Solubility of AgCl in pure water:

$$AgCl(s) + 2\,H_2O \; \rightleftharpoons \; Ag(OH_2)_2^+ + Cl^- \tag{3.9}$$

Solubility of AgCl in dilute aqueous NH_3:

$$AgCl(s) + 2\,NH_3 \; \rightleftharpoons \; Ag(NH_3)_2^+ + Cl^- \tag{3.10}$$

Since a change in principal species will occur only if the new principal species is more stable than the old, equilibrium in a chemical reaction will always shift so as to favor the formation of the substance with the new principal species.

When principal species on both sides of the chemical equation are affected by a change in conditions, then equilibrium will shift so as to favor that side of the equation on which the stabilization of molecular species is more effective. For example, the strength of higher-valent metal ions such as cerium(IV) or copper(II) as oxidizing agents can be affected markedly by complex-ion formation.

$$Ce(IV) + e^- \; \rightleftharpoons \; Ce(III) \tag{3.11}$$

$$Cu(II) + e^- \; \rightleftharpoons \; Cu(I) \tag{3.12}$$

Even though both oxidation states of the metal ion tend to form complex ions, the higher oxidation state usually forms the stronger complexes. Addition of a complexing agent will therefore shift the half-reaction equilibrium to the left and reduce the strength of the higher oxidation state as an oxidizing agent.

For instance, consider the half-reaction for the reduction of cerium(IV) in aqueous acid. Owing to the great complexity, on the molecular level, of ceric and cerous salts in aqueous solution, it is advisable to base the discussion on the formal equation (3-11), and to consider the *formal* standard reduction potential, E_{red}°, which is computed using the formal concentrations of Ce(IV) and Ce(III) rather than the molar concentrations of free hydrated ceric and cerous ions. Values obtained for E_{red}° in various 1 N acids are as follows:

$$\text{In } 1\,N\ HClO_4\!: \quad E_{red}^{\circ} = 1.70\text{ V}$$

$$\text{In } 1\,N\ HNO_3\!: \quad E_{red}^{\circ} = 1.61\text{ V}$$

$$\text{In } 1\,N\ H_2SO_4\!: \quad E_{red}^{\circ} = 1.44\text{ V}$$

Evidently, the strength of Ce(IV) as an oxidizing agent varies substantially with the nature of the acid, being greatest in perchloric acid. Qualitatively, the sequence obtained for E_{red}° is consistent with the observation that the acid anions interact with the cerium cations: ClO_4^- has the least, perhaps even a negligible, tendency to associate with cations in aqueous solution; and SO_4^{2-} has a considerable tendency to form complex ions.

Effect of pH. Principal species diagrams

The best known method for changing the principal species of a solute is to change the pH so as to convert the solute to its conjugate acid or base. The degree of conversion can be calculated from the acid dissociation constant and the hydrogen ion concentration according to (3.13), where HA denotes the protonated form of the substrate.

$$\frac{[HA]}{[A^-]} = \frac{[H_3O^+]}{K_A} \tag{3.13}$$

The principal species will be either HA or A^-, whichever of the two has the higher concentration. On applying (3.13), we therefore derive the inequalities (3.14).

If $[H_3O^+] > K_A$, $[HA] > [A^-]$,   and HA is the principal species.

If $[H_3O^+] < K_A$, $[HA] < [A^-]$,   and A^- is the principal species.

$$\tag{3.14}$$

It is convenient to state these relationships in terms of pH and pK_A. Since $pH = -\log[H_3O^+]$ and $pK_A = -\log K_A$, the signs of the inequalities become reversed, as in (3.15).

If $pH < pK_A$, $[HA] > [A^-]$,   and HA is the principal species.

If $pH > pK_A$, $[HA] < [A^-]$,   and A^- is the principal species.

$$\tag{3.15}$$

The relationship of principal species to pH can be visualized by means of *principal species diagrams*. Thus, in Fig. 3.1, pH is plotted along the x-axis, and the pK_A value of the acid is shown as a vertical line placed at $pH = pK_A$. According to (3.15), the vertical line then divides the pH scale into two regions in which either HA or A^- is the principal species.

There are many reactions that are not primarily acid-base reactions, but in which a reactant also happens to be a Brønsted acid or base. Equilibrium in such reactions tends to be quite sensitive to conversion of that reactant to its conjugate acid or base. The conjugate base is likely to be a stronger reducing agent, a stronger Lewis base, and a better ligand in complex-ion formation. If both acid and conjugate base are anions (for example, HCO_3^- and CO_3^{2-}), the salts of the conjugate base are usually less soluble. Let us consider some examples.

Solubility of silver chromate. Figure 3.1(b) shows that the principal species of chromium(VI) in dilute aqueous solution changes from CrO_4^{2-}

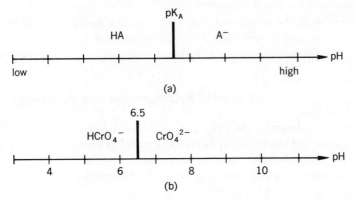

Fig. 3.1. Principal species diagrams for (a) a weak acid and (b) chromium(VI) in dilute aqueous solution: pK_A for $HCrO_4^-$ is 6.5.

to $HCrO_4^-$ at pH 6.5. Silver chromate, Ag_2CrO_4, is quite insoluble in water; $K_{SP} = 1.1 \times 10^{-12}$ (M^3). Silver hydrogen chromate, $AgHCrO_4$, on the other hand, is quite soluble, and the solid phase in equilibrium with a saturated solution continues to be Ag_2CrO_4 down to a pH of at least 5. We therefore write the following equations to represent the solubility:

$$Ag_2CrO_4(s) \;\rightleftharpoons\; 2\,Ag^+ + CrO_4^{2-}; \quad pH > 6.5 \qquad (3.16)$$

$$H_3O^+ + Ag_2CrO_4(s) \;\rightleftharpoons\; 2\,Ag^+ + HCrO_4^- + H_2O;$$
$$pH < 6.5 \text{ (valid at least to } pH \text{ 5)} \quad (3.17)$$

According to this representation, the solubility of silver chromate is independent of pH above pH 6.5 and increases with the hydrogen ion concentration below pH 6.5.

Do Eqs. (3.16) and (3.17) imply that the nature of the reaction changes abruptly at pH 6.5? No, they do not! They do imply, however, that Eq. (3.16) is the better approximation to the truth above pH 6.5, while Eq. (3.17) is the better approximation below pH 6.5. For a still better approximation, we must consider both reactions simultaneously. Just how that is done will be described in Chapter 7.

Oxidation of arsenic(III). For our next example, we wish to consider the oxidation of arsenic(III) to arsenic(V), as a function of pH. Arsenic(III) in water is a dibasic acid with the following properties.

Molecular species of arsenic(III)	K_A	pK_A	pH range in which this is the principal species
H_3AsO_3	6×10^{-10}	9.2	Below pH 9.2
$H_2AsO_3^-$	3×10^{-14}	13.5	Between pH 9.2 and 13.5
$HAsO_3^{2-}$	—	—	Above pH 13.5

Arsenic(V) is a tribasic acid in water, with the following properties.

Molecular species of arsenic(V)	K_A	pK_A	pH range in which this is the principal species
H_3AsO_4	6×10^{-3}	2.2	Below pH 2.2
$H_2AsO_4^-$	1.0×10^{-7}	7.0	Between pH 2.2 and 7.0
$HAsO_4^{2-}$	3×10^{-12}	11.5	Between pH 7.0 and 11.5
AsO_4^{3-}	—	—	Above pH 11.5

These properties are displayed in the principal species diagrams shown in Fig. 3.2. The diagrams are analogous to Fig. 3.1 but contain several vertical lines on the pH-axis, one line for each acidic hydrogen in the molecule. The vertical lines are placed at $pH = pK_{A1}, pK_{A2}, \ldots, pK_{An}$, where $K_{A1}, K_{A2}, \ldots, K_{An}$ are the successive acid-dissociation constants.

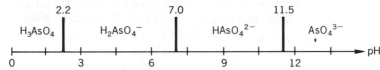

Principal-species diagram for arsenic(V)

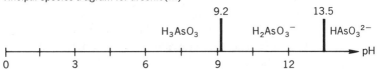

Principal-species diagram for arsenic(III)

Fig. 3.2. Acid-base behavior of arsenic(V) and arsenic(III) in aqueous solution.

Using Fig. 3.2, we can tell at a glance which are the principal species involved in the oxidation of arsenic(III) to arsenic(V) at a given pH. Thus, the balanced equations for the half-reaction at pH 4, 8, and 12 are as follows:

At pH 4: $H_3AsO_3 + 4\,H_2O \rightleftharpoons H_2AsO_4^- + 3\,H_3O^+ + 2\,e^-$ (3.18)

At pH 8: $H_3AsO_3 + 5\,H_2O \rightleftharpoons HAsO_4^{2-} + 4\,H_3O^+ + 2\,e^-$ (3.19)

At pH 12: $H_2AsO_3^- + 5\,H_2O \rightleftharpoons AsO_4^{3-} + 4\,H_3O^+ + 2\,e^-$ (3.20)

COMPLEXITY ON THE STOICHIOMETRIC LEVEL

In the day-by-day practice of quantitative analysis, the most challenging and difficult tasks often are those involving the analysis of mixtures. Ideally, one would like to analyze the original, intact mixture directly, because separation of the components tends to be laborious and entails loss of accuracy. One would also like to have specific methods of analysis for individual components, and some redundancy of information to prove the accuracy of the final result. (For instance, will the measured percentages add up to 100 percent?)

A common difficulty is that the mixture may contain two or more components whose chemical properties are so similar that the compounds all react with the same titrant. For instance, the mixture might contain several acids that react with sodium hydroxide, or several reducing agents that react with cerium(IV). We must then distinguish between *selective* titration and *nonselective* titration.

Selective titration is carried out so that the measurement is specific for either A or B. Selective titration is possible only if *one substrate has a much greater affinity for the titrant than the other*. Suppose that A has the greater affinity. Then, as titrant is added and the mixture comes to equilibrium after each addition, reaction will take place essentially in two stoichiometric steps. At first the titrant will react almost exclusively with the much more reactive substrate A, until the concentration of A is quite small. We shall call this the first step in the titration. After A has all but disappeared, what remains is essentially a solution of B, which may then be titrated in the second step of the titration. If the endpoints of both steps can be rendered visible, we can titrate the mixture selectively.

Nonselective titration of two components, A and B, is carried out under such conditions that *both* components react. At the equivalence point in the nonselective titration, the number of equivalents of titrant is equal to the sum of the equivalents of A and B. Nonselective titration results when the reactions of two or more substrates with the titrant have similar equilibrium constants. For instance, acetic acid ($pK_A = 4.76$) can be titrated with sodium hydroxide, and so can benzoic acid ($pK_A = 4.20$). As a result, any mixture of acetic and benzoic acids will be titrated nonselectively with sodium hydroxide, to measure the total number of equivalents of both acids. The reason for this will become clearer as we proceed.

In the following, we shall analyze the conditions under which selective, or stepwise, titration is feasible. (We shall use the terms "selective" and "stepwise" interchangeably for titrations with a single titrant.) For definiteness we shall select examples from the field of acid-base titration, but it should be realized that the underlying logic is of general validity. A more general treatment will be given in Chapter 8.

Conversion diagrams

The necessary conditions for stepwise titration can be shown graphically by means of conversion diagrams. Let us consider the reaction of a mixture of acetic acid ($pK_A = 4.75$) and phenol (carbolic acid; $pK_A = 10.00$) with sodium hydroxide in aqueous solution. To find out whether stepwise titration is feasible, we begin with some general ideas concerning the conversion of a weak acid HA to its conjugate base.

Let α denote the *degree of conversion* of a weak acid to its conjugate base, as defined in (3.21).

$$\alpha = \frac{[A^-]}{c_{HA}} = \frac{[A^-]}{[HA] + [A^-]} \tag{3.21}$$

Then, on dividing both numerator and denominator by $[A^-]$ and applying (3.13), we obtain (3.22).

$$\alpha = \frac{1}{1 + ([H_3O^+]/K_A)} \tag{3.22}$$

Equation (3.22), which expresses α as a function of $[H_3O^+]$ and K_A, is particularly useful for mixtures of acids. The hydrogen ion concentration can be measured conveniently with a pH meter, and the measured pH applies equally to the entire mixture: If several weak acids are present, all will be subject to the same pH. On the other hand, K_A is a characteristic constant for a given acid, regardless of whether or not other acids are present. For instance, K_A for acetic acid has the same value, 1.75×10^{-5} at 25°C, in pure water as in water containing 0.1 F phenol. Thus, if K_A is known, the simple measurement of pH will suffice to determine α.

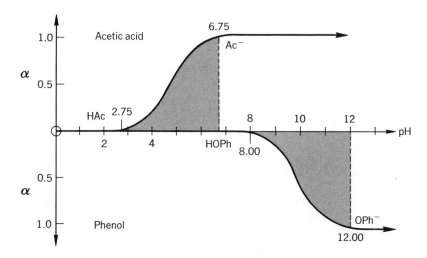

Fig. 3.3. Conversion diagram for a mixture of acetic acid and phenol.

Figure 3.3 shows graphs of α versus pH, constructed by solving Eq. (3.22) for a number of hydrogen ion concentrations and plotting the results. The curve for acetic acid is plotted in the conventional way, with α increasing above the pH-axis. However, for greater clarity, the curve for phenol (calculated with $K_A = 1.00 \times 10^{-10}$) is plotted so that α increases *below* the pH-axis. It is clear from the preceding discussion that these curves will apply not only to seperate aqueous solutions of the pure acids, but also to aqueous solutions of their mixtures.

Conversion diagrams such as those in Fig. 3.3 are useful for displaying the pH interval in which the weak acid is converted to its conjugate base. If we aim at 1 percent accuracy, we may say that conversion is negligible if $\alpha < 0.01$, and that conversion is practically complete if $\alpha > 0.99$. The *conversion interval* then is defined as the pH range in which α increases from 0.01 to 0.99. For acetic acid, this change takes place between pH 2.75 and 6.75, as shown by the shaded area in the upper part of Fig. 3.3. For phenol, this change takes place between pH 8.00 and 12.00, as shown by the shaded area in the lower part of Fig. 3.3. Note that the portion of the pH scale between 6.75 and 8.00 belongs to neither conversion interval.

Now let us suppose that we are monitoring the pH during the titration of a mixture of acetic acid and phenol with sodium hydroxide. As titrant is added a few drops at a time, the pH of the solution at equilibrium increases after each addition, sometimes only slightly, sometimes substantially, but moving always towards higher values until the titration is stopped. The titration is therefore tantamount to a trip across the conversion diagram, from left to right. At first the trip will go across the conversion interval for acetic acid. Here the pH will change rather slowly as the sodium hydroxide reacts with the acetic acid; you will recall that a solution containing considerable quantities of both acetic acid and acetate ion acts as a pH buffer. Next, the trip will go very quickly across the region (pH 6.75–8.00) between the two conversion intervals, where very little reaction takes place. And finally it will go across the conversion interval for phenol. If the titration is stopped between the two conversion intervals, that is, between pH 6.75 and 8.00, the amount of sodium hydroxide consumed will be equivalent to the amount of acetic acid present in the mixture.

Conditions for stepwise titration of mixtures

As a general rule, selective titration of the stronger of two acids (or bases) is possible only if the conversion intervals of the two substrates are well resolved. This condition can be shown, simply and clearly, by means of principal species diagrams, if we modify the diagrams used earlier (for instance, Figs. 3.1 and 3.2) to include conversion intervals.

On solving Eq. (3.22), we find that conversion intervals begin and end as follows:

$$\alpha = 0.01 \quad \text{when} \quad [H_3O^+] = 99 \cdot K_A$$

$$\alpha = 0.99 \quad \text{when} \quad [H_3O^+] = \frac{K_A}{99} \qquad (3.23)$$

On taking logarithms and expressing the result in units of pH and pK_A, we transform (3.23) into (3.24).

$$\alpha = 0.01 \quad \text{when} \quad pH = pK_A - 1.996$$

$$\alpha = 0.99 \quad \text{when} \quad pH = pK_A + 1.996 \tag{3.24}$$

Thus, in round numbers, the conversion interval of an acid extends on the pH scale from $(pK_A - 2)$ to $(pK_A + 2)$.

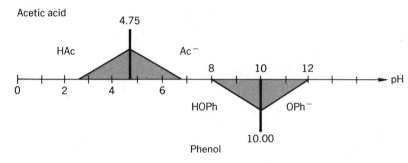

Fig. 3.4. Principal species diagram, showing conversion intervals, for acetic acid and phenol.

Figure 3.4 shows a principal species diagram for acetic acid (above the pH-axis; $pK_A = 4.75$) and phenol (below the pH-axis; $pK_A = 10.00$). As before, the changeover of principal species from acid to conjugate base is indicated by a vertical line placed at $pH = pK_A$. In addition, the conversion interval is indicated by an isosceles triangle whose apex is at pK_A, and which terminates 2 pH units on either side of pK_A. By means of this representation it becomes very clear that the two conversion intervals are well resolved, and that selective titration of acetic acid in the presence of phenol will be possible if the endpoint is taken in the pH region between the conversion intervals.

Figure 3.5 shows a similar principal species diagram for acetic acid and

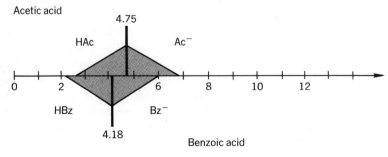

Fig. 3.5. Principal species diagram, showing conversion intervals, for acetic acid and benzoic acid.

benzoic acid. This time the conversion intervals overlap, showing that a mixture of these acids will not react with sodium hydroxide in discrete, well resolved steps. Selective titration is evidently not possible.

Mathematical analysis of selective titration, which is given explicitly in Chapter 8, shows that this simple picture is substantially correct. However, in addition to considering the overlapping, or nonoverlapping, of conversion intervals, it is necessary to consider also the relative concentrations of the two acids. Let us indicate the result: Let HX and HY denote two acids, let c_{HX} and c_{HY} denote their formal concentrations in the sample to be analyzed, and let K_A(for HX) be greater than K_A(for HY). The concentration c_{HX} of the stronger acid HX can then be determined by selective titration if the following condition (3.25) is satisfied.

Titration of HX in the presence of HY is possible if:

$$\frac{c_{HX}K_A\text{(for HX)}}{c_{HY}K_A\text{(for HY)}} \geqslant 3 \times 10^4 \qquad \text{(for 1\% accuracy)}$$

$$\geqslant 5 \times 10^5 \qquad \text{(for 0.25\% accuracy)} \qquad (3.25)$$

SUBSTRATES WITH SEVERAL REACTIVE SITES

When a substrate has two or more reactive sites, its reaction may follow either of two patterns. In the more common pattern, the sites appear to react in succession. For example, when phosphoric acid is titrated with sodium hydroxide, the principal phosphate species changes from H_3PO_4, to $H_2PO_4^-$, to HPO_4^{2-}, and (if the solution becomes sufficiently alkaline) to PO_4^{3-}. In the less common *cooperative* pattern, the reactive sites in the molecule appear to react together. For example, when silver ion in aqueous solution is titrated with ammonia, the principal silver species changes directly from $Ag(OH_2)_2^+$ to $Ag(NH_3)_2^+$. There is evidence that the intermediate complex, H_2O—Ag—NH_3^+, is formed during the titration, but it never becomes the principal species.

Reflecting (or governing) the two patterns of reaction are the stepwise equilibrium constants for the consecutive reaction steps. When the reactive sites in the substrate appear to react in succession, the stepwise equilibrium constants become progressively smaller. For example, the stepwise acid dissociation constants of phosphoric acid are in the sequence, $K_{A1} \gg K_{A2} \gg K_{A3}$. When the sites react cooperatively, the stepwise equilibrium constants become progressively greater, so that each reaction step enhances the tendency for the molecule to react further. For example, the equilibrium constant for the conversion of $Ag(OH_2)_2^+$ to H_2O—Ag—NH_3^+ is 2.0×10^3, while that for the subsequent conversion of H_2O—Ag—NH_3^+ to $Ag(NH_3)_2^+$ is 6.9×10^3.

When the reactive sites interact cooperatively, the only possible kind of titration is nonselective titration of the entire molecule. However, when the sites react in succession, titration may be either nonselective for the entire molecule or selective for a specific site or sites, the problem being very similar, logically, to that of titrating a mixture of substrates. We shall discuss it here briefly, using again some examples from the field of acid-base reactions. More detailed discussion will be given in Chapter 8.

Titration of polybasic acids

A polybasic acid can be titrated *nonselectively* if each of the stepwise acid dissociation constants is of such a magnitude as to permit titration. To derive the necessary conditions for *selective* or *stepwise* titration, it is again helpful to examine principal species diagrams.

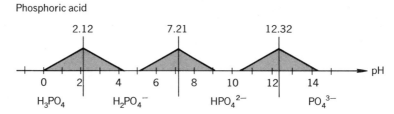

Fig. 3.6. Principal species diagram and conversion intervals for phosphoric acid in water.

For definiteness, Fig. 3.6 shows the principal species diagram for phosphoric acid in water. The pK_A values are, respectively, 2.12 for pK_{A1}, 7.21 for pK_{A2}, and 12.32 for pK_{A3}. We define the degree of conversion for the successive steps as follows.

$$\alpha_1 = \frac{[H_2PO_4^-]}{([H_3PO_4] + [H_2PO_4^-])}$$

$$\alpha_2 = \frac{[HPO_4^{2-}]}{([H_2PO_4^-] + [HPO_4^{2-}])}$$

$$\alpha_3 = \frac{[PO_4^{3-}]}{([HPO_4^{2-}] + [PO_4^{3-}])}$$

On introducing the defining equations for K_{A1}, K_{A2}, and K_{A3} [see Table 2.1 and the discussion following Eq. (2.13)], we then derive Eqs. (3.26) through (3.28), which are entirely analogous to (3.22).

$$\alpha_1 = \frac{1}{(1 + [H_3O^+]/K_{A1})} \qquad (3.26)$$

$$\alpha_2 = \frac{1}{(1 + [H_3O^+]/K_{A2})} \tag{3.27}$$

$$\alpha_3 = \frac{1}{(1 + [H_3O^+]/K_{A3})} \tag{3.28}$$

In general, for the ith step in the conversion of a polybasic acid to its conjugate base, we may write (3.29).

$$\alpha_i = \frac{1}{(1 + [H_3O^+]/K_{Ai})} \tag{3.29}$$

Next we define the conversion interval for the ith step as the pH interval in which α_i increases from 0.01 to 0.99. By analogy with Eqs. (3.23) and (3.24), we then find that the conversion interval for the ith step extends from $(pK_{Ai} - 2)$ to $(pK_{Ai} + 2)$. Conversion intervals obtained in this way are represented as isosceles triangles in Fig. 3.6.

The interpretation of Fig. 3.6 is entirely analogous to the earlier interpretation of Fig. 3.4 for a mixture of acids: Titration of phosphoric acid with sodium hydroxide is tantamount to a trip across the principal species diagram, from left to right. Since the three conversion intervals are well resolved, we conclude that proton removal from H_3PO_4 will take place in discrete stoichiometric steps, and titration can be stopped selectively after removal of one, or of two, protons. To stop the titration after removal of one proton, the pH at the endpoint should be between 4.12 and 5.21, that is, in the region where the conversion of H_3PO_4 to $H_2PO_4^-$ is already nearly complete but that of $H_2PO_4^-$ to HPO_4^{2-} has not really begun. Similarly, to stop the titration after removal of two protons, the pH at the endpoint should be between 9.21 and 10.32. Although Fig. 3.6 indicates that we should be able to titrate the third proton, the endpoint cannot be detected in aqueous solution since K_W/K_{A3}, which is the equilibrium constant for that titration, is too small.

To provide an example in which selective titration is not possible, Fig. 3.7 shows the principal species diagram for tartaric acid, a dibasic acid, in water. In this case, $pK_{A1} = 3.03$, $pK_{A2} = 4.54$, and the conversion intervals for removal of the first and second proton overlap strongly. We conclude, therefore, that proton removal from tartaric acid does *not* proceed

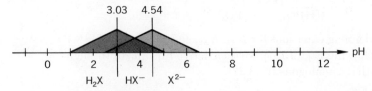

Fig. 3.7. Principal species diagrams and conversion intervals for tartaric acid in water.

in discrete stoichiometric steps. As the pH increases, during a titration with sodium hydroxide, removal of the second proton will become stoichiometrically significant above pH 2.54, well before removal of the first proton is even approximately complete.

The impression given by the principal species diagrams, that selective titration is possible only if the stepwise acid dissociation constants differ by at least four orders of magnitude, is consistent with mathematical analysis. The result, obtained in Chapter 8, is as follows.

Titration of a dibasic acid to remove selectively
> one proton is possible if:
> $pK_{A2} - pK_{A1} > 4.5$ (for 1% accuracy)
> $pK_{A2} - pK_{A1} > 5.7$ (for 0.25% accuracy)

Titration of a polybasic acid to remove selectively
> i protons is possible if:
> $pK_{A(i+1)} - pK_{Ai} > 4.5$ (for 1% accuracy)
> $pK_{A(i+1)} - pK_{Ai} > 5.7$ (for 0.25% accuracy)   (3.30)

CHEMICAL EQUATION FOR SELECTIVE TITRATION

In the earlier discussion of the selective titration of acetic acid in the presence of phenol (Figs. 3.3 and 3.4), it was necessary to consider the reaction of sodium hydroxide with both solutes. We now ask the question: Do we therefore need two chemical equations, (3.31) and (3.32), to describe that titration, or can we convey the essential information by means of a single equation?

$$HAc + OH^- \rightleftharpoons Ac^- + HOH \qquad (3.31)$$
$$HOPh + OH^- \rightleftharpoons OPh^- + HOH \qquad (3.32)$$

To answer that question, we must first consider the meaning of the term "titrant." In (3.31) and (3.32), the titrant is shown to be hydroxide ion, which is consistent with the fact that sodium hydroxide is actually used to titrate the solution. However, perhaps that approach is too naive. There are in chemistry many examples in which a substance A, on being added to a solution, generates a new substance B, and B in turn reacts with the substrate. Thus, even though A is actually added, we write an equation that shows the substrate as reacting with B.

In this section we shall give a stoichiometric criterion that enables us to identify the effective titrant when the stoichiometry is complex. We shall find that the effective titrant need not be the substance actually used to titrate the solution, and that selective titration can indeed be represented by a single chemical equation. In particular, we shall find that the titration of acetic acid with NaOH in the presence of phenol is represented by (3.33).

$$HAc + OPh^- \rightleftharpoons Ac^- + HOPh \tag{3.33}$$

Effective titrant

We begin with a somewhat idealized example. Let the titration be described by Eq. (3.34), which shall be accurate both on the molecular and on the stoichiometric levels.

$$Substrate + Titrant \rightleftharpoons Product \tag{3.34}$$

Assume further that reaction between substrate and titrant is quantitative, and that any dilution during the titration may be neglected. Figure 3.8

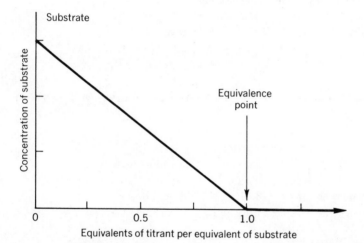

Fig. 3.8. Variation of substrate concentration during an idealized, quantitative, titration.

then shows the substrate concentration during the titration, and Fig. 3.9 shows the corresponding relationship for the titrant. These relationships are quite characteristic and serve to identify the effective substrate and titrant, as follows:

1. The *effective substrate* is that component whose concentration decreases continuously before the equivalence point, reaches practically zero* at the equivalence point, and remains at practically zero past the equivalence point.

---

* The term "practically zero" as used here is not the same as "exactly zero." It means merely that the concentration is too small to be distinguished from zero on the linear scale of titration curves such as Fig. 3.8 or Fig. 3.9.

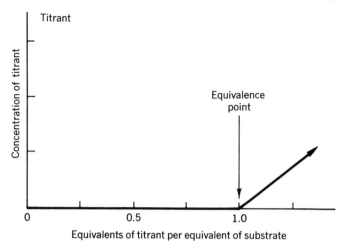

Fig. 3.9. Variation of titrant concentration during an idealized, quantitative, titration.

2. The *effective titrant* is that component whose concentration is practically zero up to and including the equivalence point, and then increases markedly.

3. The *effective equation for the titration* is the chemical equation for the reaction of the effective substrate with the effective titrant.

In any successful real titration there is always one solute whose concentration varies in the characteristic manner for the effective substrate, and a second solute whose concentration varies in the characteristic manner for the effective titrant. The effective titrant, so defined, need not be the reactant added from the buret. Let us explore the meaning of these definitions through a few examples.

Acetic acid plus sodium hydroxide

When a solution of pure acetic acid in pure water is titrated with sodium hydroxide, the hydroxide concentration varies during the titration as shown in Fig. 3.10: Before the equivalence point, the added hydroxide is almost completely destroyed by reaction with acetic acid; past the equivalence point, any further hydroxide is in excess, and the hydroxide concentration builds up. The close similarity of Fig. 3.10 to Fig. 3.8 shows that hydroxide ion is the effective titrant. The equation for the titration is therefore the familiar Eq. (3.35).

$$\text{HAc} + \text{OH}^- \;\rightleftharpoons\; \text{Ac}^- + \text{HOH} \qquad (3.35)$$

On the other hand, when a solution containing both acetic acid and phenol is titrated with sodium hydroxide, there are two Brønsted acids, HAc and HOPh, both of which are capable of reacting with hydroxide ion.

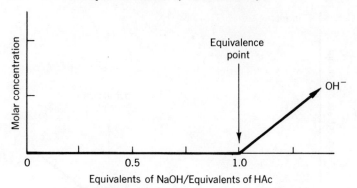

Fig. 3.10. Titration of pure acetic acid in pure water with aqueous sodium hydroxide.

We have seen that reaction takes place in two well resolved steps: HAc disappears first, and HOPh reacts next. As a result, the concentration of the two Brønsted bases, OH⁻ and OPh⁻, varies during the titration as shown in Fig. 3.11. The OH⁻ concentration is practically zero both before

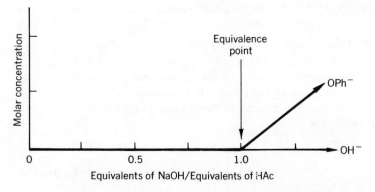

Fig. 3.11. Selective titration of acetic acid with sodium hydroxide in the presence of phenol.

and past the equivalence point, as hydroxide ion is destroyed, first by reaction with acetic acid, and then by reaction with phenol. By contrast, the concentration of OPh⁻ shows the characteristic shape by which we identify the effective titrant. It is practically zero up to the equivalence point and increases sharply thereafter. The effective equation for the selective titration of acetic acid in the presence of phenol is therefore (3.36), which is identical to (3.33).

$$HAc + OPh^- \ \rightleftharpoons \ Ac^- + HOPh \qquad (3.36)$$

As a corollary, *the equilibrium constant for this titration is precisely the same as if phenoxide ion instead of hydroxide ion had been added directly from the buret.*

Stepwise titration of phosphoric acid with sodium hydroxide

Reaction of H_3PO_4 with hydroxide ion takes place in well resolved steps, as indicated in Fig. 3.6. Accordingly, Fig. 3.12 shows the concentrations

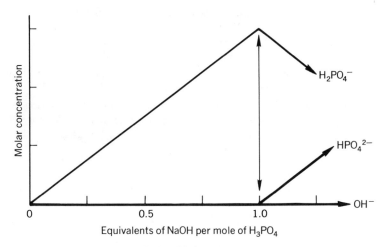

Fig. 3.12. Titration of phosphoric acid with sodium hydroxide in water.

$H_2PO_4^-$, HPO_4^{2-}, and OH^- during the first step, and during part of the second step, of the titration. Of the three relationships, only that for HPO_4^{2-} has the characteristic shape for the effective titrant around the first equivalence point. As a result, the effective equation for the selective titration of H_3PO_4 to the first equivalence point is (3.37).

$$H_3PO_4 + HPO_4^{2-} \rightleftharpoons 2\,H_2PO_4^- \qquad (3.37)$$

The equilibrium constant for this titration has already been discussed: It is given in (2.13) and is equal to K_{A1}/K_{A2}.

Effective titrant rule

The preceding examples may be generalized as follows:

If the reaction of a mixture of substrates (or of a substrate with two or more reactive sites) with a titrant takes place in two well resolved steps, then the effective titrant in the selective titration of the first step is identical to the product of the second step. (3.38)

This generalization is called the *effective titrant rule.* You should convince yourself that it is consistent with Eqs. (3.36) and (3.37), and practice applying it by working Problem 10 at the end of the chapter.

PROBLEMS

(*Note:* Values of K_A and K_B are given in Table A.1.)

1. Suppose you were given a solid unknown containing a mixture of the four electrolytes AX, AY, BX, and BY. Show that chemical analysis for the four ions A^+, B^+, X^-, and Y^- *does not* provide sufficient information to calculate the amounts of the four compounds in the solid sample.

2. Distinguish between *stoichiometric* and *molecular* chemical equations. Under what circumstances is the use of each to be preferred?

3. Write the correct *molecular equations* for the following (aqueous) acid-base titration reactions. You may want to refer to Table A.1 for help.

	Substrate	*Titrant*
(a)	Benzoic acid	Sodium hydroxide
(b)	Nitrous acid	Sodium hydroxide
(c)	Hydrochloric acid	Ammonia
(d)	Ammonia	Acetic acid
(e)	Triethylamine	Nitric acid
(f)	Phenol	Sodium hydroxide
(g)	Nitrous acid	Triethylamine
(h)	Aniline	Perchloric acid

4. Calculate equilibrium constants for each of the titrations listed in Problem 3. Indicate for each whether the titration is likely to be successful. (What are your criteria?)

5. Explain why the addition of dilute acid causes a tremendous increase in the solubility of Ag_3PO_4.

6. The reaction between cerous ion and permanganate ion

$$MnO_4^- + 5\ Ce(III) + 8\ H_3O^+ \longrightarrow Mn^{2+} + 5\ Ce(IV) + 12\ H_2O$$

does not occur to any appreciable extent in 1 N perchloric acid. The reaction does, however, proceed as written in 1 N H_2SO_4. Why? Calculate the equilibrium constant for the reaction under both sets of conditions. (Pertinent reduction potentials are listed in Table A.5.)

7. Draw principal species diagrams for the following substances in aqueous solution as the pH varies from 1 to 13. Indicate the conversion intervals. Which of the substances would you expect to undergo reactions other than simple acid-base reactions in this pH range? How would such additional reactions be indicated in the principal species diagrams?

(a)	Diethylamine	(c)	Citric acid
(b)	Tin(II)	(d)	Benzidine
		(e)	Boric acid

8. Calculate the degree of conversion, α, of nitrous acid to nitrite ion at $pH = 2.47$. What are the molar concentrations of HNO_2 and NO_2^- in a 0.200 F HNO_2 solution at this pH? At what pH is $[NO_2^-] = 2 [HNO_2]$?

9. Would you expect selective titration to be feasible for the following? What titrant would you use? How many well resolved equivalence points would result? (Assume 0.1 F solutions of each substrate and the titrant.)

(a) Carbonic acid
(b) Citric acid
(c) A mixture of benzoic acid and boric acid
(d) A mixture of phenol and oxalic acid
(e) A mixture of ethylamine, diethylamine, and triethylamine

10. Consider the following sets of titrations (assume that the solutions are 0.1 N in each reactant).

Substrate	Titrant
(a) Hydrazoic acid	NaOH
(b) Phosphorous acid	NaOH
(c) Ammonia + pyridine	HNO_3
(d) Formic acid + hydrogen sulfide	NaOH

For each titration construct diagrams analogous to Figs. 3.10, 3.11, and 3.12 in which you show *all* equivalence points. Indicate the *effective titrant* and write the correct molecular equation for each step. Calculate equilibrium constants for each reaction.

chapter 4

Volumetric Analysis
in Aqueous Solution

Until fairly recently, volumetric analysis—the science of chemical analysis by titration—was one of the major areas of chemistry. The chemical and pharmaceutical industries alone employed thousands of chemists to monitor the composition of reactant mixes and products, and many other analytical chemists were engaged in the development of better methods. Today, much of the routine analysis and product control is done by automatic machinery, and the function of volumetric analysis has changed accordingly: Although the science is still responsive to the needs of industry, its primary application is in chemical research.

There is, of course, a marked difference in the ways volumetric analysis is used in industry and in research. In industrial control, random samples of the reactants and products are analyzed at regular intervals, and one needs rapid, highly precise methods in order that significant deviations from the norm can be detected and quickly corrected. One also needs industry-wide standards of analysis so that the products of different factories can be compared on a uniform scale. By agreement, in the United States such standards are prescribed by the American Society for Testing Materials (ASTM).

In chemical research one encounters samples covering a wide range of often unpredictable composition. One therefore needs flexible methods whose underlying chemistry is well understood, and which can be adapted to a wide variety of substrates and conditions. One also needs methods of proven and verifiable accuracy for the chemical characterization of pure substances and the determination of equivalent weights.

In this chapter we shall describe a few of the most useful methods of volumetric analysis, and how they work in aqueous solution. In later chapters we shall broaden our horizon and consider nonaqueous solutions as well.

ACID-BASE TITRATION IN AQUEOUS SOLUTION

Acid-base reactions are so familiar that we need hardly review their chemistry here. Instead, we shall examine the "inner workings" of acid-base titrations, the built-in features that make titration feasible. Our purpose is to gain a preliminary perspective of the problem. Detailed mathematical analysis and derivation of all but the simplest relationships will be postponed to later chapters.

Acid-base strength and innate titration error

It is convenient to classify acid-base titrations in aqueous solution as follows.

1. Strong acid + strong base
2. Weak acid + strong base
   Weak base + strong acid
3. Weak acid + weak base

With rare exceptions, the reagent actually used to titrate the aqueous solution will be a strong base or a strong acid. Nonselective titrations therefore usually fit into classes (1) or (2). In selective titrations, on the other hand, the *effective* titrant will be a weak base or a weak acid, and such titrations often fit into class (3). For our present purpose, it will be sufficient if we consider titrations of class (1) and (2) only. Titrations belonging to class (3) are basically similar and will be considered in Chapter 8.

The reaction of a strong acid with a strong base in aqueous solution is essentially that of hydrogen ion with hydroxide ion, Eq. (4.1), and the equilibrium constant for titration is therefore equal to $1/K_W$, or 1.0×10^{14} (M^{-2}) at 25°C.

$$H_3O^+ + OH^- \rightleftharpoons 2\,H_2O \tag{4.1}$$

Equation (4.1) is of the correct stoichiometry to fit case 3(c) in Chapter 2 (p. 38), and the innate titration error can be estimated from (2.19c). We thus find that the innate error decreases with increasing substrate concentration and becomes less than 0.25 percent when $c_S > 0.7 \times 10^{-4}\,F$. Since titration is rarely attempted if c_S is much below $10^{-3}\,F$, this condition is almost certain to exist. In a more typical case, with $c_S \approx 0.03\,F$, the innate error is only 0.0006 percent.

The reaction of a weak acid HA with strong base is shown in (4.2), and that of a weak base B with strong acid is shown in (4.3).

$$HA + OH^- \rightleftharpoons A^- + HOH \tag{4.2}$$

$$B + H_3O^+ \rightleftharpoons BH^+ + HOH \tag{4.3}$$

We show general symbols HA and B instead of chemical formulas for specific substrates because mathematical analysis proves that the feasibility of titration depends only on the dissociation constant and the concentration c_S, and not on the specific nature of the weak Brønsted acid or base.

Equations (4.2) and (4.3) are both of the correct stoichiometry to fit case 3(b) in Chapter 2 (p. 38). Both innate titration errors can therefore be estimated from (2.19b). Indeed, the similarity goes even further: Both titrations (weak acid + strong base, and weak base + strong acid) involve precisely the same mathematics. We shall therefore discuss in detail only the titration of weak acids with strong base. To arrive at the corresponding relationships for the titration of weak bases with strong acid, we need only write K_B instead of K_A and $[H^+]$ instead of $[OH^-]$.

The equilibrium constant for titration according to (4.2) is as follows.

$$K = \frac{[A^-]}{[HA][OH^-]} = \frac{[H^+][A^-]}{[HA]} \cdot \frac{1}{[H^+][OH^-]}$$

$$= \frac{K_A}{K_W} = 10^{14}\,K_A \qquad \text{at 25°C} \tag{4.4}$$

On applying (2.19b), we then find that the innate titration error for titration of a weak acid with strong base is

$$<1\% \text{ if } K_A c_S > 3 \times 10^{-10}$$
$$<0.25\% \text{ if } K_A c_S > 5 \times 10^{-9} \tag{4.5}$$

A typical value for c_S is $0.03\,F$. On substituting this value in (4.5), we find that for 1 percent accuracy, K_A must be greater than 1.0×10^{-8}, and for 0.25 percent accuracy, K_A must be greater than 1.6×10^{-7}. These results, and comparable conclusions for other titrations, are summarized in Table 4.1. The table clearly shows that the innate error in acid-base titration increases as the strength of either the acid or the base decreases.

INNATE ERROR IN ACID-BASE TITRATION
IN AQUEOUS SOLUTION

Table 4.1

Titration	Innate error, %
Strong acid + strong base; $c_S = 0.03\ F$	0.0006
Weak acid + strong base; $c_S = 0.03\ F$	
$\quad K_A$ of acid = 1.6×10^{-7}	0.25
$\quad K_A$ of acid = 1.0×10^{-8}	1.0
Weak base + strong acid; $c_S = 0.03\ F$	
$\quad K_B$ of base = 1.6×10^{-7}	0.25
$\quad K_B$ of base = 1.0×10^{-8}	1.0

*p*H during acid-base titration

To understand why it is more difficult to titrate a weaker acid, we must consider the tactics by which the equivalence point can be detected. One approach is to monitor the *p*H.

As we shall find in Chapter 7, theory and experiment are in accord that the *p*H during an acid-base titration varies in the characteristic manner shown in Fig. 4.1. Graphs such as these, which show the variation of some

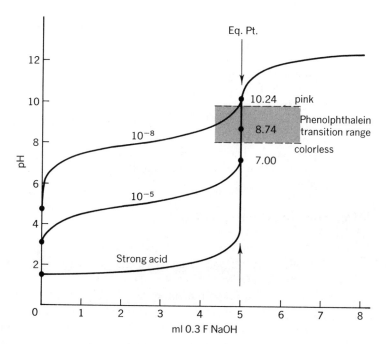

Fig. 4.1. Variation of *p*H during the titration of 50.00 ml of 0.03 *F* acids with 0.3 *F* NaOH. Lower curve, strong acid; middle curve, weak acid with K_A of 10^{-5}; upper curve, weak acid with K_A of 10^{-8}.

property of the solution during a titration, are called *titration curves*. Let us begin by considering the lowest curve, which represents the titration of strong acid with strong base. At first the pH increases rather slowly, but just before the equivalence point the slope becomes suddenly very steep, and in the figure the pH jumps by more than 5 units as the volume of NaOH increases from 4.99 ml to 5.01 ml. Finally, past the equivalence point, the slope is again quite small. The point of maximum slope (which is an inflection point) coincides with the equivalence point at pH 7.00. We therefore have two independent criteria for recognizing the equivalence point: We can look for the point of maximum slope, or we can look for the point of pH 7.00. Moreover, so steep is the slope near the equivalence point that we need not apply these criteria with high precision: If we take the titration endpoint anywhere between pH 5 and 9, the endpoint error will be less than 0.2 percent.

The second titration curve in Fig. 4.1, which represents the titration of a weak acid with $K_A = 10^{-5}$, is essentially similar. Again, there is a sharp "break," centered at the equivalence point at pH 8.74; and again the equivalence point is readily identified. Note, however, that the "break" is not quite as sharp as for the titration of strong acid with strong base.

Finally, the upper titration curve in Fig. 4.1 represents the titration of a weak acid with $K_A = 10^{-8}$, for which the innate titration error is estimated at 1 percent. The equivalence point, at pH 10.24, is again an inflection point on the titration curve. However, the slope at the equivalence point is now much smaller, and the definition of the inflection point is correspondingly less precise. In an actual titration, any error in the measurement of pH will therefore cause a relatively large error in the placement of the equivalence point.

In summary, accurate titration is feasible only if there is a sharp "break" in the pH at the equivalence point, and this condition will exist only if K_A for the weak acid (or K_B if the substrate is a weak base) is not too small: The practical lower limit for titration in aqueous solution is a dissociation constant of about 10^{-8}.

pH at the equivalence point

The most common method of detecting the equivalence point is to add a pH indicator that produces a characteristic color signal. Let us therefore consider how to predict the pH at the equivalence point.

It is helpful to remember that a solution at the equivalence point need not be prepared by taking equivalent amounts of the reactants; an identical solution can be prepared by taking the appropriate amounts of the pure products. Thus, in the titration of a weak acid HA with sodium hydroxide, the solution at the equivalence point is identical to a solution prepared from an equivalent amount of the pure sodium salt, NaA, in water. Let us find the pH of such a solution.

A solution of pure NaA in water will be slightly alkaline because the A^- ion is a Brønsted base [Eq. (4.6)]. The hydroxide ion concentration at equilibrium is calculated as follows.

$$A^- \quad + \quad HOH \; \rightleftharpoons \; HA + OH^- \qquad (4.6)$$
$$(c_S - x) \approx c_S \qquad\qquad\quad (x) \qquad (x)$$

Let c_S denote the formal concentration of NaA and x the molar concentration of OH^-. The self-ionization of water will be neglected. Therefore, the molar concentration of HA is also equal to x, and that of A^- is $(c_S - x)$. The expression for the base-dissociation constant thus takes the form (4.7).

$$\frac{x^2}{(c_S - x)} = K_B(\text{for } A^-) \qquad (4.7)$$

On applying (2.8), we find that $K_B(\text{for } A^-) = K_W/K_A$, where K_A is the acid-dissociation constant of HA.* It is also true that whenever titration is feasible, $c_S \gg x$, so that $(c_S - x) \approx c_S$. Substituting in (4.7) and solving for x, we then obtain (4.8).

$$x = [OH^-] = \sqrt{\frac{K_W c_S}{K_A}} \qquad (4.8)$$

Finally, we calculate the hydrogen ion concentration by introducing the equation, $[H_3O^+] = K_W/[OH^-]$. The result is (4.9).

Weak acid + strong base, at equivalence point,

$$[H_3O^+] = \sqrt{\frac{K_A K_W}{c_S}} \qquad (4.9)$$

For the titration of a weak base with strong acid, we consider (analogously) the dissociation of the titration product BH^+ as a Brønsted acid in water [Eq. (4.10)].

$$BH^+ \quad + \quad H_2O \; \rightleftharpoons \; B + H_3O^+ \qquad (4.10)$$
$$(c_S - x) \approx c_S \qquad\qquad\quad (x) \qquad (x)$$

The hydrogen ion concentration of the acidic solution is then given by (4.11), where K_B is the dissociation constant of the weak base.

Weak base + strong acid, at equivalence point,

$$[H_3O^+] = \sqrt{\frac{K_W c_S}{K_B}} \qquad (4.11)$$

---

* The constant $K_B(\text{for } A^-)$ is often referred to as the hydrolysis constant K_h for A^- since Eq. (4.6) represents an hydrolysis reaction, that is, a process in which a water molecule is split.

To choose a color endpoint indicator for a particular titration, we solve Eq. (4.9) or (4.11) to predict the pH at the equivalence point, and then take an indicator that will give a distinctive color signal when that pH is reached or passed. Figure 4.2 shows typical results of such pH calculations; c_S is assumed to be 0.03 F throughout, and pK_A or pK_B of the weak acid or base is plotted along the x-axis.

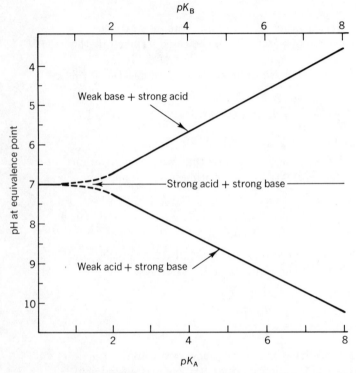

Fig. 4.2. pH at the equivalence point in acid-base titrations.

Color endpoint indicators

A color pH indicator is a weak acid whose conversion to its conjugate base produces a color change in the solution. For example, conversion of the weak acid methyl orange to its conjugate base in aqueous solution causes a color change from red to orange; and conversion of the weak acid, phenolphthalein, causes a change from colorless to red-purple. Ideally, the color of the indicator is visible at such low concentrations that the addition of the indicator will have no significant effect on the pH of the solution.

In Chapter 3 we defined the conversion interval of a weak acid as the pH range in which the degree of conversion α advances from 0.01 to 0.99.

We found [Eq. (3.24)] that the conversion interval for a weak acid extends on the pH scale from $(pK_A - 2)$ to $(pK_A + 2)$. However, for pH measurement with color indicators, only about half of the conversion interval is really useful. This pH range, in which the solution color is a sensitive function of pH, is called the *transition range* of the indicator. Transition ranges and characteristic colors of three common pH indicators are listed below. A more complete list is given in Table 9.2.

Methyl orange: $pK_A = 3.45$; pH of transition range—3.1 (red) to 4.5 (orange).

Bromothymol blue: $pK_A = 7.3$; pH of transition range—6.0 (yellow) to 7.6 (blue); the solution color changes from yellow to green to turquoise to blue.

Phenolphthalein: $pK_A = 9.3$; pH of transition range—8.0 (colorless) to 9.8 (red-purple).

The transition range of phenolphthalein (8.0 to 9.8) is on the alkaline side of the pH scale. Phenolphthalein therefore makes a good endpoint indicator for the titration of weak acids with strong base. We can see (from Fig. 4.2, for example) that the pH range 8.0 to 9.8 includes the equivalence point if the pK_A of the weak acid is between 3.5 and 7, and that it is only slightly more alkaline than the equivalence point for stronger acids. The change from colorless to fully-developed red-purple therefore brackets the equivalence point, or occurs just past it. This color change will be sharp (one drop of titrant will be sufficient to produce it) if, and *only if*, the pH changes rapidly at the equivalence point, that is, if the titration curve shows a sharp break. Thus Fig. 4.1 shows that phenolphthalein will produce a sharp color signal, practically at the equivalence point, in the titration with strong base of a strong acid, or of a weak acid if $K_A = 10^{-5}$. On the other hand, if $K_A = 10^{-8}$, the equivalence point is outside the transition range of phenolphthalein and, even if the chosen indicator were more suitable, the pH change is slow enough so that at least several drops of titrant will be needed to produce the full color change. The titration endpoint will be correspondingly uncertain. Of course, that is precisely what we expect because the innate titration error is about 1 percent.

On the basis of similar arguments we conclude that methyl orange will be a good endpoint indicator for the titration of weak bases with strong acid, since the transition range includes, or comes just after, the equivalence point in the titration. Bromothymol blue is a good indicator for the near-neutral pH range, although some people object to its many colors.

In the titration of strong acid with strong base, the break in pH at the equivalence point is so sharp that methyl orange, bromothymol blue, and phenolphthalein all give practically identical endpoints. In actual practice, however, solutions of sodium hydroxide often contain a little sodium carbonate, and this will introduce discrepancies. At the phenolphthalein endpoint, carbonate is converted to bicarbonate ($CO_3^{2-} \rightarrow HCO_3^-$) and thus acts as a monoacid base, while at the methyl orange endpoint it is converted all the way to CO_2 ($CO_3^{2-} \rightarrow H_2CO_3 \rightleftharpoons CO_2$) and thus acts as a diacid base.*

Acidimetric standards

Substances used as chemical standards must be easily purified to a high state of purity, must be stable and nonhygroscopic, and must react with simple, well understood stoichiometry. For convenience of handling they should also be solid at room temperature and have high equivalent weights. The following substances are often used as standards in acid-base work.

Weak acids. There are a number of weak organic acids that make excellent standards for measuring the normality of strong aqueous base to the phenolphthalein endpoint. Best known among them are: *Potassium acid phthalate*, $KHC_8H_4O_4$, a water-soluble acid salt which reacts as a monobasic acid with pK_A 5.41; equivalent weight = 204.22. *Benzoic acid*, $HC_7H_5O_2$, a weak organic acid with pK_A 4.18; equivalent weight = 122.12. Benzoic acid is not very soluble in water, but the salt formed during titration is usually soluble. *Oxalic acid dihydrate*, equivalent weight = 63.03. *Trinitrobenzoic acid*, equivalent weight = 257.12.

Weak bases. There is a surprising shortage of really satisfactory base standards. Pure sodium carbonate and natural calcite ($CaCO_3$) are sometimes used, the former after heating at 270–300° to decompose any sodium bicarbonate, but both have rather low equivalent weights. The water-soluble organic base, *tris*-hydroxymethyl methylamine, $(HOCH_2)_3CNH_2$ (TRIS for short) is available as a highly pure solid and is gaining in popularity. The pK_B of TRIS is 5.92; equivalent weight = 121.14. Weak bases are useful for standardizing strong acids to the methyl orange endpoint.

Constant-boiling hydrochloric acid. A strong standard acid is sometimes more useful than a weak one because it may be used to standardize strong base to either the methyl orange or phenolphthalein endpoint. Constant-boiling hydrochloric acid is an approximately 6.10 F solution of HCl in water, with a density of 1.096 at 25°C and a boiling point of 108.6°C at 760 mm of Hg. Ordinarily one does not like to use liquid solutions as standards because the composition may change unless precautions are taken to prevent evaporation. However, it happens that the binary system, HCl—H_2O, has a "constant-boiling" composition at which the composition of the vapor is identical to that of the boiling liquid. Hydro-

---

* Alkaline aqueous solutions readily absorb carbon dioxide from the atmosphere. This phenomenon causes error in acid-base titrations to the phenolphthalein endpoint, because $CO_2(H_2CO_3)$ becomes converted to HCO_3^- and thus consumes one equivalent of base per mole of CO_2 absorbed. It is therefore prudent to approach the equivalence point from low pH, that is, to place the base in the buret and the acid in the titration flask.

chloric acid of the constant-boiling composition is readily prepared and is a reliable standard, because its composition is known with high accuracy,* and because it can be handled and stored at room temperature with little risk that it will change in composition.

To prepare constant-boiling hydrochloric acid, one places hydrochloric acid of specific gravity 1.10 into a clean, all glass distilling apparatus and distills slowly at atmospheric pressure. After about one-half of the original solution has gone over and been discarded, the third quarter, which is essentially the constant-boiling liquid, is collected. The composition of the constant-boiling acid varies slightly with the atmospheric pressure p and may be calculated from (4.12).

For constant-boiling hydrochloric acid,

$$\text{Wt. } \% \text{ HCl} = 20.221 - 0.0024(p - 760 \text{ mm Hg})$$

$$\text{Equiv. wt. (when weighed in air)} = 180.193 + 0.0214(p - 760 \text{ mm Hg})$$

$$(4.12)$$

PRECIPITATION OF INSOLUBLE SILVER SALTS

Interest in the quantitative analysis of the halide ions dates back to the very beginning of modern chemistry. The halogens, and especially chlorine, are among the most useful of chemical reagents and form an almost endless number and variety of compounds. Common table salt was of great political importance during the time of Lavoisier, being subject to an exorbitant and much-hated royal tax. In modern medicine, many drugs are manufactured as hydrochlorides, which makes analysis for chloride ion a good method for testing their purity.

Chloride, bromide, iodide, and numerous other anions that form insoluble silver salts can be determined by titration with aqueous silver nitrate. The method depends on the quantitative precipitation of the silver salt. The silver salt must be so insoluble that the concentration of free (unprecipitated) silver ion in the solution remains very small until the equivalence point, so that there is a sudden sharp increase in the silver ion concentration at precisely the point where the added silver nitrate becomes in excess. This sudden change is then harnessed to produce a visible endpoint signal.

Since the method depends on the silver salt being highly insoluble, the accuracy improves with decreasing solubility. Indeed, we shall find in Chapter 8 that the innate titration error is approximately 1.8 times the

* C. W. Foulk and M. Hollingsworth, *Journal of the American Chemical Society*, **45**, 1223 (1923).

solubility of the pure silver salt in pure water. Some relevant solubility data are listed in Table 4.2. For the salts shown, solubility in water varies in the sequence Ag_2CrO_4 > $AgCl$ > $AgSCN$ > $AgBr$ > AgI, and the innate titration error varies in the same sequence. In practice, silver chloride is near the upper limit of solubility at which titration with quantitative accuracy is still feasible.

Table 4.2   TITRATION TO PRODUCE INSOLUBLE SILVER SALTS AT 25°C

Salt	K_{SP} in water	Solubility in pure water (F)[a]	Innate titration error (approx.)
$AgCl$	1.8×10^{-10}	1.3×10^{-5}	$\pm 2 \times 10^{-5}\,F$
$AgBr$	7.7×10^{-13}	8.8×10^{-7}	$\pm 2 \times 10^{-6}\,F$
AgI	8.3×10^{-17}	9.1×10^{-9}	$\pm 2 \times 10^{-8}\,F$
$AgSCN$	1.1×10^{-12}	1.0×10^{-6}	$\pm 2 \times 10^{-6}\,F$
Ag_2CrO_4	1.1×10^{-12}	7×10^{-5}	—[b]

[a] Calculated from K_{SP}. Actual solubilities may be somewhat higher owing to ion-pair and/or complex formation. See Chapter 5.
[b] Usually not titrated directly with aqueous silver nitrate.

Titration with silver nitrate can be done over a wide range of aqueous pH, from strong acid to about pH 11. Above pH 11, the precipitation of dark brown silver oxide may cause trouble. The equivalence point in the titration can be detected potentiometrically, as described in Chapter 10, or with a color indicator. However, there is unfortunately no known single color indicator that will give accurate results over the entire useful pH range.

Mohr's method for chloride and bromide

Mohr's method is one of the oldest volumetric methods that still finds occasional use today. Its usefulness is limited to the determination of chloride and bromide ion, and to the pH range 6.5 to 10.5.

In Mohr's method one adds enough sodium chromate to the titration mixture so that the concentration of Na_2CrO_4 near the equivalence point is $0.002 - 0.01$ F. The pH must be above 6.5 to avoid the conversion of CrO_4^{2-} to $HCrO_4^{-}$ [see discussion after Eq. (3.17)]. The reaction with silver ion then takes place in two well resolved steps, shown for chloride ion in (4.13) and (4.14). The titration endpoint is taken as the first appearance of the orange-red precipitate of silver chromate.

$$Cl^- + Ag^+ \rightleftharpoons AgCl \quad \text{(grayish-white precipitate)} \quad (4.13)$$

$$CrO_4^{2-} + 2\,Ag^+ \rightleftharpoons Ag_2CrO_4 \quad \text{(orange-red precipitate)} \quad (4.14)$$

Under the stated conditions, the precipitate of silver chromate will appear slightly past the equivalence point, but not enough so as to cause serious error. In theory, one can cause the precipitation of silver chromate to coincide precisely with the equivalence point, simply by increasing the concentration of Na_2CrO_4 and thus shifting the equilibrium in (4.14) to the right. In practice this is not possible, however, because the yellow chromate color then becomes so strong as to mask the color of the precipitate.

Volhard's method

If the unknown sample contains bases such as ammonia that form stable complex ions with Ag^+, the titration must be done in acid to minimize the effect that such bases might have on the solubility of the precipitated silver salt. For instance, if Cl^- is titrated in the presence of ammonia (below pH 9, where NH_4^+ is the principal species), there action that must be suppressed by the addition of acid is shown in (4.15).

$$AgCl(s) + 2\,NH_4^+ + 2\,H_2O \; \rightleftharpoons \; Ag(NH_3)_2^+ + Cl^- + 2\,H_3O^+ \quad (4.15)$$

Volhard's method is designed to give accurate results below pH 1, that is, in moderately strong acid. The determination of the amount of a substrate X^- is done in two steps: In the first step, the solution of the substrate is acidified with nitric acid and a known number of equivalents of silver ion is added. The equivalents of Ag^+ must be greater than those of X^- so that precipitation of AgX, according to (4.16), is quantitative.

$$X^- + (\text{excess})\,Ag^+ \; \rightleftharpoons \; AgX(s) \qquad (4.16)$$

In the second step, the excess amount of Ag^+ is determined by back-titration with thiocyanate, Eq. (4.17).

$$Ag^+ + SCN^- \; \rightleftharpoons \; AgSCN(s) \qquad (4.17)$$

(*Back-titration* is the name given to the titration of the unreacted portion of a reagent that was added in known amount in an earlier step in the analysis.) To display the equivalence point in (4.17), ferric ammonium sulfate is added to the titration flask at $pH < 1$, to produce an iron(III) concentration of 0.01–$0.02\ F$. As thiocyanate is added, reaction with Ag^+ and Fe^{3+} takes place in two well resolved steps: The first reaction is the formation of a white precipitate of silver thiocyanate according to (4.17); after that reaction is complete, the solution undergoes a sudden color change from greenish-yellow to rust-orange as further thiocyanate reacts with Fe^{3+} according to (4.18).

$$\underset{\textbf{(Greenish-yellow)}}{Fe^{3+}} + SCN^- \; \rightleftharpoons \; \underset{\textbf{(Rust-orange)}}{FeSCN^{2+}} \qquad (4.18)$$

When the sudden color change occurs, the concentration of $FeSCN^{2+}$ is still very small, so that the color change is a satisfactory endpoint signal for

(4.17). Past the endpoint, as still more thiocyanate is added, the rust-orange hue changes slowly to dark red-brown as $FeSCN^{2+}$ is converted to $Fe(SCN)_2^+$ and perhaps to still higher ferric thiocyanate complexes.

A potential source of error in Volhard's method is that the precipitate of AgX, produced in the first step of the analysis, may be partially decomposed as thiocyanate is added in the second step of the analysis. The equation for the decomposition is shown in (4.19).

$$SCN^-(aq) + AgX(s) \rightleftharpoons AgSCN(s) + X^-(aq) \qquad (4.19)$$

However, it can be shown that the error due to (4.19) will be negligible if the solubility of AgX is less than that of AgSCN. When that condition is not satisfied, the precipitate of AgX may have to be separated prior to back-titration.

Since AgBr and AgI are less soluble than AgSCN (as shown in Table 4.2), Br^- and I^- can be titrated by the following procedure: One adds (1) nitric acid, (2) a known amount of Ag^+, (3) ferric ammonium sulfate, and then titrates directly with a standard solution of ammonium or sodium thiocyanate to the rust-orange endpoint. This sequence of addition of the reagents must be observed especially when iodide is present, because I^- [but not AgI(s)] is oxidized by iron(III) to molecular iodine.

AgCl is *more* soluble than AgSCN, but fortunately there is a simple procedure for removing solid silver chloride from effective contact with an aqueous solution. Just before the back-titration with thiocyanate, one adds about 1 ml of nitrobenzene. This sweet smelling organic liquid is insoluble in water and surrounds the precipitate of AgCl with a tenacious oily coating that is practically impervious to thiocyanate ions in aqueous solution. After adding the nitrobenzene and swirling vigorously, one then proceeds with the back-titration in the normal way.

The Volhard method has been used also for the analysis of basic anions such as AsO_4^{3-}, CNO^-, CN^-, CO_3^{2-}, CrO_4^{2-}, $C_2O_4^{2-}$, PO_4^{3-} and S^{2-}, all of which form insoluble silver salts at sufficiently high pH. After such a substrate has been treated with a known amount of Ag^+ at the appropriate pH, the precipitated silver salt is separated (by filtration, if necessary), and the excess of Ag^+ is determined by back-titration at low pH.

Silver, in the absence of interfering anions, can be determined by direct titration with thiocyanate, using Volhard's method. This is especially convenient for the analysis of silver alloys, since nitric acid is used to dissolve the sample:

$$Ag(\text{in alloy}) + NO_3^- + 2\,H_3O^+ \longrightarrow Ag^+ + NO_2 + 3\,H_2O$$

After dissolution of the sample, the solution must be boiled to free it from the red-brown nitrogen dioxide (NO_2) whose presence would interfere with the titration.

Fajans' method

When a solid titration product precipitates in the form of many small crystals, it is sometimes possible to produce a sharp color change at the equivalence point by surface-dying the crystals with an adsorption indicator. This stratagem is used in Fajans' method. For example, in the determination of bromide ion by titration with silver nitrate according to Fajans' method, the phthalein dye, eosin, is added to the aqueous titration mixture at 10^{-5} F concentration and the color of the silver bromide precipitate is observed. As silver nitrate is added, the precipitate will have the characteristic light yellow color of pure silver bromide until the equivalence point is reached, where the color changes suddenly to red. The color change results from the sudden adsorption of eosin anions from solution.

Solid silver bromide shares with other silver salts, and with heavy-metal salts in general, the property of being a highly specific adsorbent. Of course, any pure crystal will show a marked preference for adsorbing its own ions— that, after all, is how the pure crystal grows—and solid silver bromide preferentially adsorbs bromide ions. However, in the absence of bromide ions, the surface will show a strong affinity for eosin anions, and a much weaker affinity for such oxy-anions as nitrate, perchlorate, or sulfate. Thus, at the equivalence point, where the concentration of bromide ions in solution suddenly becomes very small, the silver bromide precipitate suddenly becomes free to adsorb eosin ions and the surface color becomes red. That the red color is due to the adsorption of eosin can be shown by the following observations: The precipitate of silver bromide does not become red in the absence of eosin; and there is no sign of a red precipitate when silver nitrate is added to 10^{-5} F eosin in plain water.

Indicator	Substrates titrated	pH
ADSORPTION INDICATORS FOR TITRATIONS WITH SILVER ION IN AQUEOUS SOLUTION		Table 4.3
Dichlorofluorescein	Cl^-, Br^-, I^-, SCN^-	4–10
Eosin	Br^-, I^-, SCN^-	2–10 (sometimes 1–10); optimum 4–5
Rhodamine 6G	Ag^+ (titrate with Br^-)	minimum 0.5

Table 4.3 summarizes the most important adsorption indicators for titrations with silver ion, and the conditions under which each may be used. Control of the pH is generally necessary because these indicators are weak organic acids, and if the pH is too low, too much of the indicator is con-

verted to an inactive conjugate acid. It is sometimes desirable to add a "dispersing agent" such as dextrin or methanol to the titration flask to stabilize and promote the formation of small crystals and thus to accentuate the color change.

Argentimetric standards

Silver nitrate, and sodium and potassium halides, of analytical-reagent grade from reliable sources are pure enough to serve as chemical standards, except when accuracy of better than 0.1–0.2 percent is required. It is good practice to crush the crystals with an agate mortar and pestle, to remove occluded moisture, and to dry them at 110°C for about an hour prior to use. Old samples of silver nitrate are best dried by fusion at 220–250°C for a few minutes. (The melting point of $AgNO_3$ is 208°C.)

In work requiring high accuracy, or as a check on the reagent, pure sodium chloride or carefully prepared constant-boiling hydrochloric acid can be recommended as standards. The latter reagent has the advantage that double-checking of its composition is possible: The chloride normality can be determined gravimetrically, by precipitation of AgCl; and the acid normality can be determined volumetrically, by comparison with other acidimetric standards.

OXIDATION-REDUCTION

The chemistry involved in redox analysis is, on the whole, more intricate than that involved in acid-base titration or the precipitation of insoluble silver salts. The redox titration itself is often only the final step of a procedure that may include preliminary treatment of the sample (to make sure that the substrate is in the desired oxidation state), control of the pH, expulsion of atmospheric oxygen, and addition of a catalyst. Nevertheless, redox analysis ranks among the principal tools of analytical chemistry. Redox reactions are useful not only for the analysis of metals and ores in mining and metallurgy, but also for the analysis of a host of metal ions and organic and biochemical substrates in the research laboratory. They are particularly well suited to potentiometric methods of titration, as we shall see in Chapter 10, and also provide us with some of the best color endpoints in chemistry.

The most common oxidizing and reducing agents of analytical chemistry are listed in Tables 4.4 and 4.5. Before going on with this section, spend a few minutes memorizing their names and formulas.

COMMON OXIDIZING AGENTS

Table 4.4

Reagent	Medium	Oxidized state	Reduced state	Equivalents per formula weight	E°_{red} (V)
Permanganate	Acid	MnO_4^-	Mn^{2+}	$5(MnO_4^-)$	$+1.51$
Dichromate	Acid	$Cr_2O_7^{2-}$	Cr^{3+}	$6(Cr_2O_7^{2-})$	$+1.33$
Ceric	Acid	Ce^{4+}	Ce^{3+}	$1(Ce^{4+})$	$+1.44$[a]
Iodate[b]	Acid	IO_3^-	I^-	$6(IO_3^-)$	$+1.195$
Iodine	Acid or neutral	I_2	I^-	$2(I_2)$	$+0.54$
Hydrogen peroxide	Acid	H_2O_2	H_2O	$2(H_2O_2)$	$+1.77$
Ferric iron	Acid	Fe^{3+}	Fe^{2+}	$1(Fe^{3+})$	$+0.77$
Lead dioxide	Acid	$PbO_2(s)$	Pb^{2+}	$2(PbO_2)$	$+1.455$
Mercuric chloride	Acid	$HgCl_2$	$Hg_2Cl_2(s)$	$1(HgCl_2)$	—
Permanganate	Basic	MnO_4^-	$MnO_2(s)$	$3(MnO_4^-)$	$+0.588$

[a] In 1 F H_2SO_4.
[b] This reduction is normally carried out in two steps. E°_{red} is for the reduction of IO_3^- to I_2. (See p. 89)

COMMON REDUCING AGENTS

Table 4.5

Reagent	Medium	Reduced state	Oxidized state	Equivalents per formula weight	E°_{ox} (V)
Iodide	Acid or neutral	I^-	$I_2(I_3^-)$	$1(I^-)$	-0.54
Oxalate	Acid	$C_2O_4^{2-}$	CO_2	$2(C_2O_4^{2-})$	$+0.49$
Arsenious acid	pH 5–9	H_3AsO_3	$H_2AsO_4^-$	$2(H_3AsO_3)$	-0.56
Thiosulfate	Acid or neutral	$S_2O_3^{2-}$	$S_4O_6^{2-}$	$1(S_2O_3^{2-})$	-0.08
Stannous tin	Acid	Sn^{2+}	Sn^{4+}	$2(Sn^{2+})$	-0.15
Zinc	Acid	$Zn(s)$	Zn^{2+}	$2(Zn)$	-0.76
Ferrous iron	Acid	Fe^{2+}	Fe^{3+}	$1(Fe^{2+})$	-0.77

Effect of acidity

Redox equilibria are remarkable for the enormous range and variety of their response to changes in pH. To illustrate this point, let us examine a few half-reactions for reduction. For greater clarity in the following, we shall represent the hydronium ion by $H^+(aq)$, or simply by H^+.

Some redox half-reactions, such as the reduction of a solvated ion to a lower oxidation state, involve neither hydrogen ions nor hydroxide ions and therefore show no direct dependence on acidity. An example is (4.20).

$$Co^{3+}(aq) + e^- \rightleftharpoons Co^{2+}(aq) \qquad (4.20)$$

To be sure, if studied over a sufficiently wide pH range, such half-reactions will show some perturbation because the oxidized and reduced forms of the reagent will react as Brønsted acids and bases, and complex formation with the anion of the pH buffer (or added strong acid) may also produce an effect.*

Another kind of redox half-reaction, which is often found in organic chemistry and biochemistry, involves the gain of hydrogen atoms by the substrate. Such half-reactions show a direct dependence on acidity since stoichiometry requires that one hydrogen ion reacts for each hydrogen atom gained. An example, prominent in the biochemistry of cell respiration, is the reduction of the coenzyme diphosphopyridine nucleotide (**DPN⁺**, for short) according to (4.21).

$$H^+(aq) + \textbf{DPN}^+ + 2\,e^- \; \rightleftharpoons \; \textbf{DPNH} \tag{4.21}$$

The greatest dependence on acidity is shown by half-reactions in which reduction assumes its ancient meaning of "reduction in the oxygen content." For example, the reduction of lead dioxide to Pb^{2+} entails the loss of two oxygen atoms and requires four hydrogen ions, as shown in (4.22).

$$4\,H^+(aq) + PbO_2(s) + 2\,e^- \; \rightleftharpoons \; Pb^{2+} + 2\,H_2O \tag{4.22}$$

Similarly, the reduction of $Cr_2O_7^{2-}$ to Cr^{3+} entails the loss of seven oxygen atoms and requires *14* hydrogen ions, as shown in (4.23).

$$14\,H^+(aq) + Cr_2O_7^{2-} + 6\,e^- \; \rightleftharpoons \; 2\,Cr^{3+}(aq) + 7\,H_2O \tag{4.23}$$

In the half-reactions shown, the reduction in the oxygen content of the reagent leads to the formation of water molecules. Stoichiometry then requires that two hydrogen ions react for each oxygen atom removed, and an almost unimaginably high sensitivity to pH results. Thus Eq. (4.23) indicates that the mass-action effect of $Cr_2O_7^{2-}$ will vary as the fourteenth power of the hydrogen ion concentration! No wonder, then, that dichromate cleaning solution (a solution of potassium dichromate in concentrated sulfuric acid) is so effective at removing grease and oxidizable dirt from glass surfaces: At the high acidity of concentrated sulfuric acid, dichromate is quite an oxidizing agent.

On the basis of these examples we may state that the greatest sensitivity to pH will be shown by oxidizing ions whose name ends in -*ate*, such as nitrate, iodate, arsenate, permanganate, or perchlorate. Such ions lose oxygen atoms on being reduced, and their strength as oxidizing agents therefore increases with increasing acidity—often quite spectacularly so.

Potassium permanganate and ceric sulfate

The two most common quantitative oxidizing agents are potassium permanganate and ceric sulfate. Both substances are strong oxidizing agents and enjoy a wide range of applicability. Potassium permanganate

---

* Precipitation of insoluble hydroxides prevents many metal ions from being titrated directly at high pH.

has been in use since the early days of quantitative analysis, while ceric sulfate is a relative newcomer. The half-reactions of these reagents have already been given (Tables 1.2 and 4.4); they are repeated below.

$$MnO_4^- + 8\,H^+ + 5\,e^- \; \rightleftharpoons \; Mn^{2+} + 4\,H_2O$$

$$Ce^{4+} + e^- \; \rightleftharpoons \; Ce^{3+}$$

Titrations with potassium permanganate are usually done in a medium in which the hydrogen ion concentration is adjusted to 1–3 M. Permanganate ion acts as its own endpoint indicator: The appearance of the characteristic purple tint in the solution, right after the oxidation of the substrate is complete, is one of the sharpest color endpoints in chemistry. A minor disadvantage of potassium permanganate is that its solutions do not "keep" well; the reagent decomposes slowly on standing and solid MnO_2 precipitates out of solution. The decomposition is catalyzed by exposure to light, and the precipitate of MnO_2 acts as a catalyst for further decomposition. A more serious limitation in the analysis of iron ores that must be brought into solution by treatment with concentrated hydrochloric acid is that permanganate in acid solution will oxidize chloride ion to chlorine. We shall have more to say about that later.

Titrations with ceric sulfate are usually done in a medium in which the hydrogen ion concentration is at least 0.1 M. The titration endpoint can be detected potentiometrically, as described in Chapter 10, or with a color endpoint indicator. Ferroin (ferrous 1,10-phenanthroline sulfate) is most widely used as indicator; it gives a color change from red to pale light blue that is both sharp and accurate in most cases.

The most widely used "ceric sulfate" titrant is a solution of ceric ammonium sulfate in 2 N sulfuric acid. This solution is highly stable and retains its normality for many months. Although the oxidizing power of ceric sulfate is comparable to that of potassium permanganate in acid solution, the reaction rate with some substrates is inconveniently slow, and a catalyst must be added. Common catalysts are 1 drop of 0.01 F osmium tetroxide in 0.1 F H_2SO_4,* or 1 drop of 0.01 F iodine chloride. The catalyst is reduced quickly by the substrate and then is reoxidized quickly by ceric ion. The oxidation of chloride ion by ceric sulfate in aqueous acid is slow enough, even in 3 F hydrochloric acid, to cause no appreciable error in the titration of other substrates.

Scope. Because of the great oxidizing strength of permanganate and ceric sulfate in acid, the list of substrates that can be determined by quantitative oxidation is very long. Some representative examples are given on the next page.

---

* *Caution:* Pure osmium tetroxide (perosmic acid) is a dangerous, slightly volatile poison and produces severe symptoms of heavy-metal poisoning. However, the 0.01 F solution of the osmium tetroxide catalyst is quite safe if handled with normal care.

Antimony: $H_3SbO_3 + H_2O \longrightarrow H_3SbO_4 + 2\,H^+ + 2\,e^-$

Arsenic: $H_3AsO_3 + H_2O \longrightarrow H_3AsO_4 + 2\,H^+ + 2\,e^-$

Iron: $Fe^{2+} \longrightarrow Fe^{3+} + e^-$

Molybdenum: $Mo^{3+} + 4\,H_2O \longrightarrow MoO_4^{2-} + 8\,H^+ + 3\,e^-$

Nitrite: $HNO_2 + H_2O \longrightarrow NO_3^- + 3\,H^+ + 2\,e^-$

Oxalate: $H_2C_2O_4 \longrightarrow 2\,CO_2 + 2\,H^+ + 2\,e^-$

Peroxide: $H_2O_2 \longrightarrow O_2 + 2\,H^+ + 2\,e^-$

Sulfite: $H_2SO_3 + H_2O \longrightarrow HSO_4^- + 3\,H^+ + 2\,e^-$

Tin: $Sn^{2+} \longrightarrow Sn^{4+} + 2\,e^-$

Titanium: $Ti^{3+} + H_2O \longrightarrow TiO^{2+} + 2\,H^+ + e^-$

Uranium: $U^{4+} + 2\,H_2O \longrightarrow UO_2^{2+} + 4\,H^+ + 2\,e^-$

Vanadium: $VO^{2+} + 2\,H_2O \longrightarrow VO_3^- + 4\,H^+ + e^-$

Many of these substrates in their lower oxidation state react slowly with atmospheric oxygen and may have to be treated with a reducing agent, just before titration, to assure that the entire substrate is reduced. It is also desirable, in many cases, to remove atmospheric oxygen from the titration flask. Bubbling of pure nitrogen gas through the solution, or addition of about 1 g of solid sodium bicarbonate to the acid solution to replace air with CO_2, are usually effective.

When reduction of the substrate is required prior to titration, a versatile reducing agent is the active metal, zinc. Reduction is sometimes done simply by dipping a spiral of zinc metal into the flask just before titration. However, this method can be used only if the substrate is reduced quickly, in a matter of seconds, otherwise the competing reaction of zinc with hydrogen ion, with the evolution of hydrogen gas, becomes too important. The evolution of hydrogen gas is inhibited if zinc amalgam (a solid solution of zinc and mercury) is used in place of the pure metal. In this particular method, the acid solution of the substrate is allowed to pass under gentle suction through a column of zinc amalgam in a device called a *Jones reductor*.

Analysis of iron ore. To give a typical example, we shall describe the analysis of an iron ore. This analysis has been an essential routine task of iron-related industries for many years, and chemists have devoted a great deal of time and research to perfecting it. The finely ground ore sample is decomposed by heating with concentrated hydrochloric acid until only a white residue of silica (which need not be separated) remains. The iron is then reduced to the ferrous state, transferred quantitatively to a titration flask, diluted, and titrated immediately.

The reduction is usually done as follows: To the hot, concentrated solution of the ore in hydrochloric acid, one adds stannous chloride, a drop at a time, until the color of the solution changes to the characteristic light green color of ferrous iron, and then one drop in excess. This reaction is represented in (4.24).

$$2 \, Fe^{3+} + Sn^{2+} \; \rightleftharpoons \; 2 \, Fe^{2+} + Sn^{4+} \qquad (4.24)$$

To destroy excess stannous chloride, the solution is cooled to room temperature and a large excess of mercuric chloride is added.

$$Sn^{2+} + 2 \, HgCl_2 \; \rightleftharpoons \; Sn^{4+} + Hg_2Cl_2(s) + 2 \, Cl^- \qquad (4.25)$$

The precipitate of mercurous chloride in (4.25) is extremely insoluble and will not be oxidized in the subsequent titration. The ferrous iron solution, which is now about $4 \, N$ in HCl, is then transferred quantitatively to a titration flask and diluted with water until the hydrochloric acid concentration is about $0.5 \, N$.

In the titration with ceric sulfate, ferroin indicator is added and the solution is titrated in the normal manner.

In the titration with permanganate, several precautions are taken to minimize error due to the possible oxidation of Cl^- to Cl_2. In the Zimmermann-Reinhardt method one adds a "preventive solution" of manganous sulfate, phosphoric acid, and sulfuric acid in water. The manganous ion will catalyze the oxidation of ferrous ion by permanganate and, according to some reports, will inhibit the oxidation of chloride ion, thus reducing the *relative* rate of the undesired reaction with chloride ion. The phosphoric acid will convert the yellow ferric ion to a nearly colorless phosphate complex, thus sharpening the observation of the permanganate endpoint.

After the preventive solution has been added, the ferrous iron is titrated immediately. However, the permanganate titrant is added circumspectly, with efficient stirring, in such a manner as to minimize the accumulation of unreacted permanganate in the titration flask until the endpoint is reached. In case of appreciable chlorine error, the characteristic odor of chlorine gas will become noticeable over the solution.

Other oxidizing agents. Potassium dichromate

Considerations of reaction rate are more obtrusive in redox analysis than in any other general method of analysis. To appreciate this, just bear in mind that many substrates submitted for redox analysis are capable of being oxidized by atmospheric oxygen. Serious error is avoided only because the rate of air oxidation is slow.

Of course, when wide differences in reaction rate exist, the analytical chemist can exploit them to make his methods more selective. Thus, in redox analysis, there are many specific methods that are selective for one component in a mixture because the reaction rate with that component is selectively high. The frequent use of chemical kinetics, in addition to chemical equilibrium, adds an entire new dimension to redox analysis, and the list of specific redox titrants is therefore long. Indeed, most transition metals, heavy metals, and elements of groups V, VI, and VII of the periodic table are being used, in one oxidation state or another.

In this chapter we shall consider explicitly only two further oxidizing titrants: potassium or sodium dichromate and (in the next section) iodine. Dichromate is commonly used in fairly strong acid—$[H^+] \approx 3\ M$ is typical —where it is a strong oxidizing agent. The half-reaction under such conditions is represented by (4.23). Sodium diphenylamine sulfonate, whose color changes from colorless to purple, is often used as endpoint indicator. The purple tint, which develops just after dichromate becomes in excess, is intense enough to be seen over the green color due to chromic ion.

Dichromate is sometimes preferred over permanganate as titrant in the analysis of iron ores because it does not oxidize chloride ion to chlorine under the conditions of the titration, but the color change of the indicator occurs slightly past the equivalence point. Dichromate has also been used for the quantitative oxidation of organic alcohols.

The iodine-iodide couple

Iodine is a mild oxidizing agent of considerable specificity; iodide ion is a strong and versatile reducing agent. These properties make the iodine-iodide redox couple one of the most important reagents in volumetric analysis.

As an oxidizing agent, iodine is normally used in the presence of 0.1–0.5 F potassium iodide, which converts most of the iodine to the nonvolatile, soluble triiodide ion.

$$I_2 + I^- \; \rightleftharpoons \; I_3^-$$

$$K = \frac{[I_3^-]}{[I_2][I^-]} = 700 \tag{4.26}$$

Aqueous triiodide solutions have a characteristic dark brown color. The unknown substrate in its lower oxidation state is, in most cases, titrated directly. Volatile substrates, or substrates that are sensitive to air oxidation, can often be analyzed by an indirect method: The substrate is added all at once to a known amount of triiodide solution, which must be in excess over that of the substrate. After reaction with the substrate is complete, the excess amount of triiodide is determined by back-titration with sodium thiosulfate.

When iodide ion is used as a reducing agent, a large excess of potassium iodide is added to the substrate in its higher oxidation state. The substrate is reduced and the liberated iodine (mostly I_3^-) is determined by titration with sodium thiosulfate. For example, in the analysis of copper(II), a large excess of potassium iodide is added to the Cu(II) substrate at pH 3–4, and reaction (4.27) ensues.

$$2\ Cu^{2+} + 5\ I^- \; \rightleftharpoons \; Cu_2I_2(s) + I_3^- \tag{4.27}$$

After waiting two or three minutes for this reaction to go to completion, the liberated iodine is titrated *directly* with sodium thiosulfate. There is no need to separate the cuprous iodide precipitate: That is what makes the method convenient. Thiosulfate is such an effective and selective reducing agent for iodine that titration will be quantitative, with the stoichiometry shown in (4.28), in the presence of a wide variety of substances.

$$I_3^- + 2\,S_2O_3^{2-} \;\rightleftharpoons\; 3\,I^- + S_4O_6^{2-} \tag{4.28}$$

Endpoint in iodine titrations. The appearance or disappearance of the characteristic triiodide color (pale yellow in dilute solution) is a fairly satisfactory endpoint signal. However, this signal can be greatly improved by the addition of a clear suspension of starch (1 percent by weight) in water. Colorless solutions of potassium iodide remain colorless in the presence of starch, but solutions containing both iodide and iodine develop a brilliant deep blue color, owing to the formation of a starch-iodine complex. The human eye is extremely sensitive to this color, and the starch-iodine endpoint is regarded by some as the "best endpoint in chemistry."

To be effective as an indicator, the starch (which does not form true molecular solutions in water) must be well dispersed to give a clear colloidal suspension. Unfortunately, colloidal starch is readily degraded by bacterial action. The degradation product not only fails to give the deep blue color, but also consumes an unpredictable amount of iodine in an irreversible reaction that results in a reddish color. The starch indicator should therefore be recently prepared and stabilized with a little mercuric iodide or 1 percent (by weight) of boric acid to inhibit the growth of bacteria. It is also prudent not to mix the starch indicator with iodine for any longer than is really necessary. Thus, in the titration of iodine with thiosulfate, it is prudent to add the starch indicator only just before the equivalence point, when the brown triiodide color is beginning to pale.

Sodium thiosulfate is also prone to bacterial degradation in aqueous solution. However, the growth of sulfur bacteria that do the damage can be inhibited very effectively by adding a few drops of chloroform. This volatile, quite unreactive, organic liquid is only slightly soluble in water and settles conveniently to the bottom, being more dense than water.

Hydrolysis and air oxidation. Aqueous solutions of iodine undergo hydrolysis, and triiodide ion thus exists in equilibrium with hypoiodous acid.

$$I_3^- + HOH \;\rightleftharpoons\; HOI + 2\,I^- + H^+ \tag{4.29}$$

At acid and neutral pH, equilibrium in (4.29) lies on the left. At high pH, the equilibrium shifts to the right and the principal hydrolysis product becomes hypoiodite ion [Eq. (4.30)]. Iodine titrations are therefore not done at high pH.

$$I_3^- + 2\,OH^- \;\rightleftharpoons\; OI^- + 2\,I^- + HOH \tag{4.30}$$

Hypoiodous acid, in equilibrium with I_3^-, tends to decompose into iodide ion and molecular oxygen.

$$2 \text{ HOI} \;\rightleftharpoons\; 2 \text{ H}^+ + 2 \text{ I}^- + \text{O}_2 \qquad (4.31)$$

The decomposition is catalyzed by light and is reversed in acid below pH 1. Reversal of (4.31) actually generates triiodide ion, because the hypoiodous acid reacts further in acid according to the reverse of (4.29). Titrations involving iodine and KI below pH 1 are therefore done in an oxygen-depleted atmosphere. When iodine is used as an oxidizing titrant, the standard solution of iodine in aqueous KI is maintained at a neutral or slightly acidic pH, and is stored in the dark.

Iodine as oxidizing agent. Because iodine is a relatively weak oxidizing agent, the scope of oxidizing titrations with iodine is limited. Iodine will titrate arsenite (or arsenious acid, H_3AsO_3) to arsenate. Equilibrium in this reaction depends strongly on the pH—see the discussion surrounding Eq. (2.21)—and the titration is usually done in a medium buffered to $pH \approx 7$ with a $CO_2\text{-}HCO_3^-$ buffer.

Iodine is a specific titrant for substrates with N—N single bonds and will oxidize such bonds to molecular nitrogen. For example, the oxidation of hydrazine, $H_2N—NH_2$, by triiodide ion at pH 7 is shown in Eq. (4.32).

$$H_2N—NH_2 + 2\ I_3^- \xrightarrow[pH7]{} N_2 + 6\ I^- + 4\ H^+ \qquad (4.32)$$

Iodine is also a specific titrant for substrates with S—H bonds, oxidizing (for example) hydrogen sulfide to sulfur [Eq. (4.33)] and the amino acid cysteine to cystine [Eq. (4.34)].

$$H_2S + I_3^- \longrightarrow S + 3\ I^- + 2\ H^+ \qquad (4.33)$$

$$3\ I^- + 2\ H^+ + \ ^-O_2C—\overset{\displaystyle H}{\underset{\displaystyle NH_3^+}{C}}—CH_2\underbrace{—S—S—}_{\text{to form the S—S bond in cystine.}}CH_2—\overset{\displaystyle H}{\underset{\displaystyle NH_3^+}{C}}—CO_2^- \qquad (4.34)$$

Finally, and of immeasurable importance in volumetric analysis, iodine will oxidize thiosulfate to tetrathionate, as shown in (4.28). Titration of I_3^- with thiosulfate is almost always the final quantitative step when iodide ion is used as a reducing agent.

Iodide as reducing agent. Iodide ion is a fairly strong reducing agent; the standard oxidation potential for the half-reaction, $3\ I^- \;\rightleftharpoons\;$

$I_3^- + 2\ e^-$, is -0.536 V. As a result, iodide ion will reduce a wide variety of oxidizing agents. For example, iodide will reduce:

1. Permanganate to Mn^{2+} in acid. This reaction is useful for checking standard permanganate solutions.

2. A variety of higher-valent metallic oxides to their lower oxidation state in acid, for instance,

$$PbO_2(s) + 4\ H^+ + 2\ e^- \longrightarrow Pb^{2+} + 2\ H_2O$$

$$MnO_2(s) + 4\ H^+ + 2\ e^- \longrightarrow Mn^{2+} + 2\ H_2O$$

3. Dichromate to Cr^{3+} in acid. This reaction is useful for the determination of metals that form insoluble chromate salts (e.g., Pb^{2+}, Ba^{2+}, Ag^+).

4. Gold(III) to gold(I) iodide. Copper(II) to cuprous iodide.

5. Hydrogen peroxide and organic peroxides in acid. For instance, $H_2O_2 + 2\ H^+ + 2\ e^- \rightarrow 2\ H_2O$.

6. Chlorine, bromine, and their oxides, the latter in acid

7. Nitrous acid and nitrite salts in acid:

$$HNO_2 + H^+ + e^- \longrightarrow NO + H_2O$$

8. Iodate (IO_3^-) and periodate (IO_4^-) to iodine in acid.

In each case, the substrate is allowed to react with a large excess of potassium iodide, and the liberated iodine is titrated with a standard solution of sodium thiosulfate at neutral or acidic pH. Very high acidities are avoided, because iodide then becomes prone to air oxidation, and thiosulfate decomposes spontaneously in strong acid.

$$S_2O_3^{2-} + 2\ H^+ \rightleftharpoons S + SO_2 + H_2O \qquad (4.35)$$

Redox standards

In redox analysis it is more than usually important that reaction conditions and substances used in the standardization of titrants be as similar as possible to those used in the actual analysis because systematic errors are more common and less avoidable. In this way the effects of systematic errors will be similar in standardization and analysis and will tend to cancel in the final calculation of results. It is not unusual, therefore, for a laboratory to have its own carefully prepared redox standards. For example, to standardize a permanganate solution that is to be used in the analysis of an unknown iron ore, the laboratory might use an iron ore of known composition, or at least an analyzed bright iron wire in a similar procedure. For ultimate accuracy and for research there are, of course, tested primary standards that can be obtained in high purity from reliable supply houses or, at somewhat greater cost, from the National Bureau of Standards. The most important ones are the following.

Sodium oxalate, $Na_2C_2O_4$, for standardization of potassium permanganate solutions in an acid medium, and of ceric sulfate solutions with ICl as catalyst in an acid medium.

Arsenic trioxide (arsenious oxide), As_2O_3, for standardization of potassium permanganate solutions in an acid medium with I^- or IO_3^- catalyst, of ceric sulfate solutions with osmium tetroxide catalyst in an acid medium, and of iodine-potassium iodide solutions at $pH \sim 6.5$–7 maintained with a CO_2-HCO_3^- buffer.

Pure copper, for standardization of thiosulfate solutions by the following reaction series: Cu is oxidized to Cu(II); Cu(II) is then reduced with KI and the liberated iodine (I_3^-) is titrated with thiosulfate.

Potassium dichromate, $K_2Cr_2O_7$, for preparation of standard dichromate solutions. The latter may be reduced with KI and the liberated iodine (I_3^-) used to standardize thiosulfate solutions.

Potassium iodate, KIO_3, for standardization of thiosulfate solutions by the following reaction series: KIO_3 is added to a large excess of KI in mild acid and the liberated iodine (I_3^-) is titrated with thiosulfate.

COMPLEX-ION FORMATION WITH ETHYLENEDIAMINE TETRAACETATE ION

Metal ions with a single stable oxidation state, such as Mg(II), Ca(II), Zn(II), or Al(III), cannot be titrated by redox methods. However, such ions do have a tendency to form coordination complexes, especially with nitrogen, oxygen, and fluorine bases, and normally prefer a coordination number of six. The molecules of the organic complexing agent ethylenediamine tetraacetate ion (commonly known as EDTA) are custom made to form stable complexes with such ions.

The structural formula of EDTA is shown in (4.36).

$$
\begin{bmatrix}
\text{(3)} \quad {}^-\text{O--C--CH}_2 & & & \text{CH}_2\text{--C--O}^- \quad \text{(5)} \\
\qquad\qquad\;\; \overset{\text{O}}{\overset{\|}{}} & & & \qquad\overset{\text{O}}{\overset{\|}{}} \\
& \text{N--CH}_2\text{--CH}_2\text{--N} & & \\
& \text{(1)} \qquad\qquad \text{(2)} & & \\
\text{(4)} \quad {}^-\text{O--C--CH}_2 & & & \text{CH}_2\text{--C--O}^- \quad \text{(6)}
\end{bmatrix}^{4-}
$$

(4.36)

This ion, which has a charge of $4-$, is *hexa*dentate, that is, it has six coordinating sites or "teeth" numbered (1) through (6) in (4.36). These sites are separated by short, flexible carbon chains of just the right length so that all six "teeth" can coordinate with a single cation. Thus, the structure of the complex of EDTA with Mg(II) is shown in (4.37).

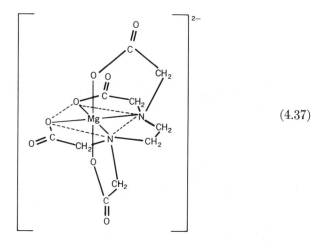

(4.37)

Stability of EDTA complexes

In this section, we shall let Y^{4-} denote the hexadentate EDTA anion whose structure is shown in (4.36), and M^{2+} any divalent metal ion,* such as Mg^{2+} or Ca^{2+}. The chemical reaction between EDTA and the metal ion is then shown in (4.38), where MY^{2-} denotes a hexadentate complex of the general structure shown in (4.37).

$$M^{2+} + Y^{4-} \rightleftharpoons MY^{2-} \tag{4.38}$$

$$K_{assoc} = \frac{[MY^{2-}]}{[M^{2+}][Y^{4-}]} \tag{4.39}$$

The equilibrium constant, expressed in (4.39), is the association constant for complex-ion formation: If K_{assoc} is large, the formation of the hexadentate complex is favored.

Numerical values of K_{assoc} for a variety of metal ions are listed in Table 4.6. The values are small for Na^+, K^+, and alkali metal ions in general, but are quite large for silver ion, the alkaline earth ions, aluminum ion, and other divalent and trivalent metal ions. Moreover, the values are quite specific, being larger for transition-metal ions than for group II or group III ions of the same charge, and for trivalent ions than for divalent ions of similar size and electronic structure.

If we are contemplating the use of EDTA as a titrant, the data in Table 4.6 are quite favorable. On applying (2.19b), we find that the innate titration error for titration of a metal ion with EDTA will be less than 0.25 percent if $c_s \cdot K_{assoc} > 5 \times 10^5$. A typical value for c_s in EDTA titrations is $10^{-2}\ F$. Hence the innate titration error will be less than 0.25 per-

---

* We use M^{2+} as a convenient symbol since most common EDTA complexes involve divalent metal ions. Complex formation for metals in other oxidation states is analogous.

**Table
4.6**   ASSOCIATION OF THE HEXADENTATE EDTA
ANION WITH VARIOUS METAL IONS

Metal ion	K_{assoc}	Lowest pH for	
		Titration with EDTA	Precipitation of hydroxide[a]
Na^+	Small[b]	—	—
K^+	Small[b]	—	—
Ag^+	2×10^7	—	10
Mg^{2+}	5×10^8	10	11
Ca^{2+}	5×10^{10}	8	>12
Sr^{2+}	4×10^8	10	>12
Ba^{2+}	6×10^7	*ca.* 10	>12
Zn^{2+}	3×10^{16}	4	6.5
Al^{3+}	1×10^{16}	4	5
Mn^{2+}	6×10^{13}	5.5	9
Fe^{2+}	2×10^{14}	5	7.5
Fe^{3+}	1×10^{25}	1	3
Cu^{2+}	6×10^{18}	3	6
Ni^{2+}	4×10^{18}	3	8
Co^{2+}	2×10^{16}	4	8
Cd^{2+}	3×10^{16}	4	8
Pb^{2+}	1×10^{18}	3.5	7
Hg^{2+}	6×10^{21}	2	9.5

[a] Or oxide.
[b] Too small to be important in EDTA titrations.

cent if $K_{assoc} > 5 \times 10^7$. According to the data in Table 4.6, this condition is satisfied by all divalent and trivalent metal ions.

If we are contemplating *selective* titration of metal ions with EDTA, the data in Table 4.6 are also encouraging. A rigorous discussion will be postponed to Chapter 8, but we may state the following result: If the two ions are present at comparable concentration, selective titration is feasible if K_{assoc} for the sought ion is at least 10^5 times greater than K_{assoc} for the other ion. According to the data in Table 4.6, it should thus be possible, for instance, to titrate Al^{3+} in the presence of Ca^{2+}, or Hg^{2+} in the presence of Mg^{2+}.

These predictions are essentially in agreement with fact. EDTA is a useful, versatile titrant for a wide variety of metal ions. Titration can be done potentiometrically, using methods to be described in Chapter 10, or with a color endpoint indicator. However, before describing the actual reaction conditions, we must first describe the acid-base properties of EDTA and of EDTA-metal complexes.

Acid-base properties

When we examine the structural formula of the EDTA ion in (4.36), we find two different kinds of basic sites: The two nitrogen atoms, (1) and (2), are chemically analogous to, say, the nitrogen atom in ammonia, while the four carboxylate groups, (3) to (6), are chemically analogous to the carboxylate group in acetate ion. And, just as ammonia is a stronger Brønsted base than acetate ion, so the nitrogen atoms in (4.36) are more strongly basic than the carboxylate groups.

In treating the acid-base properties of EDTA, it is customary to begin with the electrically uncharged ethylenediamine tetraacetic acid, H_4Y. In this tetrabasic acid, two of the protons are bonded to the nitrogen atoms, at sites (1) and (2) in (4.36), while the other two protons are bonded to carboxylate groups, probably one proton at site (3) or (4) and the other proton at site (5) or (6). The stepwise acid-dissociation constants of ethyl-

	Number of protons		pK_A	
Acid	COOH	NH	Symbol	Value
H_4Y	2	2	pK_{A1}	2.2
H_3Y^-	1	2	pK_{A2}	2.7
H_2Y^{2-}	0	2	pK_{A3}	6.2
HY^{3-}	0	1	pK_{A4}	10.0
Y^{4-}	0	0	Not an acid	

STEPWISE ACID-DISSOCIATION CONSTANTS **Table**
OF ETHYLENEDIAMINE TETRAACETIC ACID **4.7**

enediamine tetraacetic acid are given in Table 4.7 and allow us to construct the principal species diagram shown in Fig. 4.3. Note that the hexadentate Y^{4-} ion is the principal species only above pH 10.

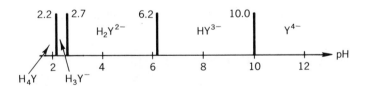

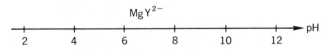

Fig. 4.3. Principal species diagram of ethylenediamine tetraacetic acid, H_4Y (top), and of the magnesium-EDTA complex, MgY^{2-} (bottom).

By contrast with EDTA, the metal-EDTA complex is quite inert as an acid or base because all six active sites shown in (4.36) are now tied to the metal ion, as shown in (4.37). The hexadentate complex therefore remains the principal species over the entire aqueous pH range; that is, conversion of MY^{2-} to HMY^- is insignificant and may be neglected on a principal species diagram such as Fig. 4.3. Accordingly, the chemical equation (in terms of principal species) for the formation of the metal-EDTA complex will vary with pH as follows:

$$M^{2+} + Y^{4-} \rightleftharpoons MY^{2-}; \qquad pH > 10.0 \qquad\qquad (4.40)$$

$$M^{2+} + HY^{3-} + H_2O \rightleftharpoons MY^{2-} + H_3O^+; \qquad 10.0 > pH > 6.2 \qquad (4.41)$$

$$M^{2+} + H_2Y^{2-} + 2\,H_2O \rightleftharpoons MY^{2-} + 2\,H_3O^+; \qquad 6.2 > pH > 2.7 \qquad (4.42)$$

Although in each case the reaction product is MY^{2-}, Eqs. (4.41) and (4.42) show that below pH 10 the formation of MY^{2-} is opposed by the mass-action effect of hydrogen ion.

Let us consider the significance of Eqs. (4.40) through (4.42) for EDTA titrations. In general, whenever there is a decrease in the tendency of a substrate and titrant to form a titration product, the innate titration error will increase. Equations (4.40) through (4.42) show that the tendency to form the MY^{2-} ion decreases progressively with decreasing pH below pH 10. Consequently, the innate error of EDTA titrations will increase progressively with decreasing pH below pH 10.

This prediction is consistent with observation. For each metal ion there is a minimum pH value below which the error becomes so great that titration is no longer satisfactory. These values are listed in Table 4.7, and they vary widely. In general, the greater the association constant for the metal-EDTA complex, the lower will be the pH at which EDTA titration is still satisfactory. Table 4.7 shows also that for most ions, titration can be done below the pH range in which precipitation of the metal oxide or hydroxide might become a problem.

Schwarzenbach's magnesium-EDTA titration

Most metal ions that form EDTA complexes also form less stable complex ions, of varying stability, with many common solutes. The formation of complex ions with solutes such as NH_3, Cl^-, or $SO_4{}^{2-}$ reduces the tendency towards precipitation of hydroxides, but it produces two undesirable effects in EDTA titrations: It raises the innate titration error (although usually within tolerable limits), and it makes it more difficult to find a color endpoint indicator. For instance, an endpoint indicator that is quite suitable for the titration of Pb^{2+} in a chloride-free medium may not be suitable for the titration of $PbCl_2$, which forms when chloride ion is present.

Among the ions that can be titrated with EDTA, magnesium ion stands out as being particularly insensitive to the presence of complex-forming solutes, with the result that all magnesium titrations can be done with a single indicator. In Schwarzenbach's method of titrating magnesium with EDTA, the weak organic acid, Eriochrome Black T (Erio T), is used as color indicator. The titration medium is adjusted to pH 10 by addition of ammonia-ammonium chloride buffer, and about 0.2 g of ascorbic acid (vitamin C) is added to remove oxygen from the solution and thus to prevent air oxidation of the indicator. The color due to Erio T at pH 10 is red in the presence of magnesium, magenta just before the equivalence point, and blue when EDTA becomes present in excess.

Schwarzenbach's method is so satisfactory that the analysis of other ions is often arranged so that the final titration is essentially one of Mg(II) with EDTA. For example, in the analysis of calcium(II), which by itself does not give a good Erio T endpoint, one deliberately adds a small amount of Mg-EDTA complex and titrates nonselectively with EDTA to the Erio T endpoint, using the same reaction conditions as if magnesium alone were the substrate. This procedure is known as a *substitution titration* because the added Mg-EDTA complex reacts according to (4.43) and Mg^{2+} is thus substituted for Ca^{2+} in the titratable substrate. As titrant is added,

$$MgY^{2-} + Ca^{2+} \rightleftharpoons CaY^{2-} + Mg^{2+} \tag{4.43}$$

reaction takes place approximately in two steps: first with Ca(II), which has the greater association constant (as shown in Table 4.6), and then with Mg(II). The titration endpoint will be as sharp as if Mg(II) had been titrated alone.

The nonselective titration of a metal ion, after addition of a known quantity of Mg(II), will succeed if the association constant of the metal ion with EDTA is greater than that of Mg(II) with EDTA, and if the metal ion has no unusual affinity for Erio T. Many of the metal ions in Table 4.6 satisfy these conditions and can therefore be determined by Schwarzenbach's method, in the presence of Mg(II), by nonselective titration to the Erio T endpoint at pH 10. Possible variants of the technique are:

1. Add a known amount of Mg(II) to the unknown substrate and titrate with a standard solution of EDTA at pH 10 to the Erio T endpoint.

2. Add a known amount of EDTA, in excess to the amount of the unknown substrate, and back-titrate with a standard solution of Mg(II) to the Erio T endpoint. The EDTA may be added at a low pH, where the substrate remains in solution; the back-titration is done at pH 10.

3. Prepare a magnesium-doped EDTA titrant containing about 0.05 equivalent of Mg(II) per equivalent of EDTA. Determine the normality of free EDTA in the titrant. Then titrate the unknown substrate with the magnesium-doped titrant at pH 10 to the Erio T endpoint.

The usefulness of EDTA titrations in quantitative analysis goes far beyond these variations in the magnesium titration. In Chapter 15 we shall briefly survey the use of EDTA for the analysis of nonmetals and weakly complexing metals, and for the direct titration of strongly complexing metals.

Chemical standards

The disodium salt of EDTA forms a dihydrate that is nonhygroscopic and stable at room temperature and ordinary humidity. Chemical formula, $Na_2C_{10}H_{14}O_8N_2 \cdot 2 H_2O$, mol. wt. 372.2. This substance ($Na_2H_2Y \cdot 2 H_2O$ for short) is available in analytical-reagent quality. The reagent may be purified further by solution in water (9–10 percent by weight) and reprecipitation with pure ethyl alcohol. $Na_2H_2Y \cdot 2 H_2O$ is dried effectively by heating at atmospheric pressure in a drying oven at 80°C for four days. Higher temperatures than 80°C should be avoided as the dihydrate then tends to decompose to the anhydrous salt. If a metal-ion standard is needed, pure reagent-grade calcium carbonate or pure natural calcite will be satisfactory.

REFERENCES

W. J. Blaedel and V. W. Meloche, *Elementary Quantitative Analysis*, 2nd Ed., Harper and Row, Publishers, New York, 1963.

I. M. Kolthoff, E. B. Sandell, E. J. Meehan, and S. Bruckenstein, *Quantitative Chemical Analysis*, 4th Ed., The Macmillan Company, New York, 1969.

I. M. Kolthoff and V. A. Stenger, *Volumetric Analysis*, 2nd Ed., John Wiley & Sons, Inc., New York, 1942–1957 (three volumes).

H. A. Laitinen, *Chemical Analysis*, McGraw-Hill Book Company, New York, 1960.

L. Meites, Ed., *Handbook of Analytical Chemistry*, McGraw-Hill Book Company, New York, 1963.

A. Ringbom, *Complexation in Analytical Chemistry*, Interscience Publishers, New York, 1963.

W. W. Scott (N. H. Furman, Ed.), *Standard Methods of Chemical Analysis*, 5th Ed., D. Van Nostrand Co., Inc., New York, 1939.

G. Schwarzenbach and H. Flaschka, *Complexometric Titrations*, Translated by H. M. N. H. Irving, Methuen & Co., Ltd., London, 1969.

D. A. Skoog and D. M. West, *Fundamentals of Analytical Chemistry*, 2nd Ed., Holt, Rinehart & Winston, Inc., New York, 1969.

PROBLEMS

(*Note:* Refer to the Appendix for values of equilibrium constants and electrode potentials.)

1. Starting with Eq. (4.3), derive the expression for the equilibrium constant for the titration of a weak base with a strong acid in terms of K_W and K_B.

2. Show that for a weak Brønsted acid HA and its conjugate base A⁻, the expression

$$pK_A(\text{for HA}) + pK_B(\text{for A}^-) = pK_W$$

always holds.

3. Express Eq. (4.9) in terms of pH rather than $[H_3O^+]$. Do the same for Eq. (4.11). You now have simple, direct relationships for calculating the pH at the equivalence point in the titration of weak acids with strong base and of weak bases with strong acid.

4. Calculate the pH at the equivalence point for each of the following titrations. Be sure when you determine formal concentrations that you have considered the dilution factor introduced when titrant a isdded. (You may assume a specific volume of substrate if you wish.) Use Eqs. (4.9) and (4.11) (as written in Problem 3, if you wish) or, if they do not apply, see if you can derive an expression for the pH at the equivalence point. If necessary, you may skip ahead to Chapter 9, Eqs. (9.25) through (9.30), for help. Then consult Table 9.2 and pick a suitable color endpoint indictor for each titration. Would you expect a successful analysis?

  (a) Titration of 0.1000 N HClO₄ with 0.1000 N NaOH
  (b) Titration of 0.05000 F chloroacetic acid with 0.1000 N NaOH
  (c) Titration of 0.2000 F NH₃ with 0.1000 N HCl
  (d) Titration of 0.1000 F HOCl with 0.05000 N NaOH
  (e) Titration of 0.1000 F NaN₃ with 0.1000 F HAc
  (f) Titration of 0.05000 F Na₂CO₃ to the first equivalence point with 0.1000 N HCl
  (g) Titration of 0.05000 F Na₂CO₃ to the second equivalence point with 0.1000 N HCl

5. A NaOH solution is to be standardized using solid potassium acid phthalate (KHP: equivalent weight 204.22). Three (dry) KHP samples weighing 0.9530, 1.0011, and 0.9723 g are dissolved in CO₂-free water and titrated to a phenolphthalein endpoint. If the three titrations require 42.10, 44.35, and 43.00 ml of base, respectively, what is the normality of the NaOH solution? (*Hint:* How many *milliequivalents* of KHP are being titrated in each case? A milliequivalent is 10^{-3} equivalent.)

6. A solution of a weak acid HA, with $K_A = 1.0 \times 10^{-4}$ is to be titrated with 0.1000 N NaOH. Approximately what pK should the indicator used in this titration have? In three trials, it is found that 50.00 ml of the acid solution require 46.25 ± 0.15 ml of NaOH for neutralization. What is the concentration of the acid?

7. You now know *four* methods for the quantitative analysis of halide ions. Briefly describe each. If given a solid chloride sample for analysis, which procedure would you choose? Why?

8. Using the appropriate data from Table 4.2, demonstrate the validity of Mohr's method, i.e., show that there will be no Ag₂CrO₄ precipitate until after the equivalence point in a chloride titration. Under common experimental conditions a concentration of 0.05 g K₂CrO₄ per 100 ml of solution is used. How many ml of 0.1 N AgNO₃ solution must be added beyond the equivalence point before Ag₂CrO₄ will begin to precipitate under these conditions? How would you correct for the error involved in producing a *detectable* endpoint? Remember that the Ag₂CrO₄ color must be seen in the presence of a large amount of silver chloride precipitate.

9. (a) What is the reason for adding acid before a Volhard titration?

(b) Calculate the solubility of silver chloride in a 0.05000 F ammonia solution? [The formation constant for $Ag(NH_3)_2{}^+$ is 1.4×10^7.]

10. Propose some analytical problems which cannot be solved by direct titration but where back titration can be employed. You may consult outside sources if you have trouble.

11. In some situations, even when direct titration is feasible, it is more convenient to deliver the substrate and titrant from two burets, rather than use a pipet for the substrate and a buret for the titrant. Describe how this procedure might be put to use. What are its advantages?

12. What is a "primary standard"? What characteristics should a reagent have in order to be useful as a primary standard? Give some examples of substances that meet these criteria and describe their specific applications.

13. What factors, other than the $E°$ value for the reaction, influence the feasibility of a redox titration?

14. Arsenious oxide, As_2O_3, is often used as a primary standard for permanganate, ceric, and triiodide solutions. The two former titrations are carried out in acid solutions while for the latter a nearly neutral medium is used. Write balanced equations and calculate the $E°$ value for each of these reactions. Why do you suppose it is necessary to do the iodide standardization in a solution buffered at a nearly neutral pH? (Note: As_2O_3 in acid becomes H_3AsO_3)

15. How many grams of the following reagents are required to prepare 250 ml of 0.1000 N solutions of the following oxidizing agents? The appropriate half-reactions are given either in the text or in Table A.5.

(a) KIO_3 (see Table 4.4)

(b) $Ce(HSO_4)_4$

(c) $KMnO_4$

(d) $K_2Cr_2O_7$

16. Hexacyanoferrate(II) ion (ferrocyanide) may be conveniently analyzed by titration with standard ceric sulfate in HCl or H_2SO_4 to a ferroin endpoint. What is the concentration of $Fe(CN)_6{}^{4-}$ in a solution that required 47.23 ml of 0.06250 N Ce(IV) for titration of a 50.00 ml aliquot? How many milliliters of 0.02105 F $KMnO_4$ solution would be required for the titration?

17. A 50.00 ml aliquot of a solution containing both permanganic acid ($HMnO_4$) and sulfuric acid required 42.50 ml of 0.1000 N NaOH for neutralization. A second 50.00 ml sample was titrated with ferrous ion which reduces $MnO_4{}^-$ to Mn^{2+}. If 25.00 ml of 0.1000 N Fe^{2+} solution was needed for the titration, what are the concentrations of the two acids?

18. The role of a redox indicator is quite analogous to the function of the chromate ion in the Mohr chloride analysis. How do you suppose a substance such as ferroin serves to indicate the endpoint in a redox titration? Try *not* to look ahead to Chapter 9.

19. Calculate the (pseudo-) equilibrium constant for the following reactions in 0.5000 M H^+ and at pH 5. In which cases would you expect successful titration?

(a) Ceric ion plus stannous ion (in perchlorate media)

(b) Permanganate ion plus ferrous ion

(c) Dichromate ion plus arsenious acid
(d) Triiodide ion plus arsenious acid

20. Describe how you would analyze a copper ore containing about 60 percent Cu so that the final step is the titration of iodine with thiosulfate. What quantities of each reagent would be required? (Assume that the thiosulfate solution is 0.1000 N and that a 50 ml buret is used.) What precautions might be necessary? Give all pertinent chemical equations and estimate the possible errors involved in the determination.

21. Write the half-reaction for the following:
(a) The oxidation of hydrazine (H_2N—NH_2) to nitrogen
(b) The oxidation of cysteine to cystine [See Eq. (4.34) for molecular structures.]

22. How might one use the iodide-dichromate reaction in the determination of Pb^{+2}, Ag^+, or Ba^{2+}?

23. Consult Table 4.6 and list some pairs of ions for which *selective* EDTA titration might be possible. On what criteria are your answers based?

24. Cobalt(II) forms an EDTA complex of high stability, but, because of the formation of insoluble hydroxides, Co^{2+} cannot be titrated at the pH 10 required for the Eriochrome Black T indicator. One way to get around this difficulty is to form the cobalt-EDTA complex at low pH, then buffer the solution at pH 10 with an ammonia-ammonium chloride buffer, and back-titrate the excess EDTA with standard Mg^{2+} to the Erio T endpoint.
(a) Write molecular equations for all reactions pertinent to this analysis including reactions of the indicator.
(b) Calculate the relative concentrations of NH_3 and NH_4Cl needed to produce a pH 10 buffer.
(c) In an actual experiment 50.00 ml of 0.01056 F EDTA was added to 25.00 ml of an unknown Co^{2+} solution. After buffering at pH 10, the solution required 17.25 ml of 0.01132 F Mg^{2+} solution for titration. What was the concentration of the Co^{2+} solution?

25. EDTA titrations are often used to determine water hardness. In a typical analytical procedure, the EDTA is first standardized using pure $CaCO_3$ and the *titer* expressed as the number of milligrams of calcium carbonate that are equivalent to 1.00 ml of EDTA. A hardness determination then yields information which is easily expressed as parts per million (p.p.m.) of calcium carbonate (by weight). 50.00 ml of a standard calcium solution whose concentration is 0.4052 g of $CaCO_3$ per liter requires 27.30 ml of magnesium-doped EDTA for titration to the Erio T endpoint. What is the titer of the EDTA in mg of $CaCO_3$ per ml? A 500.0 ml water sample takes 21.05 ml of this EDTA solution for titration. Express the hardness of the water sample as parts per million of calcium carbonate.

26. Not all complexometric titrations involve EDTA. For instance, mercury(II) may be determined quantitatively by titration with standard thiocyanate.

$$Hg^{2+} + 2\ SCN^- \rightleftharpoons Hg(SCN)_2$$

The endpoint is detected by adding a bit of ferric ion which reacts with SCN^- just past the equivalence point to form the rust-orange $FeSCN^{2+}$ complex. How is this method similar to the Mohr determination of chloride ion? to the Volhard determination? What are the differences? By consulting Tables A.3 and A.4 see if you can come up with some new (to you) precipitation or complexometric methods of analysis for specific ions.

Effect of the Solvent
on the Chemistry of Ions

It is no accident that the great majority of chemical reactions are carried out in liquid solution. The liquid state offers a better opportunity to vary reaction conditions and control reactivity than any other state of matter. We have seen that by changing the $p\text{H}$, one can convert specific solutes to their conjugate acids or bases; by using buffers, one can maintain specific solutes at a very low concentration; by adding appropriate reagents, one can cause the formation of complex ions or of insoluble precipitates, or reverse their formation. In these procedures, the solvent is kept constant.

We shall now introduce a new dimension by allowing the solvent to vary—by considering solvents other than water. In the brief space available we shall consider only the two most dramatic effects of the solvent: How it affects the behavior of dissolved electrolytes, and how it reacts as an acid or base. It is helpful to discuss these phenomena separately, and in this chapter we shall examine the behavior of electrolytes.

Our intention is to acquaint you with general ideas of solution chemistry

101

so that chemical analysis in nonaqueous solvents will seem less strange. We shall find that the chemistry of ionic solutes in nonaqueous solvents is often peculiarly different from that in water. However, the differences are, in large measure, predictable and an analytical chemist must be prepared to cope with them.

CHARGED-SPHERE MODEL OF AN ION AND THE DIELECTRIC CONSTANT

Many features of the effect of the solvent can be understood if we use a model in which an ion in solution is likened to an electrified sphere in a perfectly continuous fluid medium. It is a consequence of Coulomb's law that physical work is required to place an electrical charge on an initially uncharged sphere. As a result, the electrified sphere will have a higher energy, and thus will be less stable, than the same sphere without the electric charge. In the sense that an ion in liquid solution resembles a charged sphere, we may say therefore, that *the ionic charge is an inherent source of instability*.

On applying this model further, we find, however, that the extent of instability varies greatly with the solvent. When a charged sphere is transferred from a vacuum to a nonconducting fluid, the instability associated with the electrical charge is reduced by a mechanism called *polarization of the medium*. As shown in Fig. 5.1, the charged sphere induces a polarization

$\dot{P}$

Fig. 5.1. Electrical polarization of the medium around a charged sphere. The charge on the sphere, $+q$, induces a polarization charge of opposite sign. The net charge of sphere plus polarized shell sensed at point P is q/ε.

charge of opposite sign in a thin layer of the fluid material around it. A test charge placed at a point P outside that layer will "sense" only the *net* charge, that is, it will sense the charge on the sphere minus the polarization charge. The *force* acting on the test charge will be proportional to the *net* charge and will therefore be smaller than it would be if the charges were interacting in a vacuum. The *ratio* of the force acting on the test charge in a vacuum to that acting in the medium, at the same distance of separation, is called the *dielectric* constant of the medium, and is denoted by ε. By this definition, the dielectric constant of a vacuum is unity.

Returning to an ion in a liquid solvent, we may conclude, according to this model, that the ionic charge is partially compensated by polarization of the surrounding liquid. The *net* charge of ion plus polarized solvent shell, is less than the full ionic charge by a factor of $1/\varepsilon$. As the *net* charge is reduced, the instability associated with the ionic charge is reduced correspondingly. A liquid with a high dielectric constant is therefore expected to be a relatively good solvent for free ions. We shall find that this conclusion is in agreement with fact. But first, let us become familiar with the more common nonaqueous solvents.

SOME COMMON NONAQUEOUS SOLVENTS

Some of the more common solvents of chemistry are described in Table 5.1. These liquids include substances that are miscible with water in all proportions, such as methanol and acetone; substances that are virtually insoluble in water, such as carbon tetrachloride; substances that ionize as acids in water, such as sulfuric acid and acetic acid; and substances that ionize as bases in water, such as pyridine and ammonia. With the exception of ammonia, all substances listed in Table 5.1 are liquid at room temperature.

The term *nonaqueous solvent* is applied not only to the pure anhydrous liquids but also to mixtures of liquids, including mixtures with water. Thus, most of the water-miscible liquids in Table 5.1 are also used in the form of mixed solvents with water. Especially popular are ethanol-water, dioxane-water, and dimethylsulfoxide-water mixtures. In the genuinely anhydrous domain, common mixed solvents are mixtures of carbon tetrachloride or benzene with methanol, ethanol, or chloroform.

On examining the data in Table 5.1, we find a wide variation in the dielectric constant, from greater than 100 for sulfuric acid at 25°C to 2.21 for dioxane at 25°C. The dielectric constant of water is seen to be relatively high, being 78.54 at 25°C. By comparison, dielectric constants of gases up to moderate pressures are only slightly greater than unity, and the dielectric constant of a vacuum is, by definition, exactly unity.

FORMATION OF ION PAIRS

The charged-sphere model predicts that the instability associated with the electric charge of free ions can be reduced through the formation of ionic aggregates, particularly of ion pairs. Since this concept provides the key to an understanding of much of the chemistry of ions in media of low dielectric constant, we wish to examine it briefly.

Ion-pair formation, according to the charged-sphere model, is represented by the process shown in Fig. 5.2: Two widely separated spheres, with charges $+|q|$ and $-|q'|$, come together in the liquid medium until the spheres almost touch and the charges are separated by a characteristic dis-

Table 5.1 PHYSICAL CONSTANTS OF SOME COMMON LIQUID SOLVENTS

Solvent	Chemical formula	M.p. (°C)	B.p. (°C)	Moles per liter	Solubility in water	Dielectric constant (25°C)
Anhydrous sulfuric acid	H_2SO_4	10.4	>300 (dec)	19.0	Soluble (dec)	>100
Water	HOH	0.0	100.0	55.5	Miscible	78.54
Methanol	CH_3OH	−97.8	64.7	24.5	Miscible	32.63
Ethanol	CH_3CH_2OH	−117.3	78.5	17.0	Miscible	24.30
tertiary-Butyl alcohol	$(CH_3)_3COH$	25.5	82.2	10.6	Miscible	12.4
Acetic acid	$CH_3C\overset{O}{\overset{\|}{C}}OH$	16.6	118.5	17.4	Miscible	6.3
Ammonia	NH_3	−77.7	−33.4	40.1 (−33°)	Miscible	22.4 (−33°C)
Chloroform	$HCCl_3$	−63.5	61.2	12.4	Insoluble	4.8
Dimethylsulfoxide	$CH_3\overset{O}{\overset{\|}{S}}CH_3$	6	100 (dec)	14.1	Miscible	46.4
Dimethylformamide	$H\overset{O}{\overset{\|}{C}}N(CH_3)_2$	−61	153	12.9	Miscible	36.7
Acetonitrile	$CH_3{-}C{\equiv}N$	−45.7	80.1	17.3	Miscible	36.0
Acetone	$CH_3\overset{O}{\overset{\|}{C}}CH_3$	−95.4	56.2	13.5	Miscible	20.7
Pyridine	(pyridine ring)	−42	115.5	12.4	Miscible	12.3

Solvent	Chemical formula	M.p. (°C)	B.p. (°C)	Moles per liter	Solubility in water	Dielectric constant (25°C)
Dioxane	(structure: dioxane ring)	11.8	101	11.7	Miscible	2.21
Benzene	(structure: benzene ring)	5.5	80.1	11.1	Insoluble	2.27
Carbon tetrachloride	CCl_4	−23.0	76.8	10.3	Insoluble	2.23

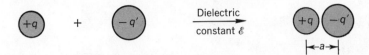

Fig. 5.2. Charged-sphere model for the formation of an ion pair. a is the center-to-center distance at closest approach.

tance a. The instability, that is, the energy associated with the electric charges of the spheres, is thus reduced by an amount W, given (on the basis of Coulomb's law) by Eq. (5.1).

$$W = -\frac{|q| \cdot |q'|}{\varepsilon a} \tag{5.1}$$

According to this theory, the greater the magnitude of W, the greater will be the tendency towards ion-pair formation. Equation (5.1) thus indicates the variables that need to be considered. They are: the ionic charges, the dielectric constant, and the characteristic distance of approach a. It turns out that a varies the least. For typical electrolytes, a is about 5 Å, and we shall not consider it further. The dielectric constant, on the other hand, may vary a great deal as the medium is changed. Since ε appears in the denominator in (5.1), ion-pair formation should be most important in solvents of low dielectric constant.

A typical electrolyte in the gas phase

We begin by considering a rather drastic change of the dielectric constant, from that of an aqueous solution ($\varepsilon = 78.5$) to that of the gas phase ($\varepsilon = 1.0$). For definiteness, we shall consider a specific solute, lithium bromide, which is a typical ionic salt. In dilute aqueous solution, lithium bromide is almost completely dissociated to lithium ions and bromide ions, which react, act, and move like independent particles. However, as a dilute gas at 1000°K, lithium bromide exists almost entirely in the form of neutral molecules. The principal molecular species is LiBr. Next, in order of probability, are $(LiBr)_2$ and $(LiBr)_3$, while the concentration of free ions in the equilibrium mixture is insignificant. These conclusions are based on direct measurements of molecular velocities in gaseous lithium bromide, which leave little doubt that the moving particles are indeed compound molecules rather than free ions.

What is the electronic structure of the LiBr molecule? Because both free ions, Li^+ and Br^-, have an inert noble gas electron configuration, it is reasonable to guess that the LiBr molecule will be similar to a lithium ion and a bromide ion in close proximity. The ions are held together, not by a covalent bond, but simply by the force of attraction that exists always between electrical charges of opposite sign. If this be granted, then it is logical to refer to the LiBr molecule as an *ion pair* and to represent its molecular formula as Li^+Br^-. There is much evidence that such a model of the LiBr molecule is substantially correct. For instance, the strong tendency in the gas phase for the LiBr molecules to associate further can be explained on the basis of attractive forces among ionic charges, as in Eqs. (5.2) and (5.3).

$$2\ Li^+Br^-(g) \;\rightleftharpoons\; \begin{vmatrix} Li^+ & Br^- \\ Br^- & Li^+ \end{vmatrix}(g) \tag{5.2}$$

$$K_{assoc} = 4.5 \times 10^5\ (M^{-1}) \qquad \text{at } 1000°K$$

$$3\ Li^+Br^-(g) \;\rightleftharpoons\; \begin{vmatrix} Li^+ & Br^- & Li^+ \\ Br^- & Li^+ & Br^- \end{vmatrix}(g) \tag{5.3}$$

$$K_{assoc} = 1.0 \times 10^{10}\ (M^{-2}) \qquad \text{at } 1000°K$$

Association of univalent ions in liquid solution

The predictions of the charged-sphere model are in substantial agreement with fact for many univalent electrolytes in a wide range of solvents. Salts that are strong electrolytes in water generally become weak electrolytes as the dielectric constant of the solvent decreases. The solution properties indicate that the free ions are in equilibrium with ion pairs which, at high concentrations and in solvents of *very* low dielectric constant, are capable of associating further.* At this point it will be sufficient to consider only the association of free ions to ion pairs. A general equation for the formation of ion pairs from free ions is given in (5.4), and the equilibrium constant is defined in (5.5).

$$C^{m+} + A^{n-} \;\rightleftharpoons\; C^{m+}A^{n-} \tag{5.4}$$

$$K_{assoc} = \frac{[C^{m+}A^{n-}]}{[C^{m+}][A^{n-}]} \tag{5.5}$$

Typical results for univalent electrolytes in solvents covering a wide range of dielectric constants are listed in Table 5.2. Some of the association constants apply to salts with unfamiliar polyatomic organic ions; such salts were chosen for reasons of solubility, but all ions have stable electron configurations. The association constants show an unmistakable trend to increase as the dielectric constant decreases, in agreement with the charged-sphere model. Moreover, the range spanned by the association constants between water and benzene is truly enormous. At typical concentrations of 0.001 to 0.1 F, the electrolytes listed in Table 5.2 will be almost completely dissociated to free ions in *water*, but their free-ion fractions will be quite small in acetic acid and very small in benzene.

---

* Ion association constants are measured by conductivity, potentiometry, solubility, spectrophotometry, and a host of other methods. See, for example, C. W. Davies, *Ion Association* (Washington: Butterworth & Co., Ltd., 1962).

Table 5.2   EXPERIMENTAL ION-ASSOCIATION CONSTANTS FOR UNIVALENT ELECTROLYTES AT 25°C

Solvent	ε	Cation	Anion	K_{assoc}
Water	78.5	Na^+	OH^-	0.2
	78.5	K^+	IO_3^-	0.5
Acetonitrile	36.0	$(n\text{-}C_4H_9)_4N^+$	Picrate$^-$	14
Nitrobenzene	34.8	$(n\text{-}C_4H_9)_4N^+$	Br^-	50
Methanol	32.6	$(n\text{-}C_4H_9)_4N^+$	Br^-	20
Ammonia ($-33.3°C$)	22.4	K^+ or Na^+	$(C_6H_5)_2N^-$	200
Acetone:				
12.48% water	25.7	Li^+	Br^-	105
5.35% water	22.5	Li^+	Br^-	560
0.45% water	20.7	Li^+	Br^-	3800
Pyridine	12.3	K^+	Picrate$^-$	1.0×10^4
Acetic acid	6.3	K^+	Cl^-	7.7×10^6
	6.3	$C_5H_5NH^+$	Ac^-	0.8×10^6
Benzene (30°C)	2.27	$(n\text{-}C_4H_9)_4N^+$	Cl^-	1.0×10^{17}
	2.27	$(n\text{-}C_4H_9)_4N^+$	ClO_4^-	0.46×10^{17}

Association of divalent and higher-valent ions

Table 5.3 lists some experimental association constants for ions of higher chargetype than one-to-one. According to Eq. (5.1), W (whose magnitude measures the tendency towards ion-pair formation) is proportional to the product of the ionic charges, and we therefore expect that K_{assoc} will in-

Table 5.3   EXPERIMENTAL ION-ASSOCIATION CONSTANTS FOR A VARIETY OF CHARGE TYPES IN WATER AT 25°C

Charge type $(|z| \cdot |z'|)$	Cation	Anion	K_{assoc}
2	Sr^{2+}	IO_3^-	6
2	Ba^{2+}	OH^-	5
2	Ca^{2+}	OH^-	25
2	Mg^{2+}	OH^-	300
2	K^+	$S_2O_3^{2-}$	5
2	Na^+	SO_4^{2-}	5
4	Mg^{2+}	$S_2O_3^{2-}$	70
4	Sr^{2+}	$S_2O_3^{2-}$	110
4	Ba^{2+}	$S_2O_3^{2-}$	210
4	Zn^{2+}	$S_2O_3^{2-}$	240
4	Mg^{2+}	SO_4^{2-}	160
4	Ca^{2+}	SO_4^{2-}	190
4	Ca^{2+}	$C_2O_4^{2-}$	1000
4	Tl^+	$Fe(CN)_6^{4-}$	1700
6	Mg^{2+}	$P_3O_9^{3-}$	2000
6	Ba^{2+}	$P_3O_9^{3-}$	2000

crease with $|z| \cdot |z'|$, where z and z' denote the charge number of the free cation and anion, respectively. The data in Table 5.3, which are all for aqueous solutions at 25°C, show a trend in the expected direction. However, there are marked specific differences within each charge type, and interactions other than those between simple charged spheres are apparently not quite negligible.

Principal molecular species of electrolytes

When the association constant is known, the principal species of the electrolyte can be found by solving the appropriate mass-action expression. Assuming that the concentrations of electrolytes in typical analytical procedures are 0.001 to 0.1 F, a rough rule of thumb is to represent the principal species as an ion pair when $K_{assoc} > 1000$, and as a pair of free ions when $K_{assoc} \leqslant 1000$. When the association constant is not known, a similarly rough rule, suggested by Eq. (5.1), is to represent the principal species as an ion pair when $\varepsilon/(|z| \cdot |z'|) < 20$, and as a pair of free ions when $\varepsilon/(|z| \cdot |z'|) \geqslant 20$. Applied to univalent electrolytes, this rule states that the principal molecular species becomes an ion pair when the dielectric constant is less than 20.

CHEMICAL CONSEQUENCES OF ION-PAIR FORMATION

Most of us are so familiar with electrolytic dissociation to free ions in water that the properties of electrolytes in solvents of low dielectric constant will at first seem strange. To illustrate a few of the concepts that must be modified, we shall consider some reactions involving electrolytes in acetic acid. The dielectric constant of acetic acid is 6.3, so that we expect the concentrations of free ions to be rather small. As a matter of fact, ion-association constants in acetic acid are typically of magnitude 10^6 or greater.

Effect of the counterion

In dilute aqueous solution, the free ions produced from the dissociation of electrolytes interact only weakly and in such a manner that the interaction depends on the ionic strength rather than on the specific ions present. When we write chemical equations for reactions in water, we therefore show only the reactive ions, and we omit the counterions. This is not true, however, when the reactive species is part of an ion pair.

As an example, consider the acid-base reaction of hydrogen chloride—

a weakly ionized acid in acetic acid—with acetate ion. The latter exists in acetic acid largely in the form of ion pairs, and the equation we write therefore shows the specific ion pairs used. Thus, when potassium acetate is the reagent, we write Eq. (5.6); when lithium acetate is the reagent, we write (5.7). Both equations, and the corresponding expressions for the equilibrium constant, expressly indicate the cation.

$$HCl + K^+Ac^- \rightleftharpoons K^+Cl^- + HAc$$

$$K = \frac{[K^+Cl^-]}{[HCl][K^+Ac^-]} \tag{5.6}$$

$$HCl + Li^+Ac^- \rightleftharpoons Li^+Cl^- + HAc$$

$$K = \frac{[Li^+Cl^-]}{[HCl][Li^+Ac^-]} \tag{5.7}$$

At 25°C, the equilibrium constant for (5.6) is 4.3×10^6; that for (5.7) is 1.5×10^6. Evidently, the cation exerts a substantial specific effect.

Common-ion effect on solubility

When we write the chemical equation for the dissolution of an ionic solid in water, we are usually justified in assuming that the crystalline solid produces dissociated ions. For instance, to represent the solubility of calcium hydroxide, we write Eq. (5.8).

$$Ca(OH)_2(s) \rightleftharpoons Ca^{2+}(aq) + 2\,OH^-(aq) \tag{5.8}$$

According to (5.8), addition of salts with a common ion (such as $CaCl_2$ or NaOH) will reduce the solubility by shifting equilibrium in (5.8) to the left. A mathematical statement of this effect is that the solubility product, $[Ca^{2+}][OH^-]^2$, must remain equal to K_{SP}. On the other hand, salts without a common ion (such as KNO_3 or NaCl) are expected to have no effect.

The situation is entirely different when the ionic solid exists in equilibrium largely with ion pairs. For example, consider the solubility of potassium bromide in acetic acid. Since this salt exists in acetic acid largely in the form of ion pairs, we write Eq. (5.9).

$$KBr(s) \rightleftharpoons K^+Br^-(in\ HAc) \tag{5.9}$$

A common-ion salt such as K^+Ac^- now should have very little effect on the solubility. To be sure, a small fraction of the dissolved K^+Br^- exists in the saturated solution in the form of free ions, and that fraction will decrease when a common-ion salt is added. However, since the free-ion fraction is at most small, any change in that fraction will have at most a small effect on the solubility. In agreement with this analysis, the solubility of KBr in acetic acid changes only slightly when 0.06 *F* KAc is added, as shown in Table 5.4.

0.06 F Salt added	Solubility of KBr (F)
None	0.0184
KAc	0.019$_0$
LiAc	0.029$_5$
NaAc	0.039$_5$

On the other hand, addition of a salt *without* a common ion, such as LiAc, now will *increase* the solubility by metathesis. This reaction is shown in (5.10).

$$K^+Br^- + Li^+Ac^- \overset{\text{HAc}}{\rightleftharpoons} K^+Ac^- + Li^+Br^- \tag{5.10}$$

As Li^+Ac^- is added to the saturated solution and $K^+Ac^- + Li^+Br^-$ forms, more of the solid KBr must dissolve in order that the *molar* concentration of K^+Br^- remain at the saturation value. The solubility, that is, the *formal* concentration of KBr in the saturated solution, therefore increases. Representative data in Table 5.4 show that the increase in the solubility is substantial.

Metathesis (or ion-pair exchange) is a typical reaction of ion pairs and has no counterpart among the reactions of free ions.

Solubility and ionic association in water

We have seen that ionic association is not strictly negligible in aqueous solutions, either, especially for higher-valent electrolytes. As a result, the solubility product sometimes gives a rather poor description of the actual solubility. Consider, for example, the solubility of calcium sulfate in water. Because the saturated solution will contain free ions as well as ion pairs, we write two equations:

$$CaSO_4(s) \rightleftharpoons Ca^{2+} + SO_4^{2-}$$
$$K_{SP} = [Ca^{2+}][SO_4^{2-}] = 6 \times 10^{-5} \tag{5.11}$$
$$Ca^{2+} + SO_4^{2-} \rightleftharpoons Ca^{2+}SO_4^{2-}$$

$$K_{assoc} = \frac{[Ca^{2+}SO_4^{2-}]}{[Ca^{2+}][SO_4^{2-}]} = 190 \tag{5.12}$$

On solving (5.11) and (5.12), we find that in the saturation solution in pure water $[Ca^{2+}] = [SO_4^{2-}] = 8 \times 10^{-3}\ M$ and $[Ca^{2+}SO_4^{2-}] = 11 \times 10^{-3}\ M$. That is, the concentration of ion pairs in the saturated solution is actually *greater* than that of the free ions. The solubility of $CaSO_4$, by definition, is equal to the total number of formula weights of $CaSO_4$ per liter of solution. The solubility is therefore the sum of 8×10^{-3} plus 11×10^{-3}, or 19×10^{-3} F. If we had neglected the ion pairs, we would have underestimated the solubility by more than half.

EFFECT OF DIELECTRIC CONSTANT
ON CHEMICAL EQUILIBRIUM

According to the charged-sphere model of an ion, the ionic charge becomes progressively more unstable as the dielectric constant is reduced. Having found that this simple picture is in substantial agreement with many facts, we shall now apply it to the reactions of ions in solution. In particular, by changing the dielectric constant we may hope to change equilibrium constants in predictable ways.

A convenient way of changing the dielectric constant of an aqueous solution is to dilute it with a water-miscible organic liquid such as ethyl alcohol, dioxane, or dimethylsulfoxide. In general, the dielectric constant of a homogeneous liquid mixture is intermediate between that of the pure

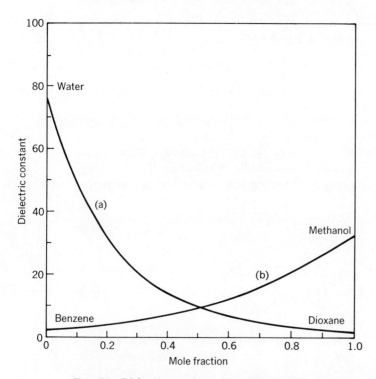

Fig. 5.3. Dielectric constant of typical liquid mixtures: (a) water-dioxane at 25°C; (b) benzene-methanol at 25°C. The mole fraction is that of the second component.

components and varies in a regular manner with composition. Two typical relationships are shown in Fig. 5.3. Thus, as dioxane ($\varepsilon = 2.21$) is added to water ($\varepsilon = 78.5$), Fig. 5.3 shows that the dielectric constant becomes progressively smaller.

To begin our discussion of equilibrium constants, let us consider a familiar acid-base reaction, that of hydronium ion with hydroxide ion in aqueous solution.

$$H_3O^+ + OH^- \rightleftharpoons 2\ HOH \qquad (5.13)$$

In this reaction, two unit ionic charges of opposite sign are destroyed. Since ionic charges become more unstable as the dielectric constant decreases, we expect that equilibrium in (5.13) will shift to the right. As shown in Table 5.5, that is precisely what happens when the dielectric

VALUES OF LOG K FOR SOME IONIC   **Table**
REACTIONS IN WATER-DIOXANE   **5.5**
MIXTURES AT 25°C[a]

Weight percent water in solvent:	100.0	80.0	55.0	30.0
Dielectric constant:	78.5	61.9	40.2	19.1
$H_3O^+ + OH^- \rightleftharpoons 2\ H_2O$[b]	14.00	14.62	15.74	17.86
$H_3O^+ + Ac^- \rightleftharpoons HAc + H_2O$	4.76	5.29	6.31	8.32
$NH_4^+ + OH^- \rightleftharpoons NH_3 + H_2O$	4.74	—	5.77	8.89
$H_3O^+ + NH_3 \rightleftharpoons NH_4^+ + H_2O$	9.26	—	8.97	8.97
$H_3O^+ + (CH_3)_3N \rightleftharpoons (CH_3)_3NH^+ + H_2O$	9.81	—	9.31	8.90

[a] Equilibrium constants are stated in molar concentration units.
[b] $K = ([H_3O^+][OH^-])^{-1}$.

constant is reduced by the addition of dioxane. For reaction (5.13) in water, $\log K = 14.00$—the familiar value for pK_W. However, in 30 percent water-70 percent dioxane, $\log K = 17.85$, which represents an increase of nearly four logarithmic units, or of nearly 10^4 in K. The effect is evidently very large. Moreover, we are justified in ascribing the effect largely to the change of dielectric constant because the dioxane molecules are quite unreactive, compared to water molecules, in acid-base reactions.

Table 5.5 lists data for two further reactions in which two unit ionic charges of opposite sign are destroyed:

$$H_3O^+ + Ac^- \rightleftharpoons HAc + HOH \qquad (5.14)$$

$$NH_4^+ + OH^- \rightleftharpoons NH_3 + HOH \qquad (5.15)$$

Again, $\log K$ increases substantially as the dielectric constant decreases, thus proving that reaction (5.13) is not an isolated example. Moreover, in this case the effect of the dielectric constant is of great practical importance. In plain water it is not possible to determine acetate ion by titration with strong acid because the equilibrium constant for (5.14) is too small.

You can convince yourself of that by making the test recommended in Eq. (2.19b). However, when the dielectric constant is reduced by adding seven parts of dioxane for every three parts of the original aqueous solution (and what could be easier?), the equilibrium constant increases by nearly four orders of magnitude, the innate titration error falls below 0.25 percent, and quantitative titration becomes entirely feasible. Similarly, titration of NH_4^+ with strong base according to (5.15) is not practical in plain water but can be made quantitative by the simple addition of dioxane.

On the basis of these examples we shall make the following generalization: If the reaction is such that ionic charges of opposite sign are *destroyed*, reduction of the dielectric constant will raise the equilibrium constant. Conversely, if the reaction is such that ionic charges of opposite sign are *generated*, reduction of the dielectric constant will lower the equilibrium constant. And if the reaction is such that there is no change in the ionic charge, changing the dielectric constant will have no effect.

Equation (5.16), the dissolution of the ionic solid KCl, shows an example in which ionic charges are generated to produce a saturated solution.

$$KCl(s) \; \rightleftharpoons \; K^+ + Cl^- \tag{5.16}$$

In agreement with prediction, the solubility of potassium chloride decreases as dioxane is added to water.

Equation (5.17) shows an example in which there is no change in the ionic charge. Note that there is a single univalent cation on each side of the equation.

$$NH_4^+ + H_2O \; \rightleftharpoons \; NH_3 + H_3O^+ \tag{5.17}$$

Values of $\log K$ for (5.17) are included in Table 5.5. Although the values are not strictly constant, their variations are well within one logarithmic unit.

Since Eqs. (5.13) through (5.17) show free ions, the predictions about the effect of the dielectric constant apply strictly only to solvents of high dielectric constant where ionic association is relatively unimportant. However, in actual practice the predictions often remain correct, at least qualitatively, over a wider range.

The Δz^2 test

There is a better way of predicting the effect of the dielectric constant than merely by deciding which side of the equation has the most ions: It is to examine Δz^2 for the reaction. After writing the balanced chemical equation, we note the charge number z for each reactant and product and calculate Δz^2 according to (5.18).

$$\Delta z^2 = \Sigma \, z^2 \text{(for the products)} - \Sigma \, z^2 \text{(for the reactants)} \tag{5.18}$$

Thus $\Delta z^2 = -2$ for reactions (5.13) through (5.15); $\Delta z^2 = +2$ for (5.16);

and $\Delta z^2 = 0$ for (5.17). (Confirm this!) We therefore state the following conclusion:

When Δz^2 is negative, the equilibrium constant will increase with *decreasing* dielectric constant; when Δz^2 is positive, the equilibrium constant will increase with *increasing* dielectric constant; and when $\Delta z^2 = 0$, the dielectric constant will have no effect. (5.19)

One advantage of examining Δz^2 is that it enables us to make valid predictions for reactions of complicated charge type. A few examples follow.

$$Fe(OH_2)_5Cl^{2+} + H_2O \rightleftharpoons Fe(OH_2)_6^{3+} + Cl^-; \qquad \Delta z^2 = +6$$

$$H_2O + Hg(CN)_4^{2-} \rightleftharpoons Hg(CN)_3(OH_2)^- + CN^-; \qquad \Delta z^2 = -2$$

$$Al(OH_2)_6^{3+} + H_2O \rightleftharpoons Al(OH_2)_5OH^{2+} + H_3O^+; \qquad \Delta z^2 = -4$$

$$Mg^{2+} + SO_4^{2-} \rightleftharpoons MgSO_4; \qquad \Delta z^2 = -8$$

Another advantage is that Δz^2 allows us to estimate the relative *sensitivity* of equilibrium constants for different reactions to any change in the dielectric constant. According to the charged-sphere model, the rate of change of $\log K$ with the dielectric constant ($d \log K / d\varepsilon$) is, in rough approximation, proportional to Δz^2. In practice things tend to be more complicated, but there is nevertheless an unmistakable trend for ($d \log K / d\varepsilon$) to increase with Δz^2. Thus the preceding examples include reactions for which Δz^2 is -2, -4, and -8. In each case we expect that the equilibrium constant will increase with decreasing dielectric constant, but the rate of increase will be least when Δz^2 is -2 and greatest when Δz^2 is -8.

CHEMICAL MODELS OF IONS AND ION PAIRS IN SOLUTION

The model of ions in solution as charged spheres in a fluid medium is essentially a physical model and does not allow for specific chemical interaction between the ions and the solvent molecules. Such interaction can be important and is often strong enough to produce genuine solvation complexes in which the ion and adjacent solvent molecules are combined into a single kinetic unit. In the remainder of this chapter we shall consider the two most important modes of chemical interaction: hydrogen bonding and coordination. These modes of interaction are available to any molecule with the appropriate structure, but they are especially important in the chemistry of ions and ion pairs.

HYDROGEN BONDING

As an example of hydrogen bonding, consider a dilute solution of ammonium ion (added as NH_4ClO_4) and ammonia in the relatively inert solvent, acetonitrile. It is clear, from a variety of solution properties, that a molecular complex is formed in which a hydrogen atom of NH_4^+ is bonded weakly to the unshared electron pair of $:NH_3$.

$$H_3N: + HNH_3^+ \quad \overset{\text{in CH}_3\text{CN}}{\rightleftharpoons} \quad H_3N\cdots HNH_3^+ \qquad (5.20)$$

$$K_{assoc} = 11\ (M^{-1}) \qquad \text{at } 25°C \qquad (5.21)$$

In Eq. (5.20), the ammonium ion is the *hydrogen-bond donor* and the ammonia molecule is the *hydrogen-bond acceptor*. The hydrogen bond is indicated by a dotted line.

A given molecule is likely to be a good hydrogen-bond donor if it has a hydrogen atom bonded to a small electronegative atom. Molecules with NH and OH bonds, and the HF molecule, are particularly effective.

A given molecule is likely to be a good hydrogen-bond acceptor if it has a small electronegative atom with an unshared electron pair, such as :N, :O, or :F. Small atomic size, permitting close approach of the interacting centers, seems to be essential. Good hydrogen bonding is therefore characteristic of electronegative elements in the first row of the periodic table. Larger electronegative atoms, such as Cl or S, are less effective at hydrogen-bonding.

For first-row atoms there is a fairly good correlation between hydrogen-bonding ability and acid-base strength: strong acids tend to be good hydrogen-bond donors, and strong bases tend to be good hydrogen-bond acceptors. Thus the acetic acid molecule is a better donor than the water molecule; O—H bonds are better donors than N—H bonds; and C—H bonds tend to be inert in hydrogen bonding, except when the carbon atom is bonded to three electronegative centers, as in Cl_3CH.

Hydrogen bonding by common solvents

Most of the solvents in Table 5.1 can be divided into two groups on the basis of their hydrogen-bond donor properties.

Hydroxylic solvents, solvents whose molecules contain an OH group, are good hydrogen-bond donors. Hydroxylic solvents listed in Table 5.1 are:

Water
Alcohols: methanol, ethanol, *tertiary*-butyl alcohol
Oxygen acids: sulfuric acid, acetic acid

Aprotic solvents consist of molecules that either have no hydrogen atoms at all or have hydrogen atoms that are virtually inert in hydrogen bonding (and also as proton donors, except under drastic conditions). Aprotic solvents listed in Table 5.1 are:

Carbon tetrachloride, benzene, dioxane, pyridine, acetone, acetonitrile, dimethylformamide, dimethylsulfoxide.
Ammonia and *chloroform* are weak hydrogen-bond donors.

The hydrogen-bond acceptor properties of the common solvents are less far-ranging. In no case is the acceptor ability entirely negligible. A rough classification, approximately in order of increasing acceptor ability, is as follows.

Solvents without :N, :O, or :F: carbon tetrachloride, chloroform, benzene
Poor hydrogen-bond acceptors: sulfuric acid, acetonitrile
Fair hydrogen-bond acceptors: acetic acid, acetone, dioxane, the alcohols, water, dimethylformamide, dimethyl-sulfoxide
Good hydrogen-bond acceptors: pyridine, ammonia

Hydrogen bonding by ions and ion pairs

Rules concerning the donor and acceptor properties of ions are similar to those for uncharged molecules. Ions with OH and NH bonds, such as H_3O^+ or NH_4^+, are good hydrogen-bond donors, especially if the OH or NH proton is acidic. Oxy-anions such as hydroxide, phenoxide (PhO^-), or acetate ion are good hydrogen-bond acceptors, especially if the oxide-oxygen atom is basic. Fluoride ion is a particularly good acceptor, and chloride ion is a good one.

There is one important difference between ions and uncharged molecules, however: The ionic charge enhances whatever affinity the ionic structure might have for hydrogen bonding. In water and other hydroxylic solvents this fact is hard to demonstrate because ions in dilute solution form hydrogen bonds primarily to *solvent* molecules (which are present at high concentration and can interact by means of their OH group), and it becomes difficult to sort out the effects of hydrogen bonding from those of other modes of solvation. However, in aprotic solvents this fact becomes obvious, and the chemistry of ions and ion pairs cannot be understood unless the formation of hydrogen-bonded complexes is kept always in mind. Let us consider a few examples.

An aprotic solvent of high dielectric constant. The aprotic solvent, acetonitrile ($CH_3C\equiv N$), has a dielectric constant of 36 at room temperature. According to the charged-sphere model, association constants for ion-pair formation should be between 10 and 100, and this prediction is in agreement with experiment for many electrolytes. (An example is shown in Table 5.2.) However, when the cation and anion are joined by a hydrogen bond, the association constant becomes markedly greater.

Instructive data are available for solutions of 3,5-dinitrobenzoic acid and triethylamine in acetonitrile. Structural formulas for these substances are shown below; however, to simplify the following, we shall let HA denote the acid and B denote the base.

$$
\begin{array}{ll}
\text{O} \quad \text{O—H} & \\
\quad\diagdown\!\!\!/ & \\
\quad\text{C} & \\
\text{H}\;\diagup\text{C}\diagdown\;\text{H} & \quad CH_3\text{—}CH_2\text{—}N\text{—}CH_2\text{—}CH_3 \\
\text{C}\!=\!\text{C} & \qquad\qquad\;\; | \\
\text{C}\;\;\;\text{C} & \qquad\qquad\; CH_2 \\
O_2N\;\;\text{C}\;\;NO_2 & \qquad\qquad\;\; | \\
\quad | & \qquad\qquad\; CH_3 \\
\quad\text{H} &
\end{array}
$$

3,5-Dinitrobenzoic acid (HA)   **Triethylamine (B)**

Proton transfer from HA to B goes practically to completion. When HA is titrated with B in *aqueous* solution, the reaction is simply that shown in (5.22).

$$HA(aq) + B(aq) \;\overset{\text{in water}}{\rightleftharpoons}\; BH^+(aq) + A^-(aq) \qquad (5.22)$$

However, when HA is titrated with B in acetonitrile, the reaction is more complicated and seems to take place in two steps. The first reaction is one involving *two* molecules of HA for every molecule of B. By measuring the conductivity of the solution, it is found that the reaction product consists largely of free ions; the reaction is best represented by (5.23).

$$2\,HA + B \;\overset{\text{in CH}_3\text{CN}}{\rightleftharpoons}\; BH^+ + A^-\!\cdot HA \qquad (5.23)$$

To explain the difference between (5.22) and (5.23), we note that in water the A^- ions are able to form hydrogen-bonded complexes with water molecules. On the other hand, in acetonitrile the solvent molecules are practically inert as hydrogen-bond donors, and the A^- ions satisfy their affinity for hydrogen bonding by reacting with HA molecules. Under these conditions, the BH^+ ions probably form weak hydrogen-bonded complexes with CH_3CN molecules, which act as acceptors.

When B is added in excess of the one-to-two ratio, a further acid-base reaction takes place and the conductivity of the solution decreases, suggesting that the product is an ion pair. This reaction is best represented by (5.24).

$$B + A^-\!\cdot HA + BH^+ \;\rightleftharpoons\; 2\,BH^+\!\cdot A^- \qquad (5.24)$$

Evidently, the tendency for proton-transfer to take place is greater than that for A^- to accept a hydrogen bond from HA. Instead, the A^- ion accepts a hydrogen bond from BH^+, and an ion pair of exceptional stability results. The association constant for the reaction, $BH^+ + A^- \rightleftharpoons BH^+A^-$, has been estimated as $8 \times 10^4 (M^{-1})$ at 25°C.

An aprotic solvent of low dielectric constant. The aprotic solvent, carbon tetrachloride, has a dielectric constant of 2.23 at 25°C and is quite inert as a hydrogen-bond donor or acceptor. To give an example of the effect of hydrogen bonding under such conditions, we shall consider the acid-base reaction between triethylamine (B) and acetic acid. Acetic acid now exists largely in the form of a hydrogen-bonded dimer whose molecular structure is shown in (5.25). As in the aprotic solvent acetonitrile, reaction with triethylamine takes place in two steps. The first step, (5.25), again involves two formula weights of acid [one mole of $(HAc)_2$] for every formula weight of B. However, the product is now an ion pair that forms a hydrogen-bonded complex with an acetic acid molecule so that the molecular formula is $BH^+ \cdot Ac^- \cdot HAc$. The second step, (5.26), represents the reaction of that complex with a second molecule of triethylamine.

$$(5.25)$$

$$(5.26)$$

Reactivity and solubility. If an ion with a strong affinity for hydrogen bonding cannot satisfy that affinity, its escaping tendency from solution becomes high. As a result, the ion may become highly reactive, or the electrolyte may become highly insoluble.

A spectacular example of these effects is provided by the behavior of hydroxide ion in mixtures of water with the aprotic solvent, dimethylsulfoxide. As the water concentration in the mixed solvent becomes small, alkali hydroxides become practically insoluble. If the more soluble electrolyte $C_6H_5N(CH_3)_3{}^+OH^-$ is used instead of an alkali hydroxide, it can be shown that the reactivity of hydroxide ion as a Brønsted base becomes enormously high. Thus the base strength of the 0.05 formal hydroxide salt in 95 percent dimethylsulfoxide-5 percent water is more than a million times greater than in pure water, and this factor goes up to more than 10^{11} as the water concentration in the mixed solvent drops below 1 percent. This enormous enhancement of the base strength enables us to study the chemistry of very weak acids.

COORDINATION

It is well-known that monatomic cations associate with complexing agents or *ligands* to form coordination complexes of definite composition and molecular geometry. The ligand molecule contains an unshared electron pair that coordinates with the cation, being bound to the cation either by the attraction of opposite electrical charges or by reacting with it as a Lewis base. The best ligands are good Brønsted bases or are molecules that can be oxidized easily.

Good solvents for electrolytes generally have molecules that are capable of coordinating with cations, and the existence of coordination complexes with solvent molecules has been demonstrated experimentally in many cases. Thus aluminum(III) in water forms a stable octahedral complex, $Al(OH_2)_6^{3+}$, in which six oxygen atoms are coordinated with the aluminum ion (Eq. 5.27). Cobalt(II) in methanol similarly forms a stable octahedral

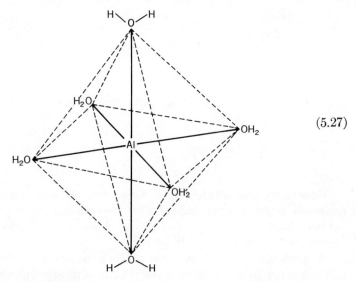

$$(5.27)$$

complex with six methanol molecules. Silver(I) in acetonitrile forms a linear complex, $CH_3CN-Ag-NCCH_3^+$. Indeed, coordination of monatomic cations with solvent molecules is so common in dilute solution that we normally assume it to happen, unless we are told otherwise.

The coordination with solvent molecules has a marked effect on the chemical properties of the ions in solution. If the solvated complex is stable, the ion will have little affinity for reaction with other ligands that might be present in solution, and the coordination chemistry will be relatively simple. Thus in liquid ammonia, monatomic cations form stable ammonia complexes and are relatively unreactive. But in acetonitrile, the coordination complexes with CH_3CN molecules are relatively unstable and solvent molecules are easily displaced from the coordination shell by other ligands. For instance, the inherently low solubility of silver chloride in acetonitrile is increased enormously when chloride ion is added to the solution, owing to the formation of $AgCl_2^-$. As a result, the silver-silver chloride electrode is not very useful for monitoring the concentration of free chloride ion in acetonitrile.

Coordination with a cation also modifies the properties of the coordinated solvent molecules. For example, hydrated cations such as $Al(OH_2)_6^{3+}$ or $Fe(OH_2)_6^{3+}$ are moderately strong Brønsted acids because the water protons in the coordination shell are repulsed by the positive charge on the cation.

$$Al(OH_2)_6^{3+} + H_2O \rightleftharpoons Al(OH_2)_5OH^{2+} + H_3O^+$$

$$K_A = 2.45 \times 10^{-5} \text{ (M)} \quad \text{in water at } 25°C \tag{5.28}$$

Hydrated cations, and particularly their conjugate bases, can also associate to larger units in which the cations are joined by oxide, or hydrous oxide, bridges. In the case of aluminum(III), association begins above pH 3 with the probable formation of $[(H_2O)_5AlOH]_2^{4+}$, whose structure is suggested in (5.29).

$$\left[\begin{array}{c} \overset{\displaystyle H \quad H}{\underset{\displaystyle }{OH \cdots O}} \\ (H_2O)_4\,Al \overset{\diagup}{\underset{\diagdown}{}} \overset{\diagdown}{\underset{\diagup}{}} Al(OH_2)_4 \\ \underset{\displaystyle H \quad H}{O \cdots HO} \end{array} \right]^{4+} \tag{5.29}$$

Above pH 5, the association of aluminum(III) produces a giant molecule that precipitates out of solution as a hydrous gel; and above pH 9, aluminum(III) again becomes soluble owing to the formation of a complex ion whose formula is usually written as $Al(OH)_4^-$.

USE OF NONAQUEOUS SOLVENTS IN ANALYTICAL CHEMISTRY

Analytical chemists are interested in the use of nonaqueous solvents both by necessity and by choice. In chemical synthesis, and especially in organic chemistry, the use of solvents such as acetone, benzene, or chloroform is commonplace, and many unknown substrates are prepared for analysis most conveniently in such solvents. (Nonaqueous solvents are not necessarily very expensive; some can be bought at prices that compete with the cost of distilled water.) Also, by use of nonaqueous solvents the scope of chemical analysis is enlarged and new reagents, such as acid anhydrides, metal hydrides, or organic Grignard reagents that decompose in water, become available. However, the most important advantage of allowing the solvent to vary is that familiar methods of analysis, especially those that we have studied in Chapter 4, then become enormously more powerful. We have seen how the dielectric constant of the solvent can be changed so as to make the titration of a substrate more accurate by increasing the equilibrium constant for the titration. By exploiting possible differences in Δz^2, the dielectric constant can be changed also so as to make the titration of a mixture of substrates more selective.

Karl Fischer titration for water

The determination of small amounts of water in salt hydrates, in hygroscopic liquids, in fibers, and in a host of other materials can be done by titration, but necessarily requires use of an anhydrous solvent. The *Karl Fischer reagent* is such a solvent, and has wide use in water determinations. The reagent looks and smells like a mad witch's brew, but consists of nothing more sinister than methanol, pyridine, sulfur dioxide, iodine, and a preservative. In carrying out the determination, the unknown is dissolved in methanol and titrated, while protected from atmospheric moisture, with approximately 0.1 N Karl Fischer reagent. The stoichiometry of the analytical reaction is shown in (5.30).

$$H_2O + SO_2 + I_2 \longrightarrow SO_3 + 2\,HI \qquad (5.30)$$

As a molecular equation, (5.30) is not quite accurate because the pyridine in the Karl Fischer reagent reacts as a Lewis base with SO_2 and SO_3 and as a Brønsted base with HI, thus reducing the volatility of these substances and shifting the equilibrium in (5.30) to the right. The endpoint is taken when the characteristic brown iodine color persists in the titration flask. The method can detect as little as 0.001 F water.

PROBLEMS

1. Describe the *charged-sphere model* for ions in solution. How close do you think this model is to physical reality? What shortcomings might this picture have?

2. Show that you understand the terms *polarization* and *dielectric constant* by giving your own working definitions of these concepts. What is the relationship between the polarization and the dielectric constant of a solvent? What is meant by the statement "the dielectric constant of dimethylsulfoxide is 46.4"?

3. The dielectric constant ε of methanol at 25°C is 32.63 while $\varepsilon = 12.4$ for *tert*-butyl alcohol at this temperature. Based on this information alone, what differences would you expect in the solvent properties of these two substances?

4. What factors influence the tendency of ions to form ion pairs in solution? What is the significance of each of these variables?

5. Consider 0.001 F solutions of the following salts in the solvents specified. Decide in each case whether the free ions will be the principal species, or whether ion-pairing will predominate. Explain your answers.

    (a) KCl in water
    (b) KCl in pyridine
    (c) $MgSO_4$ in water
    (d) K_2SO_4 in acetone

6. (a) Using the values of $K_{SP} = 6 \times 10^{-5}$ for $CaSO_4$ and $K_{assoc} = 190$ for the $Ca^{2+}SO_4^{2-}$ ion pair, show that a saturated solution of calcium sulfate in pure water has a composition of $[Ca^{2+}] = [SO_4^{2-}] = 8 \times 10^{-3} M$ and $[Ca^{2+}SO_4^{2-}] = 1.1 \times 10^{-2} M$.

    (b) If $K_{assoc} = 1 \times 10^3$ and $K_{SP} = 2 \times 10^{-9}$ for calcium oxalate, what is the total solubility of CaC_2O_4 in water?

7. Using data from Table 5.5, calculate the equilibrium constant for the reaction

$$Ac^- + H_2O \;\rightleftharpoons\; HAc + OH^-$$

in a 30 percent water-70 percent dioxane mixed solvent. (*Hint:* Refer to Problem 2 in Chapter 4.)

8. The pK_B for ethylamine ($CH_3CH_2NH_2$) in water at 25°C is listed in Table A.1 as 3.25. Assuming the equilibrium constant for

$$CH_3CH_2NH_3^+ + H_2O \;\rightleftharpoons\; CH_3CH_2NH_2 + H_3O^+$$

to be independent of changes in ε, calculate the equilibrium constant for

$$CH_3CH_2NH_2 + H_2O \;\rightleftharpoons\; CH_3CH_2NH_3^+ + OH^-$$

in a 55 percent water-45 percent dioxane mixture.

9. For each of the following reactions, use the generalizations given in (5.19) to predict the effect on the equilibrium constant of changing the solvent from pure water to a water-methanol mixture. In which case will the relative change in K be the greatest?

    (a) $Cd(CN)_3(OH_2)^- + CN^- \;\rightleftharpoons\; Cd(CN)_4^{2-} + H_2O$
    (b) $Fe(OH_2)_6^{2+} + H_2O \;\rightleftharpoons\; Fe(OH_2)_5(OH)^+ + H_3O^+$
    (c) $CaCO_3(s) \;\rightleftharpoons\; Ca^{2+} + CO_3^{2-}$
    (d) $PO_4^{3-} + H^+ \;\rightleftharpoons\; HPO_4^{2-}$
    (e) $Co^{2+} + 4\ SCN^- \;\rightleftharpoons\; Co(SCN)_4^{2-}$ (Careful!)

10. Describe hydrogen-bond formation. What characteristics must a molecule have to take part in hydrogen bonding? Why? Name some species which are good at hydrogen-bond formation and some which are not.

11. Table 5.1 lists some common solvents in order of decreasing dielectric constant. Classify each of these solvents according to its effectiveness as (a) a hydrogen-bond donor and (b) a hydrogen-bond acceptor. Do you see any correlation? If you do, see if you can offer an explanation.

12. Rewrite Eqs. (5.22), (5.23), and (5.24) using the structures for 3,5-dinitro-benzoic acid and triethylamine given in the text. Indicate the formation and breaking of hydrogen bonds and identify the path of the proton during each acid-base reaction. [Refer to Eq. (5.25).] Why do different equations apply for the three sets of conditions?

13. Lithium nitrate, which is virtually unassociated in water, can be titrated as a base with perchloric acid in an acetic acid solvent under anhydrous conditions: $Li^+NO_3^- + HClO_4(\text{in } HAc) \rightarrow Li^+ClO_4^- + HNO_3$. In one experiment, 50.00 ml of an unknown $LiNO_3$ solution required 43.15 ml of 0.1025 N $HClO_4$ to reach the potentiometrically determined equivalence point. Calculate the concentration of lithium nitrate in this solution.

14. You are given a solution of ~ 0.35 F $LiNO_3$ in water. Devise an experiment for the analysis of this solution by titration with perchloric acid. Give all details, including the volumes of reagents that you would use for each operation.

The Solvent
as an Acid or Base

Having examined the effect of the dielectric constant, we now continue our discussion of nonaqueous solvents by examining their behavior as Brønsted acids and bases. We shall find that the levels of acidity and basicity attainable in dilute aqueous solutions are limited, but that these limitations can be overcome through the use of nonaqueous solvents. We shall also find that equilibrium constants for acid-base reactions in various solvents can be predicted, at least in rough approximation. This will enable us to choose solvents for acid-base titration objectively, with a minimum of guesswork, and for a wide variety of substrates.

THE BRØNSTED DEFINITIONS

As defined by J. N. Brønsted, an acid is a proton donor, a base is a proton acceptor, and an acid-base reaction is a process in which a proton is transferred from an acid to a base. An example is the acid-base reaction of hydrogen sulfide with ammonia in aqueous solution. As shown in Eq. (6.1), hydrogen sulfide is the acid and ammonia is the base.

$$\underset{\substack{\text{Acid}_1 \rightarrow}}{H_2S} + \underset{\substack{\text{Base}_2 \rightarrow}}{NH_3} \rightleftharpoons \underset{\substack{\leftarrow \text{Base}_1}}{HS^-} + \underset{\substack{\leftarrow \text{Acid}_2}}{NH_4^+}; \qquad K = 200 \qquad (6.1)$$

According to Brønsted's definitions, the reverse reaction (NH_4^+ with HS^-) is also an acid-base reaction. NH_4^+ is now the acid, and HS^- is the base. The molecule that is produced when the proton is removed from an acid is called *the conjugate base* of the acid. Thus, HS^- is the conjugate base of H_2S, and by analogy, H_2S is the conjugate acid of HS^-.

Because the equilibrium in Eq. (6.1) lies on the right, we say that H_2S is a stronger acid than NH_4^+, and that NH_3 is a stronger base than HS^-. In general, we say that:

the stronger acid reacts with the stronger base to produce an equilibrium mixture consisting largely of the conjugate weaker base and weaker acid. $\qquad$ (6.2)

It then follows that there is an inverse relationship between the acid strength of an acid and the base strength of the conjugate base: If $Acid_1$ is stronger than $Acid_2$, then $Base_1$ (the conjugate to $Acid_1$) is weaker than $Base_2$.

Brønsted's classification of solvents

Brønsted pointed out that acid-base reactions in a given solvent depend on three independent solvent properties: the dielectric constant (which controls dissociation to free ions), the acid strength, and the base strength. By assigning a value of either "high" or "low" to each property, he then generated eight solvent classes, as shown in Table 6.1. Although the borderline between "high" and "low" for each property is not sharply defined,

Table 6.1   BRØNSTED'S EIGHT SOLVENT CLASSES

Class	Dielectric constant	Acid strength	Base strength	Common examples
1	High	High	High	Water, methanol, ethanol
2	High	High	Low	Sulfuric acid
3	High	Low	High	Ammonia, dimethylformamide, dimethylsulfoxide
4	High	Low	Low	Acetone, acetonitrile
5	Low	High	High	*tert*-butyl alcohol
6	Low	High	Low	Acetic acid
7	Low	Low	High	Pyridine, dioxane
8	Low	Low	Low	Benzene, chloroform, carbon tetrachloride

the classification is nevertheless useful as a basis for discussion. In classifying actual solvents, we shall adopt the following critera:

1. The dielectric constant is called "high" if it is greater than 20.
2. The acid strength is called "high" if it is comparable to, or greater than, that of water.
3. The base strength is called "high" if it is comparable to, or greater than, that of water. The classification of various common solvents on this basis is shown in Table 6.1.

Speed and reversibility of proton transfer

In discussing the acid or base strength of a solvent, we usually consider only those proton transfer reactions that are rapid and reversible and neglect those that are not. For this reason we distinguish between *protic solvents* and *aprotic solvents*.

Protic solvents are acidic solvents, of high or low acidity, whose molecules rapidly come into equilibrium with proton acceptors in solution. The protic solvents include the hydroxylic solvents, especially water, alcohols, and oxygen acids; solvents with N—H or S—H protons, such as ammonia, ethylenediamine ($H_2NCH_2CH_2NH_2$), or the foul-smelling butyl mercaptan (C_4H_9SH); and the liquid hydrogen halides, especially liquid hydrogen fluoride.

Aprotic solvents are either genuinely inert as proton donors or their reactions as proton donors are negligible under typical mild solution conditions. Thus, solvents such as acetone (CH_3COCH_3), in which the protons are bonded entirely to carbon atoms, are usually classified as aprotic. However, the CH protons are not *always* inert: For instance, in the presence of an alkali hydroxide, the acetone molecules react slowly to form a conjugate base whose high reactivity is responsible for a number of characteristic products. The electronic structure of the conjugate base is best described as a resonance hybrid:

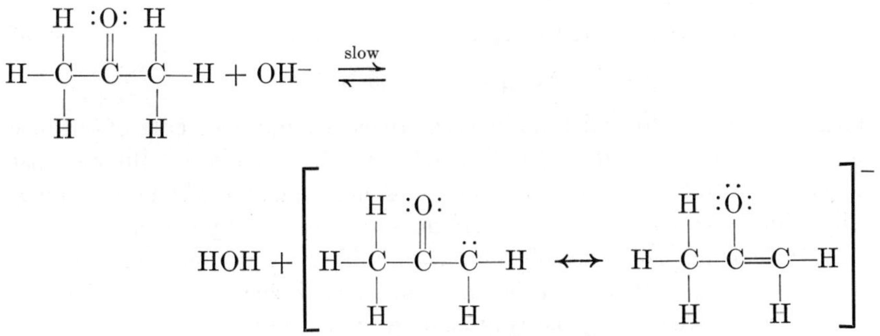

Conjugate base of acetone

The reactions of solvent molecules as proton *acceptors* are generally rapid and reversible. Strictly speaking, all solvents are at least slightly basic because all molecules have at least a small affinity for protons. Even such stable entities as the CH_4 molecule or an argon atom react with bare protons without activation energy in the gas phase to form CH_5^+ and ArH^+.

Lyonium and lyate ions

The conjugate acid of the solvent molecule is called *lyonium* ion (*-onium*, because it is a cation; *ly-*, from lysis or rupture, as in proto*lys*is and solvo-*lys*is). The conjugate base of the solvent molecule is called *lyate* ion. Thus, in aqueous solution, the lyonium ion is H_3O^+ and the lyate ion is OH^-. Structural formulas for the lyonium and lyate ions of other common solvents are shown in Table 6.2. For each protic solvent, the table shows both a lyonium and a lyate ion. However, for the aprotic solvents, the table shows only a lyonium ion for reasons stated in the preceding section.

Proton transfer among protic solvent molecules, to form a lyonium ion and a lyate ion, is of course the familiar process of self-ionization, which Brønsted called *autoprotolysis*. Thus, in the case of water, we write the familiar Eq. (6.3), in which one water molecule reacts as an acid and the other as a base.

$$HOH + HOH \;\rightleftharpoons\; H_3O^+ + OH^- \tag{6.3}$$

$$K_w = [H_3O^+][OH^-] = 1.0 \times 10^{-14}$$

In the case of methanol, the self-ionization is represented by (6.4).

$$CH_3OH + CH_3OH \;\rightleftharpoons\; CH_3OH_2^+ + CH_3O^- \tag{6.4}$$

$$K_{IP} = [CH_3OH_2^+][CH_3O^-] = 1.2 \times 10^{-17}$$

The self-ionization of anhydrous sulfuric acid is represented by (6.5), and that of pure acetic acid by (6.6).

$$H_2SO_4 + H_2SO_4 \;\rightleftharpoons\; H_3SO_4^+ + HSO_4^- \tag{6.5}$$

$$K_{IP} = [H_3SO_4^+][HSO_4^-] = 10^{-3}$$

$$HAc + HAc \;\rightleftharpoons\; H_2Ac^+ + Ac^- \tag{6.6}$$

$$K_{IP} = [H_2Ac^+][Ac^-] = 3.5 \times 10^{-15}$$

It may seem paradoxical that in these reactions one molecule of sulfuric "acid" or of acetic "acid" actually reacts as a Brønsted *base*. But we must keep in mind that these substances were named long ago, when an acid was still defined as a substance that produces an excess of hydrogen ions over hydroxide ions in aqueous solution. It is unlikely that such time-honored names as "sulfuric acid" and "acetic acid" will be changed in the foreseeable future. Instead, to avoid confusion, we put up with lengthy names for their lyonium ions, such as "the conjugate acid of acetic acid" or "the

STRUCTURAL FORMULAS FOR THE
LYONIUM AND LYATE IONS OF
SOME COMMON SOLVENTS

**Table
6.2**

Solvent	Structural Formula for		
	Solvent molecule	Lyonium ion	Lyate ion
Water	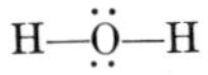 H—Ö—H	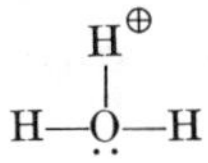 H—O—H with H⁺ above	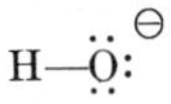 H—Ö:⁻
Methanol	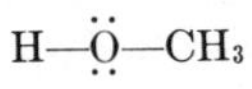 H—Ö—CH₃	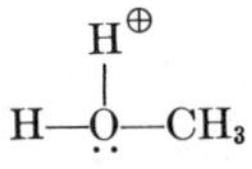 H—O—CH₃ with H⁺ above	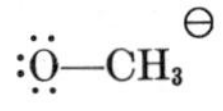 :Ö—CH₃ ⁻
Acetic acid	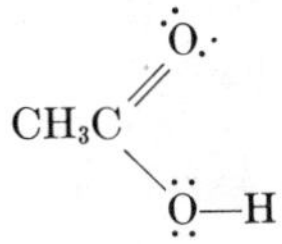 CH₃C(=Ö·)(—Ö—H)	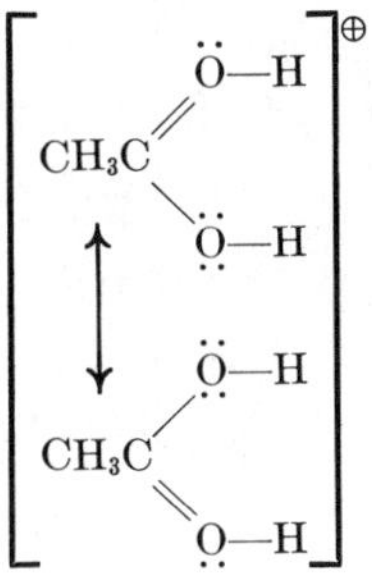 [CH₃C resonance structure]⁺	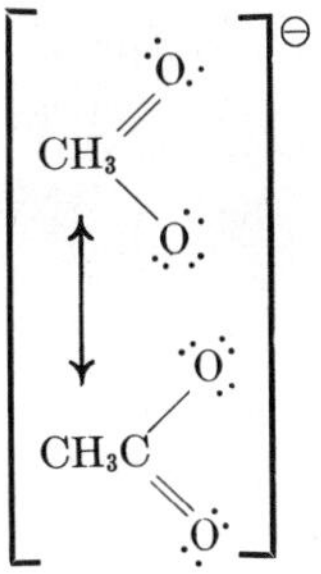 [CH₃ resonance structure]⁻
Sulfuric acid	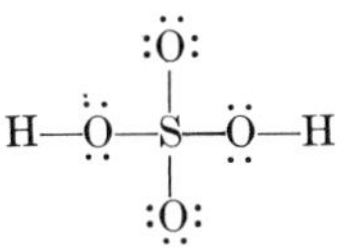 H—Ö—S(=:O:)(=:O:)—Ö—H	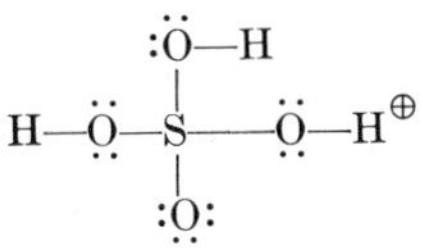 H—Ö—S(:Ö—H)(:O:)—Ö—H⁺	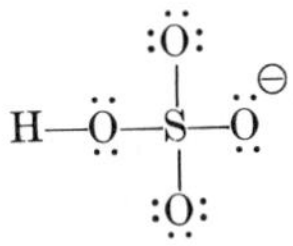 H—Ö—S(:O:)(:O:)—Ö⁻
Ammonia	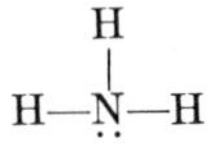 H—N(H)(H)—H	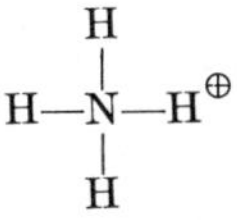 H—N(H)(H)—H⁺ (with extra H)	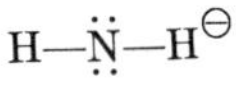 H—N̈—H⁻
Dimethylformamide	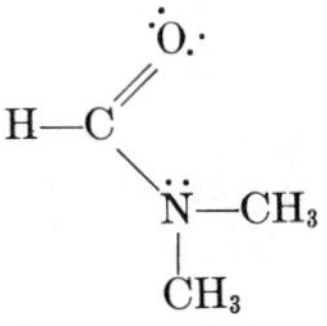 H—C(=Ö·)—N(—CH₃)(CH₃)	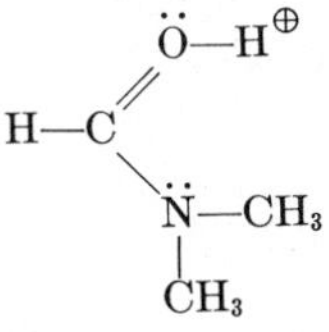 H—C(=Ö—H⁺)—N(—CH₃)(CH₃)	—ᵃ
Acetonitrile	CH₃C≡N:	CH₃C≡N—H⁺	—ᵃ
Acetone	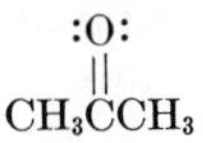 CH₃C(=:O:)CH₃	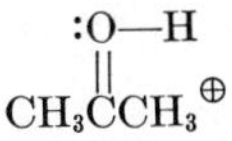 CH₃C(:O—H)CH₃⁺	—ᵃ
Dioxane	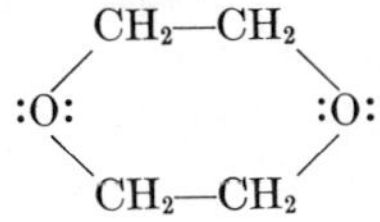 :Ö:(CH₂—CH₂)(CH₂—CH₂):Ö:	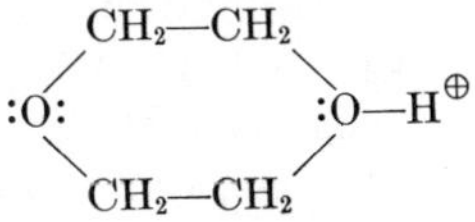 :Ö:(CH₂—CH₂)(CH₂—CH₂):Ö—H⁺	—ᵃ

ᵃ Structural formulas are not shown for the lyate ions of aprotic solvents. Reasons for this are given in the text.

lyonium ion of acetic acid." Or we simply use abbreviations, such as H_2Ac^+.

The equilibrium constant for self-ionization is expressed by the ion product K_{IP} of the solvent. Note, in Eqs. (6.3) through (6.6), that K_{IP} varies considerably with the solvent.

Acid and base dissociation, and ionization

According to Arrhenius' theory, the conductivity of an acid HA in solution results from its dissociation to $H^+ + A^-$. Brønsted argued that the underlying reaction is best described as proton transfer from HA to a solvent molecule, to produce A^- and the lyonium ion of the solvent, but he continued to call it *acid dissociation*. Thus the acid dissociation of nitric acid is represented by (6.7) in water and by (6.8) in ethanol.

$$HNO_3 + HOH \rightleftharpoons H_3O^+ + NO_3^- \tag{6.7}$$

$$K_A = \frac{[H_3O^+][NO_3^-]}{[HNO_3]} = 21 \quad \text{in water}$$

$$HNO_3 + C_2H_5OH \rightleftharpoons C_2H_5OH_2^+ + NO_3^- \tag{6.8}$$

$$K_A = \frac{[C_2H_5OH_2^+][NO_3^-]}{[HNO_3]} = 2.7 \times 10^{-4} \quad \text{in ethanol}$$

Note, therefore, that acid dissociation is a different reaction in each solvent.

The equilibrium constant for acid dissociation is expressed by the acid dissociation constant K_A, as indicated in the equations. Note that K_A may vary considerably with the solvent. The high value of K_A for nitric acid in water is consistent with the familiar assumption that the acid dissociation of HNO_3 is practically complete in dilute aqueous solution. On the other hand, the much smaller value for K_A in ethanol implies that HNO_3 behaves like a weak electrolyte in that solvent.

By analogy with acid dissociation, which he described as proton transfer *to* a solvent molecule, Brønsted described *base dissociation* as proton transfer *by* a solvent molecule. For example, the base dissociation of ammonia in methanol is shown in (6.9).

$$NH_3 + CH_3OH \rightleftharpoons NH_4^+ + CH_3O^- \tag{6.9}$$

$$K_B = \frac{[NH_4^+][CH_3O^-]}{[NH_3]} \quad \text{in methanol}$$

The products of base dissociation are the conjugate acid of the base, and the lyate ion of the solvent. The equilibrium constant is expressed by the base dissociation constant K_B, as indicated in the example.

You know, of course, that a Brønsted acid or base may be an ion, and acid or base dissociation is therefore not limited to electrically neutral molecules. For example, the base dissociation of cyanide ion in water is shown in Eq. (6.10).

$$CN^- + HOH \;\rightleftharpoons\; HCN + OH^- \qquad (6.10)$$

$$K_B = \frac{[HCN][OH^-]}{[CN^-]} \quad \text{in water}$$

In the discussion of Brønsted acids and bases in solvents of low dielectric constant, we must distinguish between *ionization* (which measures the extent of proton transfer with the solvent) and *dissociation*. For example, when hydrogen chloride gas is bubbled into pyridine, proton transfer to solvent molecules is practically complete, as shown by the fact that the vapor pressure of HCl gas over the solution is very low. However, because of the low dielectric constant, the product of proton transfer is largely an ion pair and the conductivity is that of a typical weak electrolyte. We represent this state of affairs by writing two equations, one for acid ionization to form an ion pair, and the other for dissociation of the ion pair to form free ions.

Acid ionization of HCl in pyridine (Py):

$$HCl + Py \;\rightleftharpoons\; PyH^+Cl^- \qquad (6.11)$$

$$K_i = \frac{[PyH^+Cl^-]}{[HCl]}$$

Dissociation:

$$PyH^+Cl^- \;\rightleftharpoons\; PyH^+ + Cl^- \qquad (6.12)$$

$$K_d = \frac{[PyH^+][Cl^-]}{[PyH^+Cl^-]}$$

Accordingly, we say that HCl in pyridine is strongly ionized and weakly dissociated.

The equilibrium constant for acid ionization is the ionization constant K_i, which is the equilibrium constant for the formation of the ionized ion pairs. As shown in (6.11), the expression for K_i does not include the molar concentration of the solvent. The equilibrium constant for ion-pair dissociation is the familiar dissociation constant K_d.

Base dissociation is similarly analyzed into separate steps of base ionization and ion-pair dissociation, as illustrated below for ammonia in *tert*-butyl alcohol.

Base ionization of NH₃ in *tert*-butyl alcohol (HOBu):

$$NH_3 + HOBu \;\rightleftharpoons\; NH_4^+OBu^- \qquad (6.13)$$

$$K_i = \frac{[NH_4^+OBu^-]}{[NH_3]}$$

Dissociation:

$$NH_4^+OBu^- \rightleftharpoons NH_4^+ + OBu^- \qquad (6.14)$$

$$K_d = \frac{[NH_4^+][OBu^-]}{[NH_4^+OBu^-]}$$

In the case of ammonia in *tert*-butyl alcohol, it is apparent from the low conductivity and the fairly high volatility that ammonia is both weakly ionized and weakly dissociated.

The advantage of dissecting the overall processes of acid and base dissociation into separate ionization and dissociation steps is that we thus sort out the effects due to the inherent acid and base strength of the solvent molecules from those due to the dielectric constant. A strongly basic (or acidic) solvent can be counted on to promote acid (or base) *ionization*, regardless of the dielectric constant, but it will promote dissociation only if the dielectric constant is high.

ACIDITY AND BASICITY OF SOLUTIONS

The *acidity of a solution* is measured by the ability of the solution to transfer a proton to a base. In practice, one adds a small amount of a test base (B) to the solution and evaluates the extent of proton transfer by measuring the ratio $[BH^+]/[B]$ after equilibrium has been reached. The *basicity of a solution* is measured similarly by the ability of the solution to remove a proton from an acid.

We now wish to show that *for dilute solutions in a given solvent, the acidity is proportional to the concentration of lyonium ion, and the basicity is proportional to the concentration of lyate ion*. The proof is straightforward, and we shall give it only for the case of acidity. The analogous proof for the case of basicity will be left as an exercise for the reader.

Because the solution is at equilibrium, the test system (consisting of B and BH^+) will be in equilibrium with the solvent (S) and its lyonium ion:

$$BH^+ + S \xrightarrow{K_A} B + SH^+$$

The acidity, which is measured by $[BH^+]/[B]$, is therefore directly proportional to SH^+, as shown in (6.15).

$$\frac{[BH^+]}{[B]} = \frac{[SH^+]}{K_A(\text{for } BH^+ \text{ in } S)} \qquad (6.15)$$

However, Eq. (6.15) is restricted in its validity to dilute solutions in a given solvent, because the acid dissociation constant $K_A(\text{for } BH^+ \text{ in } S)$ is a genuine constant only under those conditions. If the solvent is changed, the acidity of dilute solutions in the new solvent will become proportional to the concentration of the new lyonium ions, with a new and different constant of proportionality.

pH scale in nonaqueous solvents

For dilute aqueous solutions, the pH scale is defined conveniently so that pH $= -\log [H_3O^+]$. Analogously, for dilute solutions in a nonaqueous solvent, the pH scale is defined conveniently as shown in (6.16).

$$p\text{H(in } \mathsf{S}) = -\log [\mathsf{SH}^+] \tag{6.16}$$

In writing (6.16) we show the solvent S explicitly, because we have seen that a given concentration of lyonium ion will produce a characteristically different acidity in each solvent. [This conclusion derives from Eq. (6.15), which requires a new and characteristically different constant of proportionality for each solvent.]

Occasionally we shall wish to use the term "pH" so that it denotes the acidity of a given medium on the same pH scale as that of dilute aqueous solutions. In that case we shall call it the "effective pH", and state clearly that we mean the aqueous pH scale. Unfortunately, the effective pH (on the aqueous pH scale) for a nonaqueous solvent cannot be given a sound theoretical definition. However, in many cases it can be estimated with a fair degree of reliability.

Leveling effect of the solvent

There is an important corollary to the theorem that the acidity in a given solvent is proportional to SH^+: The solvent exerts a leveling effect on acidity.

To prove this corollary, let us consider how one might go about increasing the acidity of the given solvent. In all probability, one would add a solute that ionizes as an acid. For example, to raise the acidity of water, one might try adding a little acetic acid. However, acetic acid is a relatively weak acid, and if the resulting acidity is not high enough, one might try using a stronger acid instead.

As the strength of the added acid increases, the acidity of the resulting solution will increase also, up to a limit. Let us suppose that the dielectric constant is high, so that the lyonium ion concentration can be computed from the familiar mass-action expression for acid dissociation.

$$\underset{(c-x)}{\text{HA}} + \mathsf{S} \rightleftharpoons \underset{(x)}{\mathsf{SH}^+} + \underset{(x)}{\mathsf{A}^-} \tag{6.17}$$

$$K_\mathsf{A} = \frac{x^2}{(c-x)} \tag{6.18}$$

In (6.17) and (6.18), HA denotes the added acid, c its formal concentration, x the lyonium ion concentration (which measures the acidity), and K_A the acid dissociation constant. As the strength of the added acid increases, the value of K_A becomes greater and the acidity of the solution, measured by x, goes up. However, there is a limit to how acidic the solution can become, because x can never become greater than c.

To illustrate this point, we have solved Eq. (6.18), assuming that $c = 0.01\ F$ and that K_A varies from 10^{-4} up. The results are given in Table 6.3. The calculation shows that the acidity can be raised substantially by

Table 6.3   LEVELING EFFECT OF THE SOLVENT ON ACIDITY

K_A	$[SH^+]^a$
1×10^{-4}	0.000952
1×10^{-2}	0.00618
1	0.00990
1×10^{2}	0.0099990
1×10^{4}	0.009999990
Infinite	0.010000000

^{a} x in Eq. (6.18); $c = 0.01\ F$.

use of a stronger acid, until $K_A \approx 1$. At that point, acid dissociation is about 99 percent complete and any further increase in K_A, no matter how great, can raise the acidity at most by one percent.

On the basis of this analysis we may reach the following conclusion: *The acidity produced by an acidic solute in a basic solvent cannot be greater than that produced by an equivalent quantity of lyonium ions.* Because the acidity is thus "leveled" by proton transfer to the solvent, this phenomenon is called *the leveling effect of the solvent* on the acidity.

The strong mineral acids in water provide us with a good example of leveling of acidity. The acid dissociation constants of $HClO_4$, HBr, HCl, and HNO_3 in water are all greater than 10 and dilute aqueous solutions of these acids uniformly display an acidity characteristic of an equivalent concentration of hydronium ions. If the acidity were not leveled by proton transfer to the solvent, the same acids would show marked differences in strength, with $HClO_4$ being the strongest of the four acids and HNO_3 the weakest.

By means of a similar argument, we may reach the conclusion that proton transfer *by* the solvent molecules levels the *basicity* attainable in protic solvents. This leveling effect may be stated as follows: *The basicity produced by a basic solute in a protic solvent cannot be greater than that produced by an equivalent quantity of lyate ions.* Thus in water, the basicity attainable by adding $0.01\ F$ of any base, no matter how strong, cannot be greater than that produced by an $0.01\ M$ solution of hydroxide ion.

ACID-BASE PROPERTIES OF COMMON SOLVENTS

Because of the leveling effect of the solvent, solutions of ultra-high acidity can be prepared only if the lyonium ion is a very strong acid. This implies, according to (6.2), that the solvent itself must be a very weak base. Similarly, solutions of ultra-high basicity can be prepared only in solvents that are very weak acids. In this section we shall report some quantitative results.

Measurement of K_A and K_B for very weak acids and bases in water

A major achievement of chemists during recent decades has been to extend the aqueous pH scale beyond the limits (zero to fourteen) set by dilute aqueous solutions.* To prepare strongly basic solutions, it is convenient to use a soluble hydroxide salt [for instance, $C_6H_5N(CH_3)_3^+ OH^-$] in dimethylsulfoxide-water mixtures. As the proportion of water in this mixed solvent becomes small, the effective pH increases sharply; pH values as high as 24 can be attained. An explanation for this phenomenon has been suggested in Chapter 5 (p. 113). To prepare strongly acidic solutions, it is convenient to use sulfuric acid-water mixtures. As the proportion of water decreases, the effective pH decreases continuously, reaching a value of -11 in 100 percent sulfuric acid.

By utilizing the extended pH range, chemists have been able to measure K_A and K_B for very weak acids and bases in aqueous solution. As a result, the substances of many familiar nonaqueous solvents, such as ethanol, acetone, or acetic acid, have been studied as dilute solutes in aqueous solutions and their acid and/or base-dissociation constants have been measured. Strictly speaking, such data indicate the acid or base strength of the given substance in aqueous solution only. However, it turns out that the results apply also, in good approximation, when the given substance is used as a nonaqueous solvent.

To illustrate the method of measurement, consider the ionization of acetic acid as a Brønsted *base*. The well-known reaction of acetic acid in water is of course ionization as an acid [Eq. (6.19)], with the familiar K_A value of 1.75×10^{-5}. Ionization as a Brønsted base [Eq. (6.20)] is so minute in dilute aqueous solutions as to be negligible for practical purposes.

---

* For more information, see L. P. Hammett, *Physical Organic Chemistry* (New York: McGraw-Hill Book Company, 2nd Ed., 1969); C. L. Rochester, *Acidity Functions* (New York: Academic Press, Inc., 1970).

$$HAc + HOH \rightleftharpoons H_3O^+ + Ac^- \qquad (6.19)$$

$$HAc + HOH \rightleftharpoons H_2Ac^+ + OH^- \qquad (6.20)$$

However, in solutions containing 60 percent or more of sulfuric acid, this situation becomes reversed. Reaction (6.19) is now quite negligible, but the ultraviolet absorption spectrum and other solution properties all indicate that HAc is being converted to H_2Ac^+. In 100 percent sulfuric acid, the conversion of HAc to H_2Ac^+ is practically complete.

By measuring the ratio of [HAc] to $[H_2Ac^+]$ as a function of the effective pH, pK_A for H_2Ac^+ can be calculated by applying Eq. (6.21), which is the familiar expression for the acid dissociation constant in logarithmic form.

$$pK_A(\text{for } H_2Ac^+) = pH - \log \frac{[HAc]}{[H_2Ac^+]} \qquad (6.21)$$

It is found in this way that K_A for H_2Ac^+ in water is 1.6×10^6. K_B for the conjugate base (HAc), that is, the equilibrium constant for (6.20), is then obtained simply by applying Eq. (2.8), which in this case reduces to (6.22).

$$K_A \cdot K_B(\text{for conjugate acid-base pair}) = K_{IP} \qquad (2.8)$$

$$\therefore K_B(\text{for HAc in water}) = \frac{K_W}{K_A \text{ (for } H_2Ac^+ \text{ in water)}}$$

$$= \frac{1.0 \times 10^{-14}}{1.6 \times 10^6}$$

$$= 6 \times 10^{-21} \qquad (6.22)$$

Results

The acid-base properties of some common solvents are summarized in Table 6.4. The information given is similar to that of Table 6.1, but the three variables—dielectric constant, acid strength, and base strength—are now evaluated quantitatively. pK_A refers to the dissociation of the given substance as an acid; pK_B refers to its dissociation as a base.* In examining the pK values, recall that $pK = -\log K$ so that the smallest values indicate the highest acid and base strengths. Values preceded by a *ca.* sign may be in error by one pK unit or more.

The foremost impression conveyed by Table 6.4 is that the range of dissociation constants is enormous. K_A varies by 41 orders of magnitude;

---

* The pK_A values in Table 6.4 measure the acid strength of the *electrically neutral solvent molecules* and should not be confused with pK_A values for the corresponding conjugate acids (lyonium ions). For example, the pK_A value for acetic acid is listed as 4.75, which is the equilibrium constant for (6.19).

ACID-BASE PROPERTIES OF SOME   **Table**
COMMON SOLVENTS   **6.4**

Substance	ε (25°C) for pure liquid substance	pK_A for substance in aqueous solution	pK_B for substance in aqueous solution
Water	78.54	15.74[a]	15.74
Methanol	32.63	15.15	16
Ethanol	24.30	15.95	15.94
tert-Butyl alcohol	12.4	16	*ca.* 18
Acetic acid	6.3	4.75	20.2
Sulfuric acid	>100	*ca.* −7	*ca.* 28
Ammonia	22.4[b]	*ca.* 34	4.74
Pyridine	12.3	—	8.85
Dimethylformamide	36.7	—	14.0
Dimethylsulfoxide	46.4	—	*ca.* 14
Dioxane	2.21	—	17.2
Acetone	20.7	—	21.2
Acetonitrile	36.0	—	24

[a] K_A (for H_2O) = $[H_3O^+][OH^-]/[H_2O]$ = $K_W/[55.55]$.
   K_B (for H_2O) = $[H_3O^+][OH^-]/[H_2O]$ = $K_W/[55.55]$.
[b] At −33°C.

K_B varies by 23 orders of magnitude. There are large variations in both K_A and K_B for protic solvents, and large differences in K_B for aprotic solvents. Water and the alcohols make an interesting solvent series in which the acid-base properties are fairly constant but the dielectric constant varies widely. Among the aprotic solvents, pyridine is the strongest base, next come dimethylsulfoxide and dimethylformamide (whose base strength is slightly greater than that of water), then come dioxane, acetone, and acetonitrile. Not listed in the table are benzene, chloroform and carbon tetrachloride, whose pK_A values in water are not precisely known, but whose base strength is substantially lower than that of acetonitrile.

ACID-BASE TITRATIONS IN NONAQUEOUS SOLVENTS

The wide range and variety of acid-base properties of the common solvents gives us great flexibility in the design of acid-base titrations. The strategy, always, is to find conditions under which the equilibrium constant for acid-base titration of the substrate is high, so that the innate titration error will be small. In the rest of this chapter we shall consider a number of relevant topics. We shall consider:

1. The ion product of the solvent and why it must be small.
2. Acid-base titrations in acetic acid and some of the underlying chemistry.

3. Acid-base titrations in methanol and methanol-benzene mixtures, with remarks about specific solvation effects.

4. Prediction of acid dissociation constants in various solvents with the aid of some rough-and-ready rules.

The underlying theory of nonaqueous acid-base titrations is not different, in principle, from that of titrations in general, which will be treated in Chapters 7 through 10.

Ion product of the solvent and acid-base titration

Due to the leveling effect of the solvent, a given acid normality will produce the highest possible acidity if the acid reactant consists entirely of lyonium ions. Similarly, a given base normality (in a protic solvent) will produce the highest possible basicity if the basic reactant consists entirely of lyate ions. Consequently, the innate titration error for acid-base titration in a protic solvent is never less than that for the titration of lyonium ion with lyate ion. For example, if the solvent is methanol, the innate error for acid-base titration is never less than that for an equivalent titration of methyloxonium ion with methoxide ion.

$$CH_3OH_2^+ + CH_3O^- \;\rightleftharpoons\; 2\,CH_3OH \qquad (6.23)$$

$$K = \frac{1}{K_{IP}(\text{for methanol})}$$

The equilibrium constant for (6.23) is equal to $1/K_{IP}$, where K_{IP} is the ion product of methanol.

The innate titration error for the titration of lyonium ion with lyate ion can be predicted according to (2.19c) if the ion product of the solvent is known: It is less than 0.25 percent if $c_S^2 K = c_S^2/K_{IP} > 5 \times 10^5$. On using a typical value of 0.03 F for c_S, we find that K_{IP} must be less than 2×10^{-9}. As a result, solvents with ion products greater than about 10^{-9} are not suitable for acid-base titration. A striking example is anhydrous sulfuric acid, whose ion product, as shown in (6.5), is 10^{-3}. Sulfuric acid has many important uses in chemistry, but being a solvent for acid-base titration is not one of them. Ion products for a number of protic solvents are listed in Table 6.5.

While a high value of K_{IP} makes acid-base titration impossible, a low value of K_{IP} is of advantage. When K_{IP} is near 10^{-9}, the only really satisfactory titration is that of lyonium ion with lyate ion. However, if K_{IP} is smaller, the titration of weakly ionized acids with lyate ion, or of weakly ionized bases with lyonium ion, becomes possible. For instance, we found in Chapter 4 that in water (where $K_{IP} = K_W = 10^{-14}$) acids and bases with dissociation constants as low as 10^{-7} can be titrated with 0.25 percent

ION PRODUCTS OF SOME COMMON PROTIC   **Table**
SOLVENTS AT 25°C   **6.5**

Solvent	pK_{IP}
Water	14.00
Methanol	16.92
Ethanol	19.5
Acetic acid	14.45
Sulfuric acid	*ca.* 3
Ammonia	22 (at $-33°C$)

accuracy. By a generalization of Eqs. (4.4) and (4.5), we find that in solvents of high dielectric constant the innate titration error is less than 0.25 percent if:

$$K_A c_S > 5 \times 10^5 K_{IP}, \quad \text{for the titration of weakly ionized acids with lyate ion}$$

$$K_B c_S > 5 \times 10^5 K_{IP}, \quad \text{for the titration of weakly ionized bases with lyonium ion}$$

(6.24)

We see from (6.24) that solvents with small values for K_{IP} permit the titration of weakly ionized acids or bases with small dissociation constants.

Acid-base titrations in acetic acid

Acetic acid is a popular nonaqueous solvent for acid-base titrations. It is readily available in a high state of purity and will dissolve a wide variety of substrates. In terms of Brønsted's classification, it is a solvent of high acid strength, low base strength, and low dielectric constant.

The highest basicity attainable in acetic acid is that of solutions of acetate ion. Stronger bases are converted to their acetate salts. For example, ammonia is converted to ammonium acetate, and sodium carbonate is converted to sodium acetate.

$$NH_3 + HAc \;\rightleftharpoons\; NH_4^+Ac^- \qquad (6.25)$$

$$Na_2CO_3(s) + 2\,HAc \;\longrightarrow\; 2\,Na^+Ac^- + CO_2 + H_2O \qquad (6.26)$$

The highest acidity attainable in acetic acid is that of solutions of the lyonium ion, H_2Ac^+, which is a phenomenally strong acid. Indeed the levels of acidity attainable in acetic acid are so high that the common inorganic acids, $HClO_4$, H_2SO_4, HBr, and HCl, are found to differ markedly in strength. Perchloric acid, the strongest of the common acids and one of the strongest acids in chemistry, is largely ionized. The other acids are weaker, and the degree of ionization decreases in the sequence $HClO_4$ > H_2SO_4 > HBr > HCl. The high level of acidity also manifests itself in other ways. Sulfuric acid reacts with sodium acetate as a monobasic acid in acetic acid.

$$H_2SO_4 + Na^+Ac^- \rightleftharpoons Na^+HSO_4^- + HAc \qquad (6.27)$$

Hydrogen chloride, which is surely one of the stronger acids of chemistry, is barely strong enough to be titrated with sodium acetate (the innate titration error is about 0.5 percent), and solutions of chloride salts in acetic acid are distinctly basic compared to the pure solvent.

Because of the high level of acidity attainable, many bases that are too weak to be titrated with hydronium ion in water can be titrated with per- chloric acid in acetic acid. As indicated in Table 4.1, acid-base titration *in water* is limited to bases with $pK_B < 8$. In acetic acid this limit shifts, so that bases whose pK_B (in *water*) is as high as 12 can be titrated. Sub- stances that can thus be titrated in acetic acid include many important dyes and drugs, especially those derived from aniline and pyridine, and most of the essential amino acids. Substances with a carboxylate ($—CO_2^-$) group, such as the amino acid glycine ($^+H_3NCH_2CO_2^-$) and the acidimetric standard, potassium acid phthalate (6.28), are similar to acetate ion in base strength and can be titrated. Substances with a carboxyl ($—CO_2H$) group, such as formic acid (HCO_2H) or the carboxyl group of potassium acid phthalate, have approximately the same acid strength as acetic acid and give a "neutral" reaction.

the carboxyl group is neutral in acetic acid

$$(6.28)$$

the carboxylate group reacts as a Brønsted base in acetic acid

(Potassium acid phthalate)

The most common strong-acid titrant in acetic acid is a 0.01–0.1 F solu- tion of perchloric acid. $HClO_4$ is not only one of the strongest Brønsted acids known but also a potent oxidizing agent, and nonaqueous solutions of anhydrous $HClO_4$ have been known to explode spontaneously. For this reason the titrant is usually prepared to contain two or three moles of H_2O per equivalent of $HClO_4$ and in small lots of one liter or less. We know of no explosions under such conditions, and the acid normality is stable for weeks.

Acid-base titration in acetic acid is done in much the same way as if the solvent were water. Suitable endpoint indicators are crystal violet, which changes from yellow (in acid) through intermediate spectral colors to violet, or bromophenol blue, which changes from colorless (in acid) to yellow. Potentiometric titration is possible. Potassium acid phthalate (which reacts as a monoacid base) is a convenient chemical standard for determining the normality of perchloric acid solutions in acetic acid. The low solubility of $KClO_4$ helps sharpen this titration.

Acid-base titrations in methanol

Methanol is in the same solvent class with water, according to Brønsted's classification. It is a slightly weaker base and a slightly stronger acid than water, and the dielectric constant (32.63 at 25°C), although distinctly lower than that of water, is "high." It is also an excellent solvent for many organic acids and bases and is readily available in a high state of purity. These properties make it a popular choice when a water-like organic solvent is desired.

There are some important differences between water and methanol, however, owing largely to the difference in dielectric constant. To understand them, it is helpful to recall the Δz^2 rule (5.19).

1. If $\Delta z^2 = 0$, the dielectric constant has little or no effect on the equilibrium constant. Since the acid-base properties of water and methanol are similar, equilibrium constants and innate titration errors will also be similar. We shall consider two examples.

(a) Titration of benzoic acid with strong base.

$$HBz + OCH_3^-(OH^-) \rightleftharpoons Bz^- + HOCH_3(HOH)$$

$$\log K = 7.54 \text{ (in methanol)}, 9.82 \text{ (in water)}$$

In this example the difference between the values of $\log K$ in water and methanol is about as large as one is likely to encounter when $\Delta z^2 = 0$. Note that methoxide ion is the weaker base. Titration of benzoic acid is feasible in both solvents. Suitable endpoint indicators in methanol are phenol red, which changes from yellow to red, and bromothymol blue, which changes from yellow, through intermediate spectral colors of green and magenta, to blue.

(b) Titration of ammonia with strong acid.

$$NH_3 + CH_3OH_2^+(H_3O^+) \rightleftharpoons NH_4^+ + CH_3OH(HOH)$$

$$\log K = 10 \text{ (in methanol)}, 9.24 \text{ (in water)}$$

Note that $CH_3OH_2^+$ is the stronger acid. Titration is feasible in both solvents. Suitable endpoint indicators in methanol are thymol blue, which changes from yellow to red at the equivalence point, and bromophenol blue, which changes from blue, through green, to yellow. The strong-acid titrant must be standardized frequently because many strong acids react slowly with methanol.

2. If $\Delta z^2 = -2$, the equilibrium constant will be distinctly greater in

methanol than in water. As a result, certain weak acids and bases whose titration is difficult in water can be titrated easily in methanol. Some examples follow.

(a) Salts of carboxylic acids (such as sodium acetate) can be titrated easily with strong acid in methanol.

$$Ac^- + CH_3OH_2^+(H_3O^+) \; \rightleftharpoons \; HAc + CH_3OH(HOH)$$

$$\log K = 9.72 \text{ (in methanol)}, \; 4.76 \text{ (in water)}$$

A suitable endpoint indicator in methanol is thymol blue (at the color change from yellow to red), and bromophenol blue (at the color change from green to yellow).

(b) Many dibasic weak acids can be titrated selectively with strong base to the first equivalence point in methanol but not in water. Consider the titration of a weak acid, H_2X. The effective chemical equation for titration of the first proton [according to the selective titrant rule (3.38)] is:

$$H_2X + X^{2-} \; \rightleftharpoons \; 2\,HX^-$$

and Δz^2 is therefore -2. A good "rule of thumb" is that weak dibasic acids can be titrated to the first equivalence point with strong base in methanol if K_{A1}/K_{A2} *in water* is at least 10^2. The titration is best done potentiometrically. If a color indicator is needed, consult Chapter 9.

Methanol-benzene mixtures. The usefulness of methanol as a solvent for acid-base titrations can be enhanced by the addition of benzene. A common "recipe" calls for the addition of about 20 percent of benzene, but the exact amount is not critical and the optimum amount may vary with the substrate.

The addition of benzene to methanol causes a variety of effects, some of which are not fully understood. It causes a lowering of the dielectric constant. It also causes an increase in the base strength of methoxide ion in much the same way that the addition of dimethylsulfoxide to water causes an increase in the base strength of hydroxide ion. Finally, it causes a number of specific effects that are unrelated to the dielectric constant and which it is convenient to describe collectively as *specific solvation effects*.

In methanol-benzene mixtures, the most important of the specific solvation effects is characteristic of mixed solvents in general. When there are two or more solvent components, the solute molecules surround themselves with solvent shells of such composition as to give the greatest overall stability. The average composition of the solvent shells will thus be somewhat different from that of the bulk solvent, especially if the chemical properties of the solvent components are as widely different as those of methanol and benzene. The importance of this phenomenon varies greatly with the solute: Some solutes are stabilized more than others; some equilibrium constants are unexpectedly high, while others are unexpectedly low.

What concerns us at this point is that the specific solvation effects on the equilibrium constant can be quite different for different reactions. As a result, if a particular selective titration happens to be difficult when the solvent is pure methanol, it is quite possible that simple addition of benzene will spread the equilibrium constants apart so that selective titration becomes easy. In most cases we cannot predict, *a priori*, whether this stratagem will succeed, but the experiment is easy to do and we have little to lose by trying.

A satisfactory strong base for methanol-benzene titrations is tetra-butylammonium methoxide $[n\text{-}(C_4H_9)_4N^+OCH_3^-]$ in methanol solution. The titrations are usually done potentiometrically.

Prediction of acid dissociation constants in various solvents

The number of possible combinations of solvent and substrate is so large that, more than likely, the specific equilibrium constants we require have not been measured. We therefore conclude this chapter by describing a simple method for *predicting* equilibrium constants. In particular, we shall use acid-dissociation constants measured in water (where many measurements have been made) to predict acid-dissociation constants in other solvents. Base-dissociation constants are obtained from acid-dissociation constants for the conjugate acids and K_{IP} of the solvent by simple application of Eq. (2.8).

The accuracy of prediction is about $\pm 1\ pK$ unit, not high but sufficient for many purposes, and infinitely better than no information at all. The method is readily generalized so that a solvent other than water can become the reference solvent for prediction.

Assumptions. In predicting acid-dissociation constants, we make the following assumptions.

1. The *relative* strength of acids of the same charge type is independent of the solvent.

2. When a given nonaqueous solvent is used in place of water, the pK_A values for all acids of the same charge type shift by the same amount. The amount of shift will vary with the charge type of the acid and with the solvent.

3. For univalent cation acids, the variation of pK_A with the solvent can be predicted from the base strength of the solvent.

Univalent cation acids. The acid dissociation of a cation acid BH^+ in a solvent S is shown symbolically in (6.29).

$$BH^+ + S \ \rightleftharpoons\ B + SH^+ \tag{6.29}$$

To see how well our assumptions really work out, we begin with a specific example, the acid dissociation of a series of cation acids in the nonaqueous solvent acetonitrile. As shown in Table 6.4, acetonitrile is a much weaker base than water, differing in pK_B from water by 8 pK units. A given acid BH^+ should therefore be much less dissociated in acetonitrile.

Table 6.6

EXPERIMENTAL pK_A VALUES FOR
UNIVALENT CATION ACIDS IN
ACETONITRILE AND WATER

Acid	pK_A in acetonitrile	pK_A in water	Difference
HH_4^+	16.46	9.26	7.20
$(CH_3)_3NH^+$	17.61	9.76	7.85
$C_2H_5NH_3^+$	18.40	10.63	7.77
$(n\text{-}C_4H_9)_3NH^+$	18.09	10.89	7.20
Pyridinium ion	12.33	5.17	7.16
Morpholinium ion	16.61	8.36	8.25
$C_6H_5NH_3^+$	10.56	4.58	5.98
$2\text{-}NO_2C_6H_4NH_3^+$	4.89	−0.29	5.14
$H_2NNH_3^+$	16.61	7.96	8.65
$H_2NCH_2CH_2NH_3^+$	18.46	9.95	8.51

Mean and mean deviation: $\overline{7.4 \pm 0.8}$

Actual data are shown in Table 6.6. In agreement with expectation, K_A in acetonitrile is consistently smaller than K_A in water; that is, pK_A is consistently greater. The sequence of the pK_A values, which indicates the relative strength of the acids, is (with only a few exceptions) the same in the two solvents. All pK_A values are shifted by approximately the same amount, 7.4 ± 0.8 pK units. This shift compares favorably with the difference in basicity (8 pK units) of the two solvents.

Data obtained in other solvents follow a similar pattern. Results for a variety of univalent cation acids are summarized in Table 6.7. In each solvent, the shift of pK_A is approximately uniform for a typical sample of

Table 6.7

EFFECT OF SOLVENT ON pK_A FOR
UNIVALENT CATION ACIDS

Solvent	Shift of pK_A for BH^+ $[pK_A(\text{in S}) - pK_A(\text{in H}_2O)]$	Shift of pK_B in water $[pK_B(\text{for S}) - pK_B(\text{for H}_2O)]$
Methanol	0.4 ± 0.4^a	0^b
Ethanol	0.6 ± 0.7	0
Acetic acid	4.3 ± 0.3	4.5
Dimethylformamide	-0.7 ± 0.4	−2
Dimethylsulfoxide	$ca.\ -2.4$	$ca.\ -2$
Acetonitrile	7.4 ± 0.8	8

a Mean and mean deviation for a typical sample of univalent cation acids.
b Based on data in Table 6.4.

acids, and the mean shift of pK_A is approximately equal to the shift in pK_B of the solvent. These findings are expressed conveniently in (6.30).

$$pK_A(\text{of BH}^+ \text{ in } \mathsf{S}) - pK_A(\text{of BH}^+ \text{ in } H_2O)$$
$$= pK_B(\text{of } \mathsf{S} \text{ in } H_2O) - pK_B(\text{of } H_2O \text{ in } H_2O) \qquad \pm 1pK \text{ unit} \quad (6.30)$$

Uncharged acids. The acid dissociation of an uncharged acid is shown symbolically in (6.31).

$$\mathrm{HA} + \mathsf{S} \;\rightleftharpoons\; \mathsf{SH}^+ + \mathrm{A}^- \qquad\qquad (6.31)$$

Again, we shall assume that the shift of pK_A in any given solvent (relative to pK_A in water) is uniform for all acids of the given charge type. However, Δz^2 for acid dissociation is now $+2$, and the shift of pK_A, therefore, depends both on the basicity of the solvent and the dielectric constant. By making these assumptions, we can reproduce experimental pK_A values for a wide variety of uncharged acids to about ± 1 pK unit. Average shifts in pK_A for a number of solvents are listed in Table 6.8.

	EFFECT OF SOLVENT ON pK_A FOR UNCHARGED ACIDS	**Table 6.8**

Solvent	Shift of pK_A for HA $[pK_A(\text{in } \mathsf{S}) - pK_A(\text{in } H_2O)]$
Methanol	4.6 ± 0.4[a]
Ethanol	5.1 ± 0.5
Acetic acid	10.9 ± 0.4
Dimethylformamide	6.9 ± 0.8
Dimethylsulfoxide	5.2 ± 1.0
Acetonitrile	15.0 ± 1.2

[a] Mean and mean deviation for a typical sample of uncharged acids.

Effect of ionic association. If the ionic reactants or products in proton transfer associate to form uncharged ion pairs, Δz^2 for the reaction goes to zero, and the effect of the dielectric constant becomes much smaller than it otherwise would be. An important example is the acid ionization of an uncharged acid to form a hydrogen-bonded ion pair.

$$\mathrm{HA} + \mathsf{S} \;\rightleftharpoons\; \mathsf{SH}^+ \cdot \mathrm{A}^- \qquad\qquad (6.32)$$

$$K_i = \frac{[\mathsf{SH}^+ \mathrm{A}^-]}{[\mathrm{HA}]}$$

In (6.32), the net electrical charge of each molecule is zero; henze $\Delta z^2 = 0$. It is not clear, at this writing, whether the Δz^2 test (5.19) provides an adequate first approximation to the effect of the dielectric constant in reactions involving ion pairs. If it applies, the dielectric constant will have no effect on ionization constants (K_i). We expect, then, that the variation of pK_i

with the solvent will depend only on the base strength of the solvent, as shown in (6.33), which is analogous to (6.30).

For any uncharged acid HA and pair of solvents S_1 and S_2,

$$pK_i(\text{of HA in } S_1) - pK_i(\text{of HA in } S_2)$$
$$\approx pK_B(\text{of } S_1 \text{ in } H_2O) - pK_B(\text{of } S_2 \text{ in } H_2O) \quad (6.33)$$

Experimental data for ionization constants are quite scarce. However, the data that exist fit the predictions of Eq. (6.33) quite well.

REFERENCES

J. S. Fritz and G. S. Hammond, *Quantitative Organic Analysis*, John Wiley & Sons, Inc., New York, 1957.

W. Huber, *Titrations in Nonaqueous Solvents*, Academic Press, Inc., New York, 1967.

J. Kucharský and L. Šafařík, *Titrations in Nonaqueous Solvents*, Elsevier Publishing Company, Amsterdam, 1965.

S. R. Palit, M. N. Das, and G. R. Somayajulu, *Nonaqueous Titration*, Indian Association for the Cultivation of Science, Calcutta 32, India, 1954.

PROBLEMS

(*Note:* Use data given in this chapter and in Tables A.1 and A.2.)

1. Define or explain.
  (a) Brønsted's eight solvent classes
  (b) The difference between protic and aprotic solvents
  (c) The difference between acid ionization and acid dissociation
  (d) The leveling effect of the solvent on basicity

2. On the basis of the definition of pH given in Eq. (6.16), find the pH of the following solutions.
  (a) $0.01\ F$ acetic acid in methanol
  (b) $0.01\ F$ sodium acetate in methanol
  (c) $0.01\ F$ HCl in acetic acid
  (d) $0.01\ F$ HCN $+ 0.02\ F$ LiCN in dimethylsulfoxide

3. Suggest suitable solvents and titrants for titration of the following.
  (a) Sodium azide (NaN_3, the sodium salt of hydrazoic acid)
  (b) A mixture of sodium azide and sodium dihydrogen phosphate (NaH_2PO_4), nonselective and selective titration
  (c) The amino acid glycine ($^+H_3NCH_2CO_2{}^-$)
  (d) A mixture of glycine and lithium sulfate (Li_2SO_4), nonselective titration

4. Using data in Tables 6.4 through 6.8 and in Table A.1, predict the following.
(a) pK_A for trichloroacetic acid (Cl_3CCO_2H) in acetic acid (Compare with the experimental value in Table A.2.)
(b) pK_B for $H_2PO_4^-$ in methanol
(c) pK_B for pyridine (C_5H_5N) in ethanol
(d) pK_{A1} for citric acid ($H_3C_6H_5O_7$) in acetonitrile

5. The uncharged base, butter yellow (BY), produces a yellow color in solution, while the conjugate acid, BYH^+, produces a red color. Solutions of butter yellow in pure acetic acid are orange, owing to ionization of the uncharged base, $BY + HAc \rightleftharpoons BYH^+Ac^-$. The orange color is the result of mixing the yellow color of BY with the red color of BYH^+ (in BYH^+Ac^-). Predict how the color will change when a solution of butter yellow in acetic acid is treated with:
(a) Excess perchloric acid in acetic acid
(b) Excess sodium acetate in acetic acid

Titration Curves

Titration, broadly defined, is any procedure in which small increments of a reactant (the titrant) are added successively to a solution containing a reactive substrate, and some property of the solution is measured after each addition. The *titration curve* then is the relationship between the measured property and the total amount of added titrant. In the familiar volumetric titration with a color indicator, the measured property is the color of the solution and only one point on the titration curve, the equivalence point, is truly of interest. However, if these restrictions are relaxed and titration curves are studied in detail, they become veritable gold mines of information, informing us about the stoichiometry of the reaction, the reaction steps, equilibrium constants, and the probable molecular species in solution. Indeed, the information that can be extracted is so various and so definitive that the study of titration curves has developed into a sophisticated science.

In this chapter we shall take a quick look at titration curves, emphasizing those aspects that are most useful in quantitative chemical analysis. After examining some actual titration curves, we shall proceed to their theoretical calculation—how to set up the algebraic equations and how to solve them. We shall then make some sample calculations, noting what happens as the equilibrium constant becomes large. Finally, in the next chapter, we shall apply the theory of titration curves for deciding whether a proposed chemical analysis by titration will be feasible and for estimating the innate titration error.

A FEW EXAMPLES

The simplest titration curves are those in which the measured property is directly proportional to the concentration of a specific reactant or product. For an example, consider the reaction of *para*-nitrophenol with sodium hydroxide in aqueous solution, Eq. (7.1).*

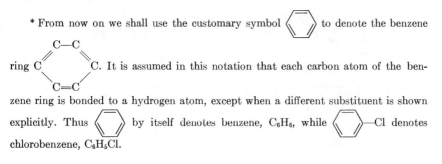

$$O_2N-\!\!\!\bigcirc\!\!\!-OH + OH^- \;\rightleftharpoons\; O_2N-\!\!\!\bigcirc\!\!\!-O^- + HOH \qquad (7.1)$$

para-Nitrophenol   **para-Nitrophenoxide ion**

$$K = 7.1 \times 10^6 \; (M^{-1}) \text{ at } 25°C$$

The reaction product, *para*-nitrophenoxide ion, absorbs in the violet, so that the color of the solution becomes yellow as titrant is added. By measuring the characteristic absorbance at a suitable wavelength and applying Beer's law (Eq. 1.2), the concentration of *para*-nitrophenoxide ion is monitored throughout the titration. The resulting *photometric titration curve* is shown in Fig. 7.1. The absorbance (and hence the *para*-nitrophenoxide concentration) increases nearly linearly with added titrant until reaction is almost complete. Then there is a sudden change of slope, and further addition of titrant causes little further reaction and only gentle dilution. The equivalence point is that point at which the curvature, or rate of change of the slope, is at a maximum. In practice, the equivalence point may be taken as the point of intersection of straight-line segments, as shown in the figure.

---

* From now on we shall use the customary symbol $\bigcirc$ to denote the benzene

ring C $\begin{smallmatrix} C-C \\ & \\ C=C \end{smallmatrix}$ C. It is assumed in this notation that each carbon atom of the ben-

zene ring is bonded to a hydrogen atom, except when a different substituent is shown explicitly. Thus $\bigcirc$ by itself denotes benzene, C_6H_6, while $\bigcirc-Cl$ denotes chlorobenzene, C_6H_5Cl.

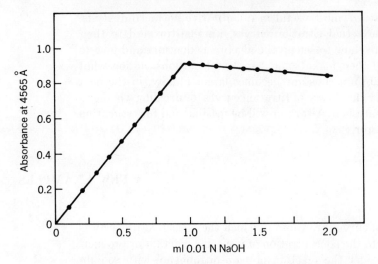

Fig. 7.1. Photometric titration of 0.001 F *para*-nitrophenol with 0.01 N sodium hydroxide in water. The absorbance is proportional to the concentration of *para*-nitrophenoxide ion.

Figure 7.2 shows the *conductometric titration curve* for the same reaction. In aqueous solution, *para*-nitrophenol is a weak electrolyte, while sodium *para*-nitrophenoxide and sodium hydroxide are strong electrolytes. The conductivity of the solution therefore increases steadily as sodium hydroxide is added. Before the equivalence point, the increase is due almost entirely to the formation of sodium *para*-nitrophenoxide; past the equivalence point, the increase is due almost entirely to the accumulation of

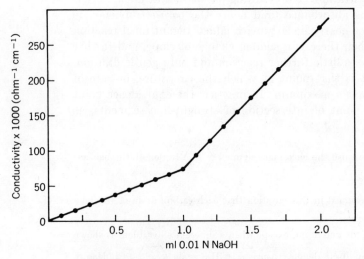

Fig. 7.2. Conductometric titration of 0.001 F *para*-nitrophenol with 0.01 N sodium hydroxide in water.

excess sodium hydroxide. The slope increases sharply at the equivalence point because the equivalent conductance of sodium hydroxide is greater than that of sodium *para*-nitrophenoxide. The equivalence point again is the point of maximum curvature and can be located by graphical construction, as shown in Fig. 7.2.

Linear versus logarithmic titration curves

Titration curves such as Figs. 7.1 and 7.2 are called *linear titration curves* because the measured property is a linear function of solute concentration, either precisely or in good approximation. The equivalence point on such curves is a point of maximum curvature.

A totally different kind of titration curve results when the measured property is a logarithmic function of concentration, such as *p*H or log ([Fe(III)/Fe(II)]). Examples of such curves had been shown in Fig. 4.1, where the *p*H is plotted versus the amount of titrant in acid-base titrations. However, the curves in Fig. 4.1 were based on theory rather than experiment. In this chapter, we shall show genuine experimental curves.

In practice, logarithmic titration curves are obtained when the titration is carried out in one compartment of an electrochemical cell and the measured property is the emf. Electrochemical cells can be used to monitor a wide variety of reactions: acid-base reactions, precipitation and complex formation, redox, reactions involving organic substrates, reactions in water and in nonaqueous solvents—in short, any reaction in which the concentration of either the substrate or the titrant can be sensed, somehow, by a suitable electrode. The logarithmic dependence of the titration curves can be derived from the Nernst equation, which states that the emf developed by an electrochemical cell varies as the logarithm of the mass-action quotient for the cell reaction. To make this statement more concrete, let us consider briefly the measurement of *p*H.

*p*H-measuring cells

The primary *p*H-measuring cell is represented in (7.2). If this notation is unfamiliar to you, look up the subject of electrochemical cells in your general chemistry textbook or in Chapter 10, or skip directly to Eq. (7.6).

$$\text{Pt-H}_2(g) \mid \text{H}_3\text{O}^+(\text{in soln. X}) \parallel \text{KCl(sat., aq)} \mid \text{Hg}_2\text{Cl}_2(s)\text{-Hg}(l) \quad (7.2)$$

The half-reactions that take place in the *p*H-measuring cell (7.2) are shown in Eqs. (7.3) and (7.4), and the complete cell reaction is shown in (7.5).

*p*H-sensing half-cell:

$$\tfrac{1}{2}\,\text{H}_2(g) + \text{H}_2\text{O} \; \rightleftharpoons \; \text{H}_3\text{O}^+(\text{in soln. X}) + \text{e}^- \qquad (7.3)$$

Reference half-cell:

$$e^- + \tfrac{1}{2}\,Hg_2Cl_2(s) \;\rightleftharpoons\; Hg(l) + Cl^-(\text{in sat. KCl}) \qquad (7.4)$$

Complete cell:

$$\tfrac{1}{2}\,H_2(g) + H_2O + \tfrac{1}{2}\,Hg_2Cl_2(s) \;\rightleftharpoons$$
$$H_3O^+(\text{in soln. X}) + Hg(l) + Cl^-(\text{in sat. KCl}); \qquad n = 1 \quad (7.5)$$

The hydrogen electrode serves as the pH-sensing electrode. The $Hg_2Cl_2(s)$ − $Hg(l)$ electrode merely serves as a convenient reference electrode. On applying the Nernst equation, we find that the emf developed by the cell is given by Eq. (7.6).

$$\mathbf{E} = \mathbf{E}° - \frac{2.303\,RT}{\mathbf{F}}\,\log \frac{[H_3O^+(\text{in soln. X})][Cl^-(\text{in sat. KCl})]}{P_{H_2}^{1/2}} \qquad (7.6)$$

The value of $[Cl^-(\text{in sat. KCl})]$ is constant because the reference half-cell is saturated with potassium chloride. If P_{H_2} is also kept constant, the cell will measure $[H_3O^+(\text{in soln. X})]$. To derive the simple relationship that applies, it is convenient to define a "cell constant" $\mathbf{E}^*$ as in (7.7).

$$\mathbf{E}^* = \mathbf{E}° - \frac{2.303\,RT}{\mathbf{F}}\,\log \frac{[Cl^-(\text{in sat. KCl})]}{P_{H_2}^{1/2}} = \text{constant} \qquad (7.7)$$

On introducing $\mathbf{E}^*$ in (7.6), we obtain (7.8); then on solving for $T = 298.2°K$ (25°C), we obtain (7.9).

$$\mathbf{E} = \mathbf{E}^* - \left(\frac{2.303\,RT}{\mathbf{F}}\right)\log\,[H_3O^+(\text{in soln. X})] \qquad (7.8)$$

$$\mathbf{E}(V) = \mathbf{E}^* + 0.059\,p\mathrm{H}\ (\text{in soln. X}) \qquad (7.9)$$

In modern practice, the more convenient glass electrode (which also measures the pH) is used in place of Pt-H_2(g), and the titration is carried out directly in the "solution X" compartment of the cell. The emf measuring instrument is called a *potentiometer*. If the potentiometer is calibrated to read pH directly, it is called a pH meter. According to Eq. (7.9), at 25°C the emf changes by 0.059 V (59 mV) per pH unit.

Potentiometric titration curves

A typical experimental pH curve for the titration of sulfuric acid with sodium hydroxide in water is shown in Fig. 7.3. As titrant is added, the pH changes rather slowly at first, until the reaction approaches close to the equivalence point. Here the change suddenly becomes precipitous as the solution changes from acidic to alkaline. The equivalence point is that point at which the slope (not the curvature) is at a maximum. It can be seen that the equivalence point is an inflection point, and that the steep region of the titration curve surrounding the equivalence point is quite

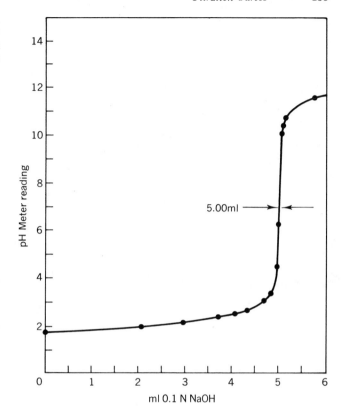

Fig. 7.3. Potentiometric titration of 10.00 ml of 0.025 F H_2SO_4 with 0.1 N NaOH in water. (Courtesy of Thomas Lasensky.)

linear. These features are in excellent agreement with those of the theoretical curves shown previously in Fig. 4.1.

When the reaction between substrate and titrant takes place in a series of well resolved steps, the potentiometric titration curve will show the characteristic precipitous change at the conclusion of each step. Thus Fig. 7.4 shows a potentiometric titration curve for the titration of sulfuric acid with a hydroxide salt in a nonaqueous solvent consisting largely of acetone. There are now two equivalence points, coinciding with the inflection points at the two "breaks" in the curve. The first equivalence point represents the reaction of $H_2SO_4 \rightarrow HSO_4^-$, the second that of $HSO_4^- \rightarrow SO_4^{2-}$.

It is instructive to compare Figs. 7.3 and 7.4 and to inquire why the change of solvent causes such a drastic change in the nature of the titration curve. The answer is that there is a marked difference in the acid-base properties of the two solvents. In water, the inherently high acid strength of H_2SO_4 is reduced by proton transfer to water molecules, and the first half of the titration curve in Fig. 7.3 thus reflects the leveling effect of the solvent on acidity described in Chapter 6. Note that the pH produced by

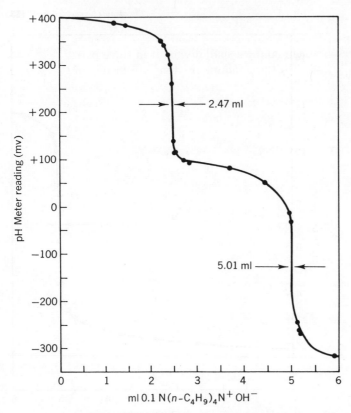

Fig. 7.4. Potentiometric titration of 10.00 ml of 0.025 F H_2SO_4 in 95 percent acetone-5 percent water with 0.1 N $(n\text{-}C_4H_9)_4N^+OH^-$ in methanol-benzene. (Courtesy of Thomas Lasensky.)

0.1 F H_2SO_4 differs by only 0.4 unit from that produced by 0.1 F $NaHSO_4$. (In Fig. 7.3, the initial pH meter reading is 1.70; after addition of one equivalent of NaOH, the pH meter reading has increased to 2.1.) Acetone, being a much weaker base than water, can sustain higher levels of acidity. Although sulfuric acid is largely ionized in acetone, the resulting lyonium ion is a much stronger acid than HSO_4^-, and the titration therefore takes place in two well resolved steps.

CALCULATIONS OF MOLAR CONCENTRATIONS FROM FORMAL CONCENTRATIONS

Solution properties vary with composition in simple, predictable ways only if the composition is stated in terms of molecular species rather than of formal components. The first step in the calculation of titration curves is

154

therefore the calculation of molar concentrations from formal concentrations. This is not a trivial problem, and we shall discuss it in three parts. In this section we shall summarize the assumptions that are made, and state their mathematical consequences. In the next section we shall consider how to set up the algebraic equations that must be solved. Finally, in the section after that, we shall consider methods for solving those equations.

In calculating molar concentrations from formal concentrations, we assume that:

    1. The solution is at equilibrium.

    2. The formal concentrations of the solute components are known.

    3. The reactions that take place in solution and the extent of those reactions, or their equilibrium constants, are known.

The distribution of the components among the molecular species is subject to constraints imposed by mass balance, chemical equilibrium, and (in the case of electrolytes) electrical neutrality. The consequences of these constraints are expressed in the form of algebraic equations whose solution leads to the desired molar concentrations. The equations are classified as follows:

Mass balance. When a solution is prepared, the amounts of the components are stated conveniently in formula weights or in formula weights per liter. For example, if we take 0.01 formula weight of acetic acid and 0.005 formula weight of sodium hydroxide and add enough pure water to make one liter of solution, the *components* will be: acetic acid, sodium hydroxide, and water; the *formal* concentrations of the components will be, $c_{HAc} = 0.01\ F$ and $c_{NaOH} = 0.005\ F$. Of course, the *molar* composition of the solution will be quite different, owing to the reactions that take place. The *equations of mass balance* express the fact that all such reactions proceed with conservation of atoms, according to the laws of stoichiometry. In general, there is one equation of mass balance for each independent component.

Chemical equilibrium. These equations relate molar concentrations to equilibrium constants. They include also the mathematical consequences of such qualitative statements as "sodium chloride in aqueous solution is dissociated to free ions." In general, there is one equation for each independent chemical equilibrium.

Electrical neutrality. These equations express the constraint that each component, as well as the whole solution, is electrically neutral. The equations are not necessary (in the mathematical sense) because electrical neutrality is inherent in, and implied by, mass balance. However, equations for electrical neutrality are useful because they are easy to set up and because they serve as a check on equations that may have been derived to express mass balance and chemical equilibrium.

SETTING UP THE ALGEBRAIC EQUATIONS

In this section we wish to show how the conditions of mass balance, chemical equilibrium, and electrical neutrality can be expressed in the language of algebra, by means of equations. The best way to do this is to work out a few typical examples in full detail.

Acetic acid in water. When acetic acid is dissolved in water, the following reactions are stoichiometrically significant:

$$HAc + HOH \; \rightleftharpoons \; H_3O^+ + Ac^- \tag{7.10}$$

$$2\,HAc \; \rightleftharpoons \; (HAc)_2 \tag{7.11}$$

By *stoichiometrically significant* we mean that the concentrations of H_3O^+, Ac^-, and $(HAc)_2$ are *not negligible* compared to the experimental error in the formal concentration of acetic acid. On the other hand, the concentration of hydroxide ion in equilibrium with H_3O^+ is too small to be stoichiometrically significant. Thus there are four unknown molar concentrations: $[HAc]$, $[H_3O^+]$, $[Ac^-]$, and $[(HAc)_2]$. To solve, we write the following four equations.

To express mass balance:

$$c_{HAc} = [HAc] + [Ac^-] + 2\,[(HAc)_2] \tag{7.12}$$

$$[H_3O^+] = [Ac^-] \tag{7.13}$$

To express chemical equilibrium:

$$\frac{[H_3O^+][Ac^-]}{[HAc]} = K_A = 1.75 \times 10^{-5}\ (M) \tag{7.14}$$

$$\frac{[(HAc)_2]}{[HAc]^2} = K_{assoc} = 0.037\ (M^{-1}) \tag{7.15}$$

Equation (7.12) expresses the conservation of acetate (Ac) groups. The concentration of $(HAc)_2$ is multiplied by two because there are two acetate groups in each molecule of $(HAc)_2$. Equation (7.13) expresses the stoichiometry of reaction (7.10): H_3O^+ and Ac^- are produced in equimolar amounts. Note, however, that Eq. (7.13) is also an expression of electrical neutrality.

Given the four independent equations, (7.12) through (7.15), we can solve for the four unknown concentrations. Let $[HAc] = x$, then from (7.15),

$$[(HAc)_2] = K_{assoc}\, x^2 \tag{7.16}$$

from (7.13) and (7.14),

$$[H_3O^+] = [Ac^-] = (K_A x)^{1/2} \tag{7.17}$$

Substituting in (7.12), we obtain (7.18).

$$c_{HAc} = x + (K_A x)^{1/2} + 2K_{assoc} x^2 \tag{7.18}$$

Equation (7.18) is a quartic equation in x. To transform it into standard form, we write (7.19) and combine terms involving equal powers of x.

$$K_A x = (c_{HAc} - x - 2K_{assoc} x^2)^2 \tag{7.19}$$

The result is (7.20).

$$4K_{assoc}^2 x^4 + 4 K_{assoc} x^3 - (4K_{assoc} c_{HAc} - 1)x^2 \\ - (2c_{HAc} + K_A)x + c_{HAc}^2 = 0 \tag{7.20}$$

Practical methods for solving (7.20) will be taken up in the next section.

Cyclohexene and bromine in carbon tetrachloride. In this example there are two solute components, and they react to form a single product [Eq. (7.21)].

$$\underset{\textbf{Cyclohexene}}{C_6H_{10}} + Br_2 \rightleftharpoons \underset{\textbf{Cyclohexene dibromide}}{C_6H_{10}Br_2} \tag{7.21}$$

The solvent, carbon tetrachloride, is chemically inert. Thus there are three unknown molar concentrations, $[C_6H_{10}]$, $[Br_2]$, and $[C_6H_{10}Br_2]$. To solve, we write the following three equations.

To express mass balance:

$$c_{C_6H_{10}} = [C_6H_{10}] + [C_6H_{10}Br_2] \tag{7.22}$$

$$c_{Br_2} = [Br_2] + [C_6H_{10}Br_2] \tag{7.23}$$

To express chemical equilibrium:

$$\frac{[C_6H_{10}Br_2]}{[Br_2][C_6H_{10}]} = K = 10^{17} \ (M^{-1}) \tag{7.24}$$

Equations (7.22) and (7.23) state that the component exists in solution, either in its original unreacted form, or in the form of an equivalent amount of reaction product.

To solve Eqs. (7.22) through (7.24), let $[Br_2] = y$. Then, using Eqs. (7.22) and (7.23), $[C_6H_{10}Br_2] = (c_{Br_2} - y)$, and $[C_6H_{10}] = (c_{C_6H_{10}} - c_{Br_2} + y)$. Substituting in (7.24), we obtain (7.25), which (you will convince yourself) is a quadratic equation that can be solved for y.

$$\frac{(c_{Br_2} - y)}{y(c_{C_6H_{10}} - c_{Br_2} + y)} = K \tag{7.25}$$

Thiosulfate and triiodide in water. In this example there is again a chemical reaction, but the stoichiometry is exceptionally intricate, and if you can work this example you can probably work any.

In the presence of an excess of iodide ion, iodine (I_2) is converted largely to triiodide (I_3^-), whose reaction with thiosulfate in aqueous solution is represented by Eq. (7.26).

$$I_3^- + 2\,S_2O_3^{2-} \rightleftharpoons S_4O_6^{2-} + 3\,I^- \qquad (7.26)$$

We shall assume that the solution is prepared from the following components: water, $Na_2S_2O_3$, I_2, and KI. We shall assume also, without serious error, that I_2 is converted to I_3^-, that the salts are completely dissociated, and that the cations are not involved in the reaction. Thus we need to find molar concentrations for four ionic species: I_3^-, $S_2O_3^{2-}$, $S_4O_6^{2-}$, and I^-. We are given the formal concentrations (c_{I_2}, c_{KI}, and $c_{Na_2S_2O_3}$) of the solute components, and the equilibrium constant for (7.26). To solve, we write the following equations.

To express conservation of iodine *atoms*:

$$c_{KI} + 2c_{I_2} = [I^-] + 3[I_3^-] \qquad (7.27)$$

To express conservation of thiosulfate groups (or of sulfur atoms):

$$c_{Na_2S_2O_3} = [S_2O_3^{2-}] + 2[S_4O_6^{2-}] \qquad (7.28)$$

To express that one formula weight of I_2 in solution produces either one mole of I_3^- or [by reacting according to (7.26)] one mole of $S_4O_6^{2-}$:

$$c_{I_2} = [I_3^-] + [S_4O_6^{2-}] \qquad (7.29)$$

To express chemical equilibrium:*

$$\frac{[S_4O_6^{2-}][I^-]^3}{[S_2O_3^{2-}]^2[I_3^-]} = K = 2.8 \times 10^{15}\ (M) \qquad (7.30)$$

Note that in (7.27), c_{I_2} is multiplied by 2 because there are two iodine atoms per formula weight of I_2. For a similar reason I_3^- is multiplied by 3 in (7.27) and [$S_4O_6^{2-}$] is multiplied by 2 in (7.28).

To calculate molar concentrations, let $[S_2O_3^{2-}] = z$. Then, from (7.28),

$$[S_4O_6^{2-}] = \tfrac{1}{2}(c_{Na_2S_2O_3} - z)$$

from (7.29),

$$[I_3^-] = c_{I_2} - \tfrac{1}{2}(c_{Na_2S_2O_3} - z)$$

from (7.27),

$$[I^-] = c_{KI} - c_{I_2} + \tfrac{3}{2}(c_{Na_2S_2O_3} - z)$$

On substituting in (7.30), we obtain (7.31), which is a quartic equation in z.

$$\frac{\tfrac{1}{2}(c_{Na_2S_2O_3} - z)(c_{KI} - c_{I_2} + \tfrac{3}{2}[c_{Na_2S_2O_3} - z])^3}{z^2(c_{I_2} - \tfrac{1}{2}[c_{Na_2S_2O_3} - z])} = K \qquad (7.31)$$

Let us summarize the intricacies that may have made this example difficult: (1) Reaction (7.26) yields more than one product; (2) the molar ratios in (7.26) are *not* simply one-to-one; (3) iodine atoms come from two independent components; (4) the final quartic equation may be tedious to solve. Similar, or analogous, intricacies arise in many other problems.

---

* The equilibrium constant was calculated on p. 31.

Sodium acetate and sodium chloride in acetic acid. In the preceding examples the stoichiometric effect of the solvent—any self-ionization or reaction with the solutes—was negligible. We now wish to consider an example in which a reaction of the solvent is significant.

In acetic acid, sodium acetate and sodium chloride exist largely in the form of ion pairs, which are in equilibrium with free ions [Eqs. (7.32) and (7.33)]. In addition, sodium chloride has a small tendency to react with acetic acid, as shown in (7.34).

$$Na^+Ac^- \rightleftharpoons Na^+ + Ac^- \tag{7.32}$$

$$Na^+Cl^- \rightleftharpoons Na^+ + Cl^- \tag{7.33}$$

$$Na^+Cl^- + HAc \rightleftharpoons HCl + Na^+Ac^- \tag{7.34}$$

HCl is a weak acid in acetic acid, and the *acetolysis* reaction (7.34) is analogous to the hydrolysis of the salt of a weak acid in water.

According to (7.32) through (7.34), there are six unknown molar concentrations: $[Na^+Ac^-]$, $[Na^+Cl^-]$, $[Na^+]$, $[Ac^-]$, $[Cl^-]$, and $[HCl]$. We may assume that the following data are available: three equilibrium constants, for reactions (7.32), (7.33), and (7.34); and two formal concentrations, c_{NaAc} and c_{NaCl}—altogether five pieces of data, and good for five equations. A sixth equation can be written to express conservation of electrical charge. The six equations are derived conveniently as follows:

Conservation of chlorine atoms:

$$c_{NaCl} = [Na^+Cl^-] + [Cl^-] + [HCl] \tag{7.35}$$

Conservation of sodium atoms:

$$c_{NaCl} + c_{NaAc} = [Na^+Ac^-] + [Na^+Cl^-] + [Na^+] \tag{7.36}$$

Conservation of electrical charge:

$$[Na^+] = [Cl^-] + [Ac^-] \tag{7.37}$$

Equilibrium-constant expressions:

$$\frac{[Na^+][Ac^-]}{[Na^+Ac^-]} = K_d \text{(for } Na^+Ac^-) \tag{7.38}$$

$$\frac{[Na^+][Cl^-]}{[Na^+Cl^-]} = K_d \text{(for } Na^+Cl^-) \tag{7.39}$$

$$\frac{[HCl][Na^+Ac^-]}{[Na^+Cl^-]} = K \text{(for acetolysis of } Na^+Cl^-) \tag{7.40}$$

SOLVING THE ALGEBRAIC EQUATIONS

After the algebraic equations have been set up, they must be solved to evaluate the molar concentrations. This task also may not be trivial.

The equations for mass balance are simple linear equations that can be solved by familiar methods. However, the equilibrium-constant expressions are usually of quadratic or higher order—the algebraic order mounts rapidly with the number of chemical equilibria that must be considered simultaneously. In this section we wish to consider the solution of such equations.

Direct calculation of roots

As is well known, the general quadratic equation $ax^2 + bx + c = 0$ has two roots, given by (7.41).

$$x = \frac{-b \pm \sqrt{b^2 - 4ac}}{2a} \tag{7.41}$$

In equilibrium problems, these roots are real numbers. If the roots turn out to be imaginary, look for a mistake, probably in an algebraic sign or a decimal point. One of the roots will be the solution of the problem; the other, extraneous, root will be physically unreasonable. The extraneous root is usually a negative number, or a positive concentration too large to be consistent with mass balance.

As a specific example, we shall consider the reaction of cyclohexene with bromine, Eq. (7.21). In particular, let us solve Eq. (7.25) to obtain the molar concentration of bromine, y, when $c_{C_6H_{10}} = 0.1000\ F$ and $c_{Br_2} = 0.0500\ F$. On rearranging (7.25), we obtain

$$y^2 + \left(c_{C_6H_{10}} - c_{Br_2} + \frac{1}{K}\right)y - \left(\frac{c_{Br_2}}{K}\right) = 0$$

On substituting the values $K = 10^{17}$, $c_{C_6H_{10}} = 0.1000$, and $c_{Br_2} = 0.0500$, we obtain

$$y^2 + 0.0500y - 5 \times 10^{-19} = 0$$

The roots, according to (7.41), are

$$y = \frac{-0.0500 \pm \sqrt{(0.05)^2 + 2 \times 10^{-18}}}{2}$$

$$= +\frac{2 \times 10^{-17}}{2}; \quad -\frac{0.1000}{2}$$

The negative root is discarded because a molar concentration must be a positive number. Hence $y = 1 \times 10^{-17}\ M$.

Roots of cubic and higher-order equations. The general method (7.41) for finding the roots of quadratic equations is attributed to the Hindus, having appeared already in the writings of Aryabhatta in the fifth century. The method did not become common knowledge among mathematicians in Europe until after the Dark Ages. It was then extended by Tartaglia and Ferrari in Italy in the sixteenth century, who derived general formulas for finding the roots of cubic and quartic equations. But that is the limit of the method. Abel and Galois proved, in the nineteenth century, that direct calculation of roots for quintic and higher-order equations, as a general method, is impossible *in principle*—it is a peculiarity of our number system that no general formula exists. In practice, the direct calculation of roots has severe limitations even for cubic and quartic equations: The arithmetic is tedious and the unavoidable rounding-off of intermediate results can introduce large errors into the final result. Indeed, as a practical method for solving algebraic equations, direct calculation of roots by a general formula is pretty much limited to quadratic equations.

Successive approximations

It may not sound scientific, but the most practical method for solving complex algebraic equations is by trial and error. If the equation is of the form $f(x) = 0$, we make a plausible guess x_1 of the correct root and calculate $f(x_1)$. If we are lucky, $f(x_1)$ will be close to zero, which means that x_1 is close to the desired root. We then repeat the calculation with other trial values $x_2, x_3, \ldots$ until we find that value of x for which $f(x)$ is precisely zero, and which is the desired solution. There are systematic procedures for expediting such calculations.*

Among the methods for solving complex algebraic equations, the *method of successive approximations* is particularly useful for the calculation of points along titration curves. The constraints of mass balance and chemical equilibrium often lead to equations of the form (7.42), where y is the molar concentration of a solute, K is an equilibrium constant, and c_1 and c_2 are formal concentrations or linear combinations of formal concentrations.

$$y = K \frac{(c_1 - y)}{(c_2 + y)} \tag{7.42}$$

A previously derived example is Eq. (7.25), which can be rewritten in the form (7.25a), where $K' = K^{-1}$.

$$y = K' \frac{(c_{Br_2} - y)}{(c_{C_6H_{10}} - c_{Br_2} + y)} \tag{7.25a}$$

Note that (7.25a) is of the same form as (7.42) since c_{Br_2} may be identified with c_1 and $(c_{C_6H_{10}} - c_{Br_2})$ with c_2.

The relationship between the molar concentration of a solute and formal concentrations of the components can *always* be written in the general form (7.43), where ϕ is a known function (or can be derived), and $c_1 \pm y$, $c_2 \pm y, \ldots, c_n \pm y$ are the independent variables.

$$y = \phi(c_1 \pm y, c_2 \pm y, \ldots, c_n \pm y) \tag{7.43}$$

---

* See, for example, H. Margenau and G. M. Murphy, *The Mathematics of Physics and Chemistry*, Chapter 13 (New York: D. Van Nostrand Co., Inc., 1943).

y is the desired molar concentration, and $c_1, c_2, \ldots, c_n$ are formal concentrations or linear combinations thereof.

If (as often happens) y is small compared to $c_1, c_2, \ldots, c_n$, then solution of (7.43) becomes convenient by successive approximations. For example, consider Eq. (7.42). Regardless of whether it is true or not, we start out by assuming that y is so small that we may set the right-hand side of (7.42) equal to $K(c_1/c_2)$. The result is denoted by y_1. We then apply y_1 to the right-hand side of (7.42) and calculate a new y, which we call y_2. This process of successive approximations is continued until successive values of y agree satisfactorily. The mathematics may be summarized as follows.

$$\text{First approximation:} \quad y_1 = K \frac{c_1}{c_2}$$

$$\text{Second approximation:} \quad y_2 = K \frac{(c_1 - y_1)}{(c_2 + y_1)}$$

$$\text{Third approximation:} \quad y_3 = K \frac{(c_1 - y_2)}{(c_2 + y_2)}$$

And so on. The correct answer is reached when $y_n = y_{n-1}$, within the desired accuracy of the calculation. In practice, the approximations converge rapidly to the correct answer if y is less than 10 percent of the smallest c. The method is especially useful for titration curves: Before the equivalence point, the concentration of unreacted *titrant* is small; past the equivalence point, the concentration of unreacted *substrate* is small.

Cyclohexene and bromine in carbon tetrachloride. To appreciate the convenience of the method, let us consider again the reaction of cyclohexene with bromine. Suppose that $c_{C_6H_{10}} = 0.1000\ F$ and $c_{Br_2} = 0.0500\ F$, so that C_6H_{10} is in excess. The concentration $y = [Br_2]$ of unreacted bromine is therefore small. Equation (7.25) is then written conveniently in the form (7.44). The factor 10^{-17} represents K^{-1}.

$$y = 10^{-17} \frac{(0.0500 - y)}{(0.1000 - 0.0500 + y)} = 10^{-17} \frac{(0.0500 - y)}{(0.0500 + y)} \quad (7.44)$$

$$\text{First approximation:} \quad y_1 = 10^{-17} \frac{0.0500}{0.0500} = 10^{-17}\ M$$

$$\text{Second approximation:} \quad y_2 = 10^{-17} \frac{0.0500 - 10^{-17}}{0.0500 + 10^{-17}} = 10^{-17}\ M$$

In this case y is so small that the first approximation is obviously sufficient. We say "obviously," because y *is clearly much smaller than the experimental error in the formal concentrations.* Assuming that the latter are expressed according to the usual convention concerning significant figures, then the numbers 0.0500 appearing in (7.44) are accurate to ± 0.0001 at best, that is, the composition of the solution is in doubt by at least $\pm 1 \times 10^{-4}\ F$. Compared to that uncertainty, the error made by neglecting the additive terms $\pm y$ on the right-hand side in (7.44) is entirely negligible. Indeed, we must conclude that the amount of unreacted bromine in the solution at equilibrium is much too small to be stoichiometrically significant. This does not mean, however, that the presence of bromine can be ignored. Because bromine exists in equilibrium with C_6H_{10} and $C_6H_{10}Br_2$, the bromine concentration is buffered and reproducible and, although small, capable of producing real chemical and electrochemical effects. In that sense a stoichiometrically insignificant species is like a shadow—it is "visible" but not substantial and moves in equilibrium with the substances that produce it.

Acetic acid in water. For our next example, we wish to apply the method of successive approximations to an inherently more complex calculation, that of the molar concentrations in a 0.1000 F solution of acetic acid in water. The basic equations are (7.12) through (7.15). The final equation (7.20) is a quartic equation. However, solution by successive approximations is not difficult. In this case we make use of the fact that $[HAc] \gg [Ac^-] + 2[(HAc)_2]$. Letting

$$y = [Ac^-] + 2[(HAc)_2]$$

we proceed as follows.

From (7.12): $[HAc] = c_{HAc} - y$

From (7.13) and (7.14): $[Ac^-] = [K_A(c_{HAc} - y)]^{1/2}$

From (7.15): $[(HAc)_2] = K_{assoc}(c_{HAc} - y)^2$

Thus: $y = [K_A(c_{HAc} - y)]^{1/2} + 2K_{assoc}(c_{HAc} - y)^2$ (7.45)

Equation (7.45) is convenient for calculation because $c_{HAc} \gg y$. On letting $c_{HAc} = 0.1000\ M$, $K_A = 1.75 \times 10^{-5}\ (M)$, and $K_{assoc} = 0.037\ (M^{-1})$, numerical results are:

First approximation: $(c_{HAc} - y) = 0.1000\ F$

Substitute in (7.45): $y_1 = (1.75 \times 10^{-6})^{1/2} + 2 \times 3.7 \times 10^{-4}\ M$

$\therefore y_1 = 1.32 \times 10^{-3} + 0.74 \times 10^{-3} = 2.06 \times 10^{-3}\ M$

Second approximation: $(c_{HAc} - y_1) = 0.09794\ M$

$\therefore y_2 = 1.31 \times 10^{-3} + 0.71 \times 10^{-3} = 2.02 \times 10^{-3}\ M$

Third approximation: $(c_{HAc} - y_2) = 0.09798\ M$

$\therefore y_3 = 1.31 \times 10^{-3} + 0.71 \times 10^{-3} = 2.02 \times 10^{-3}\ M$

Evidently the second approximation is sufficient. Final results are: $[HAc] = 0.0980\ M$ (rounded to the correct number of significant figures), $[Ac^-] = [H_3O^+] = 1.31 \times 10^{-3}\ M$, and $[(HAc)_2] = \frac{1}{2} \cdot 0.71 \times 10^{-3} = 0.36 \times 10^{-3}\ M$. We defy anyone to solve the original quartic equation by any other method in less than twice the time!

CALCULATION OF TITRATION CURVES

Definitions and equations

For definiteness we shall calculate titration curves only for the stoichiometry represented by Eq. (7.46).

$$\text{Substrate} + \text{Titrant} \rightleftharpoons \text{Product} \qquad (7.46)$$

However, the calculations will be typical and the qualitative conclusions will have general validity.

A familiar reaction whose stoichiometry conforms to (7.46) is that of cyclohexene with bromine in carbon tetrachloride. The algebra set up for that reaction [Eqs. (7.22) through (7.25)] applies directly in the general case if we imagine that C_6H_{10} is the substrate, Br_2 is the titrant, and $C_6H_{10}Br_2$ is the product. There are two solute components, the substrate and the titrant. The amount of substrate placed in the titration flask is a fixed amount, while the amount of titrant increases progressively throughout the titration. As titrant is added, the volume of the solution increases.

To measure the progress of the titration, it is convenient to introduce the titrant/substrate ratio r, defined in (7.47).

$$r \equiv \frac{\text{equivalents of titrant added to the solution}}{\text{equivalents of substrate present initially}} \qquad (7.47)$$

Thus, for any titration, $r = 0$ at the beginning, $r = 1$ at the equivalence point, and $r > 1$ when the equivalence point is exceeded. To express the amounts of substrate and titrant (the components), it is convenient to use the following notation.

n_S = formula weights of substrate in the titration flask at $r = 0$

n_T = formula weights of titrant added to the solution

V = volume of solution, in liters

$c_S = \dfrac{n_S}{V}$ = formal concentration of substrate component

$c_T = \dfrac{n_T}{V}$ = formal concentration of titrant component

In (7.46), substrate reacts with titrant in a one-to-one ratio; hence $r = c_T/c_S$. To express mass balance, we write by analogy with (7.22) and (7.23):

$$c_S = [\text{substrate}] + [\text{product}]$$
$$c_T = rc_S = [\text{titrant}] + [\text{product}]$$

In the following we shall calculate the molar concentration of unreacted titrant as a function of r. On defining the molar concentration x according to (7.48), the molar concentrations of the other species become (7.49) and (7.50), and the equilibrium constant is expressed by (7.51).

$$[\text{Titrant}] = x \tag{7.48}$$

$$[\text{Product}] = rc_S - x \tag{7.49}$$

$$[\text{Substrate}] = c_S(1 - r) + x \tag{7.50}$$

$$K = \frac{[\text{product}]}{[\text{substrate}][\text{titrant}]} = \frac{(rc_S - x)}{[c_S(1 - r) + x]x} \tag{7.51}$$

Equation (7.51) is a quadratic equation in x. On rearranging, we obtain (7.52), which is in standard quadratic form.

$$x^2 + [c_S(1 - r) + K^{-1}]x - rc_S K^{-1} = 0 \tag{7.52}$$

Linear titration curves

We now wish to use Eq. (7.52) to simulate some actual titration curves. We can (and often do) control the volume of the solution so that dilution is small; then c_S is approximately constant. We can (and usually do) carry out the titration so that the solvent composition is constant; then K, too, is constant. The titrant/substrate ratio r varies from zero to greater than one. We shall find that under such conditions, the titration curve depends primarily on the relationship between x and r.

Figure 7.5 depicts x as a function of r under the following conditions: $c_S = 0.1$ F throughout. K is either zero (no reaction), 10, 1000, 10^5, or infinite. We are looking for features by which the equivalence point can be recognized. Let us examine the extreme values of K first.

When $K = 0$, titrant merely accumulates without reacting. Because dilution is small, the plot of x versus r is essentially a straight line through the origin, with slope equal to c_S. There are no distinguishing features by which the point $r = 1$ can be recognized.

When $K = \infty$, reaction between substrate and titrant is complete. Before the equivalence point, x is zero and the relationship is a horizontal line. Past the equivalence point, excess titrant accumulates and the relationship is a different line, with slope equal to c_S, which intersects the r-axis at the equivalence point. The abrupt change in slope at $r = 1$ produces a characteristic kink by which the equivalence point can be recognized.

Intermediate values of K produce curves with intermediate features. When $K = 10$, there is a gradual and rather featureless increase in slope that offers little help in locating the equivalence point. When $K = 1000$, a point of maximum curvature can be identified, though not very accurately, and the equivalence point can thus be located, but with considerable uncertainty. When $K = 10^5$, the relationship resembles closely that for

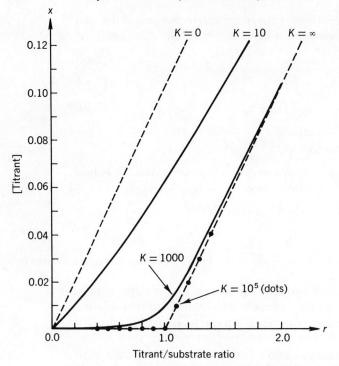

Fig. 7.5. Relationship between the molar concentration of unreacted titrant x and the titrant/substrate ratio r for various values of K, according to Eq. (7.52). $c_S = 0.1\ F$ throughout.

$K = \infty$, and recognition of the equivalence point becomes easy and accurate.

The lesson taught by Fig. 7.5 is general: If K is small, the relationship between x and r is featureless. In order for the equivalence point to be recognized, K must be large.

Finally, we wish to show that the relationship between x and r causes the curvature of linear titration curves at the equivalence point. Consider a generalized solution property $\mathcal{P}$ which, by hypothesis, will be a linear function of molar concentrations, Eq. (7.53).

$$\mathcal{P} = \mathcal{P}_0 + a_S[\text{Substrate}] + a_T[\text{Titrant}] + a_P[\text{Product}] \qquad (7.53)$$

$\mathcal{P}_0$ is the value of the property in the absence of the given solutes and a_S, a_T, and a_P are characteristic slopes. For example, $\mathcal{P}$ might be the absorbance and a the molar extinction coefficient. By applying Eqs. (7.48) through (7.50), we then express $\mathcal{P}$ as a function of r, c_S, and x. The result after collection of terms is (7.54).

$$\mathcal{P} = \mathcal{P}_0 + a_S c_S + (a_S + a_T - a_P)x + (a_P - a_S)c_S r \qquad (7.54)$$

If c_S is constant, $\mathcal{P}_0 + a_S c_S$ remain constant during the titration, and the titration curve of $\mathcal{P}$ versus r can be analyzed into two additive terms: $(a_S + a_T - a_P)x$, which is proportional to x; and $(a_P - a_S)c_S r$, which (if $c_S \approx$ constant) is proportional to r. The geometrical significance of these terms, on a plot of $\mathcal{P}$ versus r, follows.

The term "proportional to x" is essentially a plot of x versus r, with the scale of the ordinate adjusted so as to accommodate the proportionality constant. The term "proportional to r" produces no additional curvature; it has the effect of rotating the r-axis away from the horizontal. Thus, as shown in Fig. 7.6, the plot of $\mathcal{P}$ versus r is tantamount to a plot of x versus

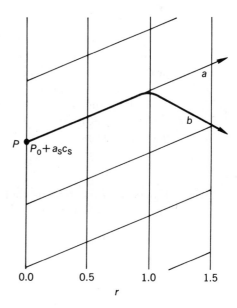

Fig. 7.6. Geometric interpretation of Eq. (7.54). Line (a) is a plot of $\mathcal{P} = \mathcal{P}_0 + a_S c_S + (a_P - a_S)c_S r$ versus r. The difference between curve (b) and line (a) is equal to $(a_S + a_T - a_P)x$. In this example, $(a_P - a_S - a_T) > (a_P - a_S) > 0$.

r in a coordinate system in which the angle between the axes is other than 90°. The equivalence point can be recognized on such a plot only if it can be recognized also on a conventional plot of x versus r. We see, therefore, that titration is feasible only if K is sufficiently large.

Molar concentrations in first approximation

If we may infer that K is always large when analytical titration is feasible, then the calculation of titration curves can be greatly simplified by the method of successive approximations. The molecular species at relatively low concentration is:

Before the equivalence point, the unreacted titrant.
Past the equivalence point, the unreacted substrate.
At the equivalence point, both titrant and substrate.

For the greater part of the titration curve the first approximation (in the series of successive approximations) will be sufficient. The algebra then becomes so simple that we can visualize the physical significance of the equations without difficulty. Let us apply the method to solve Eq. (7.51).

1. Before the equivalence point (approximately in the range $0.05 < r < 0.95$), $x \ll rc_S$ and $x \ll (1 - r)c_S$. Hence (7.51) may be simplified to (7.55).

For large K and $0.05 < r < 0.95$:

$$x = K^{-1}\frac{rc_S - x}{(1 - r)c_S + x} = K^{-1}\frac{r}{(1 - r)} \tag{7.55}$$

Equation (7.55) tells us that in a wide range before the equivalence point, centered at $r = 0.5$, x is directly proportional to $r/(1 - r)$ and inversely proportional to K. Indeed, at $r = 0.5$, $x = K^{-1}$. If x can be measured in this range, K can be deduced.

2. Past the equivalence point (approximately in the range $r > 1.05$), [Substrate] $\ll$ [Titrant], so that [Titrant] $-$ [Substrate] $\approx$ [Titrant]. On making appropriate substitutions from (7.48) and (7.50), we obtain (7.56).

For large K and $r > 1.05$:

$$x = (r - 1)c_S \tag{7.56}$$

Equation (7.56) tells us that past the equivalence point the relationship between x and r becomes independent of K: All titration curves degenerate into a single function of rc_S.

3. At the equivalence point ($r = 1$), Eq. (7.51) reduces to $Kx^2 = c_S - x$. If $x \ll c_S$, we obtain (7.57).

For large K and $r = 1$:

$$x = \sqrt{\frac{c_S}{K}} \tag{7.57}$$

Logarithmic titration curves

It is clear from Fig. 7.5 that when K is large, the concentration x of unreacted titrant before the equivalence point is very small—too small to be seen on the linear scale of that figure. Yet x, though small, varies with r in a characteristic way, stated in Eq. (7.55), and is of special interest because of its inverse dependence on K. If we wish to construct titration curves in which these features become visible, it is convenient to plot $\log x$ rather than x versus r. Such *logarithmic titration curves* are also useful because of their close connection to potentiometric titration curves. [For example, see Eq. (7.9) and the discussion that follows it.]

Typical plots of $\log x$ versus r are shown in Fig. 7.7. The calculations again are based on Eq. (7.51); $c_S = 0.1 F$ throughout and K is 10^7 or 10^5. Note that the scale of the ordinate is logarithmic (for instance, the distance between 10^{-7} and 10^{-6} is equal to that between 10^{-3} and 10^{-2}), but for convenience the graduation marks on the ordinate also show x. As expected on the basis of Eqs. (7.55) through (7.57), before the equivalence point the curve for $K = 10^7$ lies below that for $K = 10^5$ by two logarithmic units;

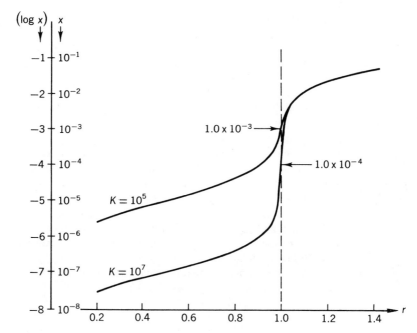

Fig. 7.7. Logarithmic titration curves calculated on the basis of Eq. (7.51). $c_S = 0.1\ F$ throughout.

at the equivalence point the difference is one unit; and past the equivalence point the curves practically coincide. In the vicinity of the equivalence point there is a characteristic precipitous change of $\log x$ with r, much like that on the experimental potentiometric titration curve in Fig. 7.3. Note, however, that the sharpness of that change increases markedly with K.

PROBLEMS

1. A solution is prepared by taking 0.0100 formula weight of H_3PO_4, 0.0150 formula weight of NaOH, and enough water to make a total volume of 0.500 liter.

(a) Find the formal concentrations, $c_{H_3PO_4}$ and c_{NaOH}, of the solute components.

(b) Assume that only the following molecular species are stoichiometrically significant: Na^+, $H_2PO_4^-$, HPO_4^{2-}. Calculate their molar concentrations.

(c) Using the results obtained in (b) and acid-dissociation constants listed in Table A.1, find the molar concentrations of the following: H_3PO_4, H_3O^+, OH^-, PO_4^{3-}. These concentrations should be small compared to those found in (b).

(d) In terms of calculation by successive approximations, the results obtained in (b) and (c) correspond to the first approximation. Now carry these calculations to the second approximation.

2. A solution is prepared by taking 0.00500 formula weight of the disodium salt of ethylenediamine tetraacetic acid (Na_2H_2Y), an exactly equivalent amount (0.00500 formula weight) of NaOH, and diluting with water to a total volume of 250.0 ml.

(a) Find the formal concentrations of the two solute components.

(b) Assume that only the following solute species are stoichiometrically significant: Na^+, H_2Y^{2-}, HY^{3-}, Y^{4-}. Calculate their molar concentrations, using data in Table A.1 as required.

(c) Find the pH of the solution and compare with the value of $(pK_{A3} + pK_{A4})/2$. Should the two be equal?

3. Apply the method of successive approximations to the calculation of the hydronium ion concentration in a 0.00500 F solution of nitrous acid (HNO_2). Assume a value of Ka = 4.50×10^{-4} (three significant figures) and carry out the approximation until your last two $[H_3O^+]$ values agree to within one percent.

4. A solution is made up from the following solute components: formic acid (HFo) 0.01000 F; benzoic acid (HBz) 0.01000 F; and NaOH 0.0100 F. Using data listed in Table A.1, find the molar concentrations of Na^+, HFo, Fe^-, HBz, Bz^-, and H_3O^+. Assume that these are the only molecular species that are stoichiometrically significant. Proceed as follows.

(a) Decide how many algebraic equations will be needed.

(b) Write equations to express mass balance.

(c) Write an equation to express electrical neutrality.

[If you wrote the complete set of equations for mass balance in (b), the equation for electrical neutrality will not be a new independent equation, but must be consistent with the equations for mass balance.]

(d) Write equations to express chemical equilibrium.

(e) Solve the set of equations.

5. Equations (7.46) through (7.52) develop the mathematics of titration curves for reactions of the type: substrate + titrant $\rightleftharpoons$ product. Using similar methods, derive an equation for the titration curve if the reaction is: substrate + titrant $\rightleftharpoons$ product P + product Q (S + T $\rightleftharpoons$ P + Q). Assume that the initial concentrations (when $r = 0$) of P and Q are zero. Apply the resulting equation to find an expression for the molar concentrations of P, Q, S, and T at the equivalence point.

6. Predict the solubility of silver chloride in 0.1 F aqueous NH_3, using data listed in Tables A.3 and A.4. Assume that the silver-ammonia complex is entirely $Ag(NH_3)_2^+$ (that is, neglect the presence of $AgNH_3^+$), and neglect all ion-pair formation.

7. Consider the titration of 0.1 F Fe^{2+} with 0.1 F Ce(IV) in 1 F H_2SO_4.

(a) Find the equilibrium constant by applying data in Table A.5.

(b) Calculate the concentrations of Fe^{2+}, Fe^{3+}, Ce(IV), and Ce(III) at the halfway point in the titration. (Remember to allow for dilution of the original concentrations during the titration.)

8. Predict the appearance of the titration curves for the following titrations in aqueous solution. For each reaction plot the experimental variable versus the titrant/substrate ratio r. Carry out the "titration" from no titrant ($r = 0$) until the titrant is 50 percent in excess. Assume that the concentration of substrate is initially 0.1 F. You may neglect the effects of dilution as the titration proceeds. Assume also that the equilibrium constants given in the Appendix remain valid throughout the titration.

(a) The conductometric titration of Ag^+ with Cl^-

(b) The amperometric (measure current at a particular applied emf, see Chapter 19) titration of Pb^{2+} with CrO_4^{2-}

(c) The potentiometric titration of Ag^+ with Br^-

(d) The potentiometric (pH) titration of benzoic acid with sodium hydroxide

(e) The photometric (measure absorbance of the product) titration of copper with EDTA

(f) The potentiometric titration of Fe^{2+} with $Ce(IV)$ in $1 \ F \ H^+$

(g) The potentiometric titration of Na_2CO_3 with HCl

9. Repeat Problems 8(a), (d), and (e), but assume that 25 ml of 0.1 F substrate is being titrated with 0.1 F titrant and consider the effect of dilution on the titration curve. Compare your result with the original result. Are there any significant differences in the two plots?

chapter 8

Innate Titration Error

In preceding chapters we made frequent use of the concept of innate titration error and gave "engineering formulas" for estimating it. Innate titration error arises, it will be recalled, because there are limits to the precision with which the "break" or abrupt change in the titration curve at the equivalence point can be recognized experimentally. If the innate titration error is less than 0.25 percent, we describe the titration as "good." If it is greater than 1 percent, we describe the titration as "difficult."

In this chapter we shall give a mathematical analysis of innate titration error. We shall consider a wide variety of problems, from the simple titration of a single substrate to selective titration of complex mixtures.

DEPENDENCE OF ERROR ON SLOPE
OF TITRATION CURVE

Innate titration error becomes serious when the equivalence point in the chemical reaction is not well defined. All methods for detecting the equivalence point are, in one way or another, logically equivalent to monitoring the concentration of unreacted titrant. The endpoint of a titration is taken as the point at which that concentration reaches some predetermined value, or some specific criterion is satisfied, such as a maximum of slope on a logarithmic titration curve, or a maximum of curvature on a linear titration curve. Examples have been shown in Figs. 7.1 to 7.4.

Theoretical titration curves such as those shown in Figs. 7.5 and 7.7 suggest that the sharpness of the titration endpoint increases with the equilibrium constant. However, we wish to introduce an even more general relationship, which states that *the sharpness of the titration endpoint increases with the slope of the logarithmic titration curve at the equivalence point.*

Titration curve near the equivalence point

For definiteness, we consider again a reaction of the simple type (8.1).

$$\text{Substrate} + \text{Titrant} \rightleftharpoons \text{Product} \tag{8.1}$$

The concentration x of unreacted titrant is given as a function of the titrant/substrate ratio r by Eq. (7.51), which is repeated below.

$$\text{Eq. (7.51):} \quad x = K^{-1} \frac{[rc_S - x]}{[(1 - r)c_S + x]} \tag{8.2}$$

If titration is feasible, K will be large, and x at the equivalence point [according to (7.57)] will be given by (8.3).

When K is large and $r = 1$,

$$x = \left(\frac{c_S}{K}\right)^{1/2} \tag{8.3}$$

Figure 8.1 shows two logarithmic titration curves near the equivalence point, one for $K = 10^7$ and the other for $K = 10^5$. These curves were calculated on the basis of Eq. (8.2) for $0.98 < r < 1.02$, and c_S is $0.1\ F$ throughout. Actually, Fig. 8.1 is simply a blowup of Fig. 7.7 around the equivalence point, but the difference in steepness of the two titration curves now becomes very striking.

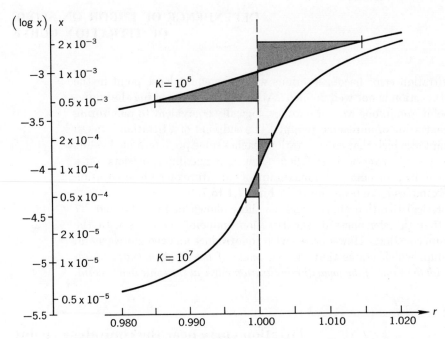

Fig. 8.1. Logarithmic titration curves near the equivalence point.

Error in locating the equivalence point

In general, there are two methods of locating the equivalence point on a logarithmic titration curve: One can stop the titration at the precalculated value of x, or one can find the point of maximum slope. Let us consider the sources of error in each.

If one tries to stop the titration at the precalculated value of x, there will be two sources of error: in the precalculation of x and in its measurement. The precalculation involves Eq. (8.3) and therefore is vulnerable to error in K and in c_S. The accuracy of available values for equilibrium constants varies widely, and c_S (or n_S/V) is never known with high accuracy: n_S must be determined by titration and is unknown beforehand, and V is also uncertain, especially if the analyst is too free with the use of his wash bottle.

The combined errors of precalculation and measurement of x are rarely less than a factor of two. Let us see what kind of an endpoint error that implies. In Fig. 8.1, when $K = 10^7$, x at the equivalence point is 1.0×10^{-4} M. Let us suppose that, because of error, the titration is actually stopped when $x = 2.0 \times 10^{-4}\ M$. In that case, r at the endpoint will be 1.0015 instead of 1.0000. That is, the endpoint error is $+0.15$ percent.

Now let us suppose we make a similar error when $K = 10^5$. As shown in the shallower curve in Fig. 8.1, x at the equivalence point is $1.0 \times 10^{-3}\ M$. If the titration is actually stopped when $x = 2.0 \times 10^{-3}\ M$, r at the endpoint will be 1.015 instead of 1.000, and the error will be 1.5 percent. We see, therefore, that a given factor of uncertainty in x (that is, a given error in $\log x$) will cause a small error when the curve is steep and a large error when the curve is shallow. Indeed, an elementary application of geometry will show that the endpoint error varies inversely as the slope.

In the other method of locating the equivalence point, by finding the point of maximum slope on the experimental logarithmic titration curve, the endpoint error also varies inversely as the slope. The argument is as follows. In order to demonstrate that the slope is at a maximum at a given point, one must show that slopes at surrounding points are smaller. Thus the endpoint error is measured by how far one must be away from the equivalence point until the slope is clearly less than maximum. Referring to Fig. 8.1, we note that for $K = 10^7$, the slope is clearly less than maximum when $r = 0.995$ or 1.005. On the other hand, for $K = 10^5$, it is not at all obvious that the slope is less than maximum even when $r = 0.99$ or 1.01, certainly not at a glance. One would need very accurate data to be sure.

Engineering formula for innate titration error

The preceding discussion indicates that the titration error varies inversely as the slope or (in the language of the calculus) as the first derivative $d \log x/dr$ at the equivalence point. Of course, we cannot expect the inverse proportionality to be exact, because such factors as sensitivity of the endpoint indicator and personal skill must also enter into the equation. However, for titrations done by thoughtful, skilled chemists with good tools, Eq. (8.4) or its equivalent (8.5)* represent a reasonably accurate summary of normal experience. All numerical estimates of innate titration errors made in previous chapters have been based on these equations.

Engineering formula for innate titration error:

$$\% \text{ error} = 40 \Big/ \left(\frac{d \log_{10} x}{dr}\right)_{r=1} \tag{8.4}$$

or

$$\% \text{ error} = 90 \Big/ \left(\frac{d \ln x}{dr}\right)_{r=1} \tag{8.5}$$

---

* Remember, $\ln x = 2.303 \log_{10} x$.

SOLUTION BY MEANS OF THE
DIFFERENTIAL CALCULUS

Explicit expressions for $(d \ln x/dr)$ must be derived from the titration curve. To show how this is done, we shall continue with the mathematical analysis of reactions of the type (8.1). Reactions of other types will be treated in the next section.

The relationship between x and r for the specific case of (8.1) has been stated in (8.2). However we are interested only *in a narrow region around the equivalence point* where the molar concentrations of substrate and titrant are small (relative to c_S) and change rapidly, while the molar concentration of product is practically equal to c_S. Since the region is narrow, such volume changes as might occur during a titration will be small, and c_S may be regarded as constant.

On letting the molar concentration of product $[rc_S - x]$ in the numerator of (8.2) be equal to c_S (recall that $r \approx 1$ and $x \ll c_S$), taking natural logarithms, and rearranging, we obtain (8.6).

$$\ln x + \ln \left[(1 - r)c_S + x\right] = \ln \left(\frac{c_S}{K}\right) \qquad (8.6)$$

Differentiating with respect to r (with c_S constant), we obtain (8.7), which is readily solved at $r = 1$ to give the desired slope at the equivalence point. The required expression for x at $r = 1$ is (8.3).

$$\frac{d \ln x}{dr} = \frac{c_S}{c_S(1 - r) + 2x} \qquad \text{when } r \approx 1 \qquad (8.7)$$

$$\frac{d \ln x}{dr} = \frac{(c_S K)^{1/2}}{2} \qquad \text{when } r = 1 \qquad (8.8)$$

Finally, on substituting in (8.5), we obtain Eq. (8.9) for the innate titration error. Note that the error varies inversely as $\sqrt{c_S K}$.

Innate titration error for a reaction of type (8.1):

$$\% \text{ error} = \frac{180}{c_S^{1/2} K^{1/2}} \qquad (8.9)$$

Inflection point and equivalence point

By a brief extension of this analysis we can show that the inflection point (or point of maximum slope) on a logarithmic titration curve coincides with the equivalence point. On taking the derivative of (8.7) with respect to r, we obtain (8.10), from which it is apparent that when $r = 1$, the curvature $(d^2 \ln x/dr^2)$ is zero.

$$\frac{d^2 \ln x}{dr^2} = \frac{c_S^3(1 - r)}{[(1 - r)c_S + 2x]^3} \qquad \text{when } r \approx 1 \qquad (8.10)$$

$$= 0 \qquad \text{when } r = 1$$

Thus the equivalence point is an inflection point on the plot of $\ln x$ versus r. Q.E.D.

Point of maximum curvature on linear titration curves

The curvature of a linear titration curve is proportional to d^2x/dr^2 (cf. Fig. 7.6).* The sharper the curvature, the more accurately the endpoint can be determined. On taking the derivative of (8.2) and regarding c_S as a constant, we obtain (8.11).† On differentiating again with respect to r, we obtain (8.12).

$$\frac{dx}{dr} = \frac{c_S x}{c_S(1 - r) + 2x} \qquad \text{if } c_S \approx \text{constant} \qquad (8.11)$$

$$\frac{d^2x}{dr^2} = \frac{2c_S^2 x[c_S(1 - r) + x]}{[c_S(1 - r) + 2x]^3} \qquad \text{if } c_S \approx \text{constant} \qquad (8.12)$$

Finally, on solving (8.12) for $r = 1$ and applying (8.3), we obtain (8.13).

At the equivalence point, for reactions of type (8.1),

$$\frac{d^2x}{dr^2} = \frac{c_S^{3/2} K^{1/2}}{4} \qquad (8.13)$$

Thus, for reactions of type (8.1), the curvature of a linear titration curve at the equivalence point varies as $K^{1/2}$, that is, a larger value of K will give a sharper endpoint. The equation shows also that the curvature varies as $c_S^{3/2}$. This is a rather strong dependence on c_S, and unnecessary dilution of the titration mixture should obviously be avoided.

Showing that the curvature (d^2x/dr^2) passes through a maximum at the equivalence point will be left as an exercise to the reader. The mathematics is straightforward but lengthy.

---

* According to the rigorous definition, the curvature of the plot of x versus r is given by:

$$\text{Curvature} = \frac{d^2x/dr^2}{[1 + (dx/dr)^2]^{3/2}}$$

However, the maximum value of (dx/dr) is c_S, and in dilute solution (where c_S is well below $1\ F$) the denominator in the rigorous expression is practically unity.

† Alternatively, Eq. (8.11) can be derived by multiplying both sides of (8.7) by x and recalling that $d \ln x/dr = x^{-1} dx/dr$.

GENERAL TREATMENT OF THE EQUIVALENCE POINT

Up to now the mathematical analysis was limited to reactions of the stoichiometric type shown in Eq. (8.1). We now wish to extend the analysis to reactions of other types and finish with a general treatment.

For mathematical purposes it is convenient to classify titration curves according to the *algebraic order* of the expression for the equilibrium constant. Any equilibrium constant is the ratio of two mass-action expressions, one for the products and the other for the reactants.

$$K = \frac{\text{mass-action expression for the products}}{\text{mass-action expression for the reactants}} \tag{8.14}$$

Thus, for the symbolic reaction

$$a\text{A} + b\text{B} \rightleftharpoons c\text{C} + d\text{D}$$

K is equal to the ratio (8.15).

$$K = \frac{[\text{C}]^c[\text{D}]^d}{[\text{A}]^a[\text{B}]^b} \tag{8.15}$$

Since [A], [B], [C], and [D] are variables, the numerator in (8.15) is an algebraic function of order $(c + d)$, and the denominator is an algebraic function of order $(a + b)$. The ratio, which is equal to the equilibrium constant, may be classified according to algebraic order as $(c + d) \mid (a + b)$. [Read: $(c + d)$ over $(a + b)$.] Similarly, the equilibrium constant for reactions of type (8.1) is of algebraic order $1 \mid 2$.

Classification according to algebraic order is useful because all titration curves belonging to the same algebraic order can be described by the same mathematics. In the following, we shall tabulate the relationships that apply at the equivalence point for several common algebraic orders. Most of the relationships will be stated without proof, but all relationships can be derived as special cases from general equations given at the end of this section. In the "typical equations," S denotes the substrate, T the titrant, and P and Q are products.

Algebraic order 2 | 2. The "typical chemical equation" for this algebraic order is given in (8.16); examples are given just below it.

$$\text{Typical equation: S} + \text{T} \rightleftharpoons \text{P} + \text{Q} \tag{8.16}$$

$$\text{Examples: HAc} + \text{NH}_3 \rightleftharpoons \text{Ac}^- + \text{NH}_4^+$$

$$\text{Fe(II)} + \text{Ce(IV)} \rightleftharpoons \text{Ce(III)} + \text{Fe(III)}$$

The equilibrium constant of order 2 | 2 is given in (8.17), and the molar concentration of titrant at the equivalence point is given in (8.18).

$$K = \frac{[\text{P}][\text{Q}]}{[\text{S}][\text{T}]} \tag{8.17}$$

$$[T] \equiv x = \left(\frac{[P][Q]}{K}\right)^{1/2} \qquad \text{when } r = 1 \qquad (8.18)$$

To obtain the most general expression for $(d \ln x/dr)$ at the equivalence point, we shall assume that some product (P and/or Q) is present in the titration flask at the start of the titration, when $r = 0$. Additional product is formed, of course, as titrant is added, owing to reaction (8.16). In Eqs. (8.19) and (8.20), [P] and [Q] denote the total molar concentrations of P and Q at the equivalence point; $c_S = n_S/V$.

$$\frac{d \ln x}{dr} = \frac{c_S}{2}\left(\frac{K}{[P][Q]}\right)^{1/2} \qquad \text{when } r = 1 \qquad (8.19)$$

$$\% \text{ error} = \frac{180}{c_S}\left(\frac{[P][Q]}{K}\right)^{1/2} \qquad \text{innate titration error} \qquad (8.20)$$

When the product is not present at the start of the titration, $[P] = [Q] \approx c_S$ at the equivalence point, and the preceding equations reduce to (8.21) and (8.22). Note that the innate titration error now becomes independent of the substrate concentration.

When no product is present at $r = 0$,

$$\frac{d \ln x}{dr} = \frac{\sqrt{K}}{2} \qquad \text{when } r = 1 \qquad (8.21)$$

$$\% \text{ error} = \frac{180}{\sqrt{K}} \qquad \text{innate titration error} \qquad (8.22)$$

Algebraic order 1 | 2. This is the special case treated earlier in this chapter. For convenience, we repeat Eq. (8.1) below and tabulate the mathematical results in the same format as above.

$$\text{Typical equations: } S + T \rightleftharpoons P \qquad (8.1)$$
$$S + T \rightleftharpoons P + \text{solvent}$$
$$S + T \rightleftharpoons P + \text{insoluble precipitate}$$
$$\text{Examples: } Mg^{2+} + (EDTA)^{4-} \rightleftharpoons Mg(EDTA)^{2-*}$$
$$HAc + OH^- \rightleftharpoons Ac^- + HOH$$

$$K = \frac{[P]}{[S][T]} \qquad (8.23)$$

$$[T] \equiv x = \left(\frac{[P]}{K}\right)^{1/2} \qquad \text{when } r = 1 \qquad (8.24)$$

If some product is present in the titration flask in addition to that produced by reaction (8.1), then [P] in the following denotes the total molar concentration of product at the equivalence point; $c_S = n_S/V$.

$$\frac{d \ln x}{dr} = \frac{c_S}{2}\left(\frac{K}{[P]}\right)^{1/2} \qquad \text{when } r = 1 \qquad (8.25)$$

* For structural formulas see Eqs. (4.36) and (4.37).

$$\% \text{ error} = \frac{180}{c_S}\left(\frac{[P]}{K}\right)^{1/2} \qquad \text{innate titration error} \qquad (8.26)$$

If the only product is that produced by reaction (8.1), $[P] \approx c_S$ at the equivalence point, and the preceding equations reduce to (8.8) and (8.9), which are repeated below. The innate titration error now varies as $c_S^{-1/2}$; that is, dilution of the solution should be kept to a minimum.

When no product is present at $r = 0$,

$$\frac{d \ln x}{dr} = \frac{(c_S K)^{1/2}}{2} \qquad \text{when } r = 1 \qquad (8.8)$$

$$\% \text{ error} = \frac{180}{(c_S K)^{1/2}} \qquad \text{innate titration error} \qquad (8.9)$$

Algebraic order 0 | 2. The algebraic order of the mass-action expression for the reaction product is zero when the reaction product consists entirely of solvent molecules or insoluble precipitates, as indicated below.

$$\text{Typical equations: } S + T \;\rightleftharpoons\; \text{solvent} \qquad (8.27)$$

$$S + T \;\rightleftharpoons\; \text{insoluble precipitate} \qquad (8.28)$$

$$\text{Examples: } H_3O^+ + OH^- \;\rightleftharpoons\; 2\,HOH$$

$$Br^- + Ag^+ \;\rightleftharpoons\; AgBr(s)$$

The results of the mathematical analysis are listed in Eqs. (8.29) through (8.32). Note that the innate titration error now varies inversely as c_S.

$$K = [S]^{-1}[T]^{-1} \qquad (8.29)$$

$$[T] \equiv x = K^{-1/2} \qquad \text{when } r = 1 \qquad (8.30)$$

$$\frac{d \ln x}{dr} = \frac{c_S K^{1/2}}{2} \qquad \text{when } r = 1 \qquad (8.31)$$

$$\% \text{ error} = \frac{180}{c_S} \cdot \frac{1}{K^{1/2}} \qquad \text{innate titration error} \qquad (8.32)$$

Any algebraic order $n \mid m$. Some very important titrations are not included in the preceding classifications—for example, the oxidation of Fe(II) by permanganate, or the reduction of I_3^- by thiosulfate.

$$MnO_4^- + 5\,Fe^{2+} + 8\,H_3O^+ \;\rightleftharpoons\; Mn^{2+} + 5\,Fe^{3+} + 12\,H_2O$$

$$I_3^- + 2\,S_2O_3^{2-} \;\rightleftharpoons\; 3\,I^- + S_4O_6^{2-}$$

All such reactions can be described by a general equation of the type (8.23)

$$S + tT + aA + bB + \ldots \;\rightleftharpoons\; pP + qQ + \ldots \qquad (8.33)$$

The mathematics for such reactions is more complex because:

1. The molar ratio of titrant to substrate is not simply one-to-one but is represented as t-to-one.

2. There are reactants (A, B, . . .) in addition to the substrate and titrant.

3. There may be more than two products.

To simplify the general expression (8.14) for the equilibrium constant, we introduce the notation (8.34) for the numerator and (8.35) for the denominator.

Mass-action expression for the products $= [P]^p[Q]^q \ldots \equiv \Pi(\text{products})$

$$(8.34)$$

Mass-action expression for the reactants $= [S][T]^t \times [A]^a[B]^b \ldots$

$$(8.35)$$

Near the equivalence point the concentrations of all reactants $(A, B, \ldots)$ except those of the substrate and titrant are practically constant. It is therefore useful to introduce a pseudo-equilibrium constant according to (8.36).

$$K[A]^a[B]^b \ldots = K_\psi \qquad (8.36)$$

Consequently, the mass-action expression for any titration, regardless of algebraic order, near the equivalence point can always be written in the form (8.37).

$$[S][T]^t = \frac{\Pi(\text{products})}{K_\psi} \qquad (8.37)$$

To derive a general titration curve for the region near the equivalence point, we must express the molar concentrations $[S]$ and $[T]$ as functions of c_S and r. We begin with Eqs. (8.38) and (8.39), where $c_T = n_T/V$.

$$[S] = c_S - \delta c_S \qquad (8.38)$$

$$[T] = c_T - \delta c_T \qquad (8.39)$$

δc_S and δc_T are the formal quantities of substrate and titrant (per liter of solution) that have been converted to product. Since one S reacts with tT [Eq. (8.33)], $\delta c_T = t\, \delta c_S$. For the same reason, the titrant/substrate ratio $r = c_T/tc_S$.

On multiplying both sides of (8.38) by t, introducing $t\, \delta c_S$ for δc_T in (8.39), and subtracting, we obtain (8.40). On letting $[T] = x$ and $c_T = rtc_S$, we obtain (8.41) and (8.42).

$$tc_S - c_T = t[S] - [T] \qquad (8.40)$$

$$[T] = x \qquad (8.41)$$

$$[S] = \frac{[tc_S(1 - r) + x]}{t} \qquad (8.42)$$

Then, on making appropriate substitutions in (8.37), we obtain (8.43).

$$\frac{x^t[tc_S(1 - r) + x]}{t} = \frac{\Pi(\text{products})}{K_\psi} \approx \text{constant} \qquad \text{when } r \approx 1 \quad (8.43)$$

From here on, the mathematics is straightforward. On solving (8.43) for $r = 1$ and rearranging, we obtain (8.44). To derive an expression for $(d \ln x/dr)$, we take logarithms of both sides of (8.43), take the derivative [assuming that $\Pi(\text{products})/K_\psi$ is constant], and solve for $r = 1$. The result is (8.45). To estimate the innate titration error, we then apply (8.5) to obtain (8.46).

$$x_{r=1} = \left[t \frac{\Pi(\text{products})}{K_\psi} \right]^{1/(1+t)} \qquad \text{when } r = 1 \qquad (8.44)$$

$$\frac{d \ln x}{dr} = \frac{t}{t+1} \cdot \frac{c_S}{x_{r=1}} \qquad \text{when } r = 1 \qquad (8.45)$$

$$\% \text{ error} = \frac{t+1}{t} \cdot \frac{90}{c_S} \cdot x_{r=1} \qquad \text{innate titration error} \qquad (8.46)$$

Equations (8.34), (8.36), (8.44), (8.45), and (8.46) characterize the equivalence point and the innate titration error for equilibrium reactions of arbitrary stoichiometric complexity.

FEASIBILITY AND DESIGN OF TITRATIONS

The engineering formulas for the innate titration error of the preceding section take much of the guesswork out of planning in volumetric analysis. If the innate error is less than 0.25 percent, accurate titration should be feasible. If the innate error is greater than 1 percent, titration might be difficult. In that case, the engineering formulas may suggest ways to improve the accuracy.

To illustrate the practical use of the engineering formulas, we shall consider some familiar reactions and common titrations. We shall find that even the most ordinary titrations—titrations that are done every day— must be controlled to some extent lest the innate error become objectionably large.

Acetic acid and sodium hydroxide in water. The equilibrium constant for this reaction is of algebraic order 1 | 2. The chemical equation is shown just before Eq. (8.23). $K = 1.8 \times 10^9$ (M^{-1}) at 25°C. The innate error is calculated from Eq. (8.26) if an acetate salt is present in the original solution, and from Eq. (8.9) otherwise.

Let us first consider the titration of acetic acid with sodium hydroxide without any extra acetate salt. If $c_S = 0.1$ F at the equivalence point, then the innate error is 0.013 percent, according to (8.9). This error varies inversely as $c^{1/2}$: If c_S is reduced to 0.001 F, the innate error increases to 0.13 percent.

The presence of an acetate salt in the original solution, which causes an increase in the concentration of reaction product (Ac^-) at the equivalence point, always has an adverse effect on the innate titration error. Suppose that $c_S = 0.001$ F and $[Ac^-] = 1.00$ M at the equivalence point. Then the titration error, instead of being 0.13 percent as above, becomes 4.2 percent, according to Eq. (8.26). Titration will be difficult at best and may even be impossible. However, the accuracy now improves in proportion to c_S.

Thus, if c_S is increased to $0.1\ F$ while $[\mathrm{Ac}^-]$ remains at $1.00\ M$, the innate error drops to 0.042 percent.

Triiodide and thiosulfate in water. The stoichiometry of this reaction is complex and the general formula (8.46) must be used. The chemical equation is shown just before Eq. (8.33): $t = 2$, $\Pi(\text{products}) = [\mathrm{I}^-]^3[\mathrm{S_4O_6}^{2-}]$, and $K_\psi = K = 2.8 \times 10^{15}$.

The titration of triiodide with thiosulfate is usually carried out in the presence of a large excess of potassium iodide. The endpoint is very sharp even at high dilution, especially if soluble starch is added just before the equivalence point to produce the intensely dark blue starch-iodine complex. Suppose that $c_S = 1.0 \times 10^{-4}\ F$ (really quite dilute!), and that $[\mathrm{I}^-] = 0.2\ M$ at the equivalence point. Then $[\mathrm{S_4O_6}^{2-}] = 1.0 \times 10^{-4}\ M$ (one formula weight of $\mathrm{I_3}^-$ produces one mole of $\mathrm{S_4O_6}^{2-}$), and $\Pi(\text{products}) = (0.2)^3 \times 1.0 \times 10^{-4} = 8 \times 10^{-7}$. Substituting in (8.44), we find that $x = 0.83 \times 10^{-7}\ M$ when $r = 1$; and finally, substituting in (8.46), we find that the innate error is 0.11 percent.

Ammonium chloride and sodium hydroxide in water. This titration is commonly regarded as not accurate enough for quantitative work. The reaction is shown in (8.47),

$$\mathrm{NH_4}^+ + \mathrm{OH}^- \;\rightleftharpoons\; \mathrm{NH_3} + \mathrm{HOH} \tag{8.47}$$

and the algebraic order is $1 \mid 2$. $K = 6 \times 10^4\ (M^{-1})$ at 25°C. If $c_S = 0.1\ F$ at the equivalence point, and if the original solution contains ammonium chloride as the only solute, the innate error is found, from Eq. (8.9), to be 2.3 percent. Endpoint detection is evidently not easy.

Reaction (8.47) is an ionic reaction in which $\Delta z^2 = -2$. Thus, on making the Δz^2 test (5.19), we find that the equilibrium constant will increase with *decreasing* dielectric constant, other things being equal. As a matter of fact, Table 5.5 shows that in dioxane-water mixtures, $\log K$ increases from 4.74 in water to 8.89 in 70 percent dioxane-30 percent water. Because of the increase in $\log K$, we expect that the innate error becomes smaller as dioxane is added to the titration medium, and this is in fact the case. If the original $0.1\ F$ aqueous solution of ammonium chloride is diluted with dioxane in the ratio of 30 parts of solution to 70 parts of dioxane, c_S will decrease from $0.1\ F$ to $0.03\ F$ at the equivalence point, K will increase from $6 \times 10^4\ (M^{-1})$ to $8 \times 10^8\ (M^{-1})$, and the innate error will decrease from 2.3 percent to 0.037 percent. Owing to the simple change of dielectric constant, the originally difficult titration becomes easy and accurate.

MIXTURES OF SUBSTRATES

The equations for the innate titration error derived in this chapter can also be applied to the titration of mixtures. To see how this is done, let us review some of the concepts introduced in Chapter 3.

The titration of a mixture may be either nonselective or selective. In a nonselective titration one determines the total normality of the entire mixture. In a selective titration one determines the normality of a specific component.

Nonselective titration of a mixture is feasible if each reactive substrate in the mixture can be titrated individually. To find out whether this is the case, we make a series of calculations of the innate titration error, one for each substrate. In each calculation we assume that the given substrate is present alone, and that its concentration is equivalent to the entire normality of the solution. The largest innate error calculated in this way will set an upper limit to the actual innate error for nonselective titration of the mixture.

Selective titration of a component A in a mixture of A and B is feasible only if A has a much greater affinity for the titrant than does B. Titration then takes place in two stoichiometric steps: The titrant reacts first with A, then with B.

To find out whether selective titration is feasible, we proceed as follows.

1. We write chemical equations, showing the expected reaction steps, and find the effective titrant according to (3.38).

2. We write an effective chemical equation for the selective titration. The reactants in that equation will be the sought-for substrate and the effective titrant. One of the reaction products will be the substance that reacts in the next titration step.

3. We formulate the mass-action expression for the effective chemical equation, find the algebraic order, and evaluate the equilibrium constant.

4. We apply the appropriate engineering formula for the innate titration error.

The method will be illustrated with a few examples.

Selective titration of acetic acid with sodium hydroxide in the presence of phenol

We begin with a familiar example that has been fully discussed in Chapter 3, a mixture of acetic acid and the weak acid phenol in aqueous solution. The principal molecular species and conversion intervals for these acids have been shown in Fig. 3.4. It is clear that reaction with sodium hydroxide takes place in two well resolved steps, with acetic acid reacting first. Figure 3.11 then shows that in the selective titration of acetic acid, phenoxide ion is the effective titrant. The chemical equation for the selective titration is therefore (3.36), which is here repeated.

Eq. (3.36): $HAc + OPh^- \rightleftharpoons HOPh + Ac^-$ (8.48)

The corresponding equilibrium constant is given in (8.49).

$$K = \frac{[Ac^-][HOPh]}{[OPh^-][HAc]} = \frac{K_A(\text{for HAc})}{K_A(\text{for HOPh})} = 1.7 \times 10^5 \quad (8.49)$$

The algebraic order of the mass-action expression (8.49) is 2 | 2. The innate titration error is therefore given by (8.20). On comparing Eq. (8.48) with the symbolic equation (8.16), we note that $S = HAc$, $T = OPh^-$, $P = HOPh$, and $Q = Ac^-$.

For definiteness, assume that the formal concentrations of acetic acid and phenol in the sample are such that $c_{HAc} = 0.05\ F$, and $c_{HOPh} = 0.10\ F$. Since acetic acid is being titrated, c_{HAc} is, of course, identical to c_S. At the equivalence point, $[Ac^-] \approx 0.05\ M$ (acetate ion is produced in the titration) and $[HOPh] \approx c_{HOPh} = 0.1\ M$ (phenol does not react appreciably until past the equivalence point). Substituting in (8.20), we then find that the innate titration error is 0.62 percent, so that selective titration is feasible.

A word about the approximations made. The effective chemical equation for stepwise titration is a reasonably accurate approximation only if the actual reaction indeed takes place in well resolved steps. Fortunately, the value obtained for the innate titration error provides a valid test. If the error is less than 1 percent, the approximation is justified and the effective chemical equation may be used with confidence to predict other properties of the titration mixture as well.

*p*H **at the equivalence point.** As a further example of the use of Eq. (8.48), we wish to calculate (1) the phenoxide concentration and (2) the *p*H at the equivalence point. Since phenoxide ion is the effective titrant, its concentration at the equivalence point x is simply calculated from Eq. (8.18). On using the same concentrations as in the preceding calculation of the innate titration error, we find that $[OPh^-] = x = 1.7 \times 10^{-4}\ M$ at the equivalence point. To find the hydrogen ion concentration, we introduce the known value (1.0×10^{-10}) of K_A for phenol.

$$[H_3O^+] = K_A(\text{for HOPh}) \frac{[HOPh]}{[OPh^-]}$$

$$= 1.0 \times 10^{-10} \frac{0.1}{1.7 \times 10^{-4}} = 5.9 \times 10^{-8}\ M$$

Thus we find that $[H_3O^+] = 5.9 \times 10^{-8}$, and the *p*H at the equivalence point is therefore 7.23.

Effect of presence of phenol. Finally we wish to compare the results of the present calculation with results obtained when phenol is absent. For the titration of 0.05 F acetic acid with sodium hydroxide in the absence of a second acid, the innate error is 0.018 percent [Eq. (8.9)], and the *p*H at the equivalence point is 8.72 [Eq. (4.9)]. The presence of phenol evidently makes the titration more difficult and has a marked effect on the *p*H at the equivalence point.

Mixture of calcium and magnesium titrated with EDTA

Ethylenediamine tetraacetate ion [EDTA, Eq. (4.36)] can be used as titrant for either calcium(II) or magnesium(II). Because calcium(II) forms the stronger complex with EDTA (see data in Table 4.6), we shall consider whether selective titration of calcium(II) in the presence of magnesium(II) is feasible. As before, we shall use the symbol Y^{4-} to denote the principal species of the EDTA titrant. The proposed titration steps are shown in (8.50) and (8.51).

$$\text{First step: } Ca^{2+} + Y^{4-} \rightleftharpoons CaY^{2-} \tag{8.50}$$

$$K_{assoc} = 5 \times 10^{10} \ (M^{-1})$$

$$\text{Second step: } Mg^{2+} + Y^{4-} \rightleftharpoons MgY^{2-} \tag{8.51}$$

$$K_{assoc} = 5 \times 10^{8} \ (M^{-1})$$

The effective titrant (that is, the species produced in the second step) is MgY^{2-}. The effective chemical equation for the proposed selective titration of Ca(II) is therefore (8.52), and the equilibrium constant is given in (8.53). The algebraic order of this mass-action expression is 2 | 2.

$$MgY^{2-} + Ca^{2+} \rightleftharpoons CaY^{2-} + Mg^{2+} \tag{8.52}$$

$$K = \frac{K_{assoc}(\text{for } Ca^{2+})}{K_{assoc}(\text{for } Mg^{2+})} = 1 \times 10^{2} \tag{8.53}$$

It turns our that the proposed titration is *not* feasible. For instance, if we assume that $c_{Ca(II)} = 0.01 \ F$ and $c_{Mg(II)} = 0.01 \ F$, then, on applying Eq. (8.20), we let $[P] = [CaY^{2-}] \approx 0.01 \ M$, $[Q] = [Mg^{2+}] \approx 0.01 \ M$, $c_S = c_{Ca(II)} = 0.01 \ F$, and $K = 1 \times 10^{2}$. The innate error is therefore calculated to be 18 percent, which is much too large for titration to be practical.

Oxidation of arsenic(III) with iodine chloride

The feasibility of selective titration in consecutive reactions is predicted by the same method. We shall consider only one (rather complicated) example: the determination of arsenic by oxidation of arsenic(III) to arsenic(V). This reaction was discussed previously under conditions where the titrant is I_3^- [Eq. (2.21)]. However, the titration is improved when iodine chloride is used in place of I_3^- because the speed of the reaction becomes greater, especially near the equivalence point. In practice, titration with iodine chloride is carried out at pH 7 in the presence of a large excess of chloride ion, which converts ICl predominantly to ICl_2^-. The reaction is then represented by (8.54).

$$H_3AsO_3 + ICl_2^- + 4 H_2O \rightleftharpoons$$
$$H_2AsO_4^- + I^- + 2 Cl^- + 3 H_3O^+ \quad (8.54)$$

Past the equivalence point, the excess amount of ICl_2^- reacts with I^- produced in (8.54) to produce iodine, as shown in (8.55).

$$I^- + ICl_2^- \rightleftharpoons I_2 + 2 Cl^- \quad (8.55)$$

Reactions (8.54) and (8.55) form a stepwise series of consecutive reactions. The effective titrant for (8.54) is therefore the reactive species produced in (8.55), that is, I_2. The effective chemical equation for the titration is therefore (8.56).

$$H_3AsO_3 + I_2 + 4 H_2O \rightleftharpoons H_2AsO_4^- + 2 I^- + 3 H_3O^+ \quad (8.56)$$

The innate titration error is calculated by applying the general method described in Eqs. (8.33) through (8.46) to the reaction given in (8.56). It may be assumed, in this calculation, that the concentration of unreacted ICl_2^- at the equivalence point is too small to be stoichiometrically significant.

WHEN THE EFFECTIVE TITRANT IS A SOLID

If we apply the effective titrant rule (3.38) consistently, there will be occasions when the substance that we identify as the effective titrant is a solid precipitate. For example, consider what happens when an aqueous solution containing ionic bromide and chloride is titrated with silver nitrate. Since silver bromide is less soluble than silver chloride, reaction takes place in steps, with silver bromide precipitating first.

$$\text{First step: } Ag^+ + Br^- \rightleftharpoons AgBr(s) \quad (8.57)$$
$$K_{SP}(\text{for AgBr}) = 0.77 \times 10^{-12} \ (M^2)$$

$$\text{Second step: } Ag^+ + Cl^- \rightleftharpoons AgCl(s) \quad (8.58)$$
$$K_{SP}(\text{for AgCl}) = 1.8 \times 10^{-10} \ (M^2)$$

If we wish to determine bromide ion only, then the titration must be stopped at the completion of the first step, and the effective titrant is $AgCl(s)$. The effective chemical equation for the selective titration of bromide ion is therefore (8.59).

$$AgCl(s) + Br^- \rightleftharpoons Cl^- + AgBr(s) \quad (8.59)$$

The equilibrium constant for the titration is therefore expressed by (8.60) and is related to solubility-product constants, as shown in (8.61).

If the solution is saturated with both AgCl(s) and AgBr(s),

$$K = \frac{[Cl^-]}{[Br^-]} \tag{8.60}$$

$$K = \frac{K_{SP}(\text{for AgCl})}{K_{SP}(\text{for AgBr})} = 2.3 \times 10^2 \tag{8.61}$$

When the effective titrant is a pure solid, the titration may be remarkably accurate because the titration curve has a peculiar truncated shape. Thus Fig. 8.2 shows theoretical titration curves for titration with silver

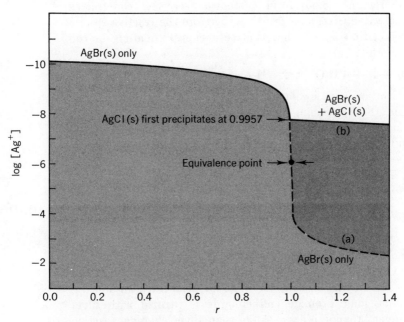

Fig. 8.2. Theoretical logarithmic titration curves: (a) 0.01 *F* bromide ion with silver nitrate in aqueous solution; (b) 0.01 *F* bromide ion with silver nitrate in the presence of 0.01 *F* chloride ion. Actual potentiometric titration curves are similar, but the point of truncation in (b) lies somewhat past the equivalence point.

nitrate of (a) 0.01 *F* bromide ion in water, and (b) 0.01 *F* bromide ion in the presence of 0.01 *F* chloride ion in water. In curve (a) the effective titrant is $Ag^+(aq)$; in curve (b) it is AgCl(s). Curve (a) has the familiar sigmoid shape of logarithmic titration curves in which the titrant is a solute in solution. Curve (b) is identical with curve (a) as long as the solid precipitate consists entirely of pure silver bromide. However, when the silver ion concentration attains a value at which the solution is also saturated with respect to silver chloride, further addition of silver nitrate results in the formation of both AgCl(s) and AgBr(s), in a definite ratio of

$2.3 \times 10^2/1$, which [according to (8.61)] is the ratio that will keep the liquid phase in equilibrium with both solid phases.

The formation of solid silver chloride has the effect of truncating the titration curve. When the first crystals of silver chloride separate from the solution, the chloride ion concentration is still practically at its original value of $0.01 \ M$, neglecting dilution during the titration. The amount of chloride in the solution is therefore relatively large and, as further small amounts of silver nitrate are added, the *fractional* decrease in the chloride ion concentration is at first quite small. Because chloride ion is in equilibrium with AgCl(s) and silver ion, if [Cl⁻] is approximately constant, so must be [Ag⁺]. That is to say, the silver ion concentration is buffered owing to the equilibrium with AgCl(s) and chloride ion. The result, in Fig. 8.2(b), is a nearly horizontal line which truncates Fig. 8.2(a) at the point where silver chloride first precipitates out of solution.

The existence of a sharp point of truncation is useful in practice. Experimentally, the logarithmic titration curve of Fig. 8.2(b) can be reproduced by potentiometric titration with a silver-silver bromide electrode. It is common experience in physical measurement that if a function undergoes a sudden change in slope, the point at which the discontinuity occurs can be located experimentally with extraordinary precision. As a result, bromide ion can be titrated potentiometrically with silver nitrate in the presence of chloride ion with quantitative accuracy even though the equilibrium constant for the titration is only 2.3×10^2 [Eq. (8.61)]. It is virtually certain that quantitative endpoints can be obtained only because of the discontinuity in slope. In the preceding example, where we considered the titration of calcium(II) with EDTA in the presence of magnesium(II) *in homogeneous solution*, the equilibrium constant was only slightly smaller yet quantitative titration was out of the question. It must be pointed out, however, that the titration of bromide ion in the presence of chloride ion becomes quantitative only if an inconvenient empirical endpoint correction is applied.*

PROBLEMS

1. Explain the significance of the "innate titration error" to volumetric analysis. Is this a determinate or indeterminate error?

2. The innate titration error for a particular titration is calculated [from Eq. (8.5)] to be 0.8 percent. Exactly what does this mean? Will results always be 0.8 percent too high? too low? How many ml of $0.1000 \ N$ titrant *would you expect* to use in this titration if a 25.00 ml aliquot of $0.1000 \ N$ substrate were taken?

---

* See, for example, D. Jacques, *J. Chem. Education*, **42**, 429 (1965); A. J. Martin, Anal. Chem., **30**, 233 (1958).

3. Under what conditions might a large innate titration error influence your decision to carry out a titration using instrumental detection rather than an indicator or vice versa? Explain.

4. Using the starting points, substitutions, and assumptions indicated in the text, give a *detailed* derivation of Eq. (8.9). Convince yourself that you understand both the mathematics and the philosophy of this approach by taking special pains to justify each step in terms of the final result.

5. What is the algebraic order (p. 178) of each of the following titration reactions?

	Substrate	Titrant
(a)	HCl	NaOH
(b)	NH_3	HCl
(c)	H_2O_2	Ce(IV)
(d)	Fe^{2+}	MnO_4^- (acid)
(e)	As_2O_3	MnO_4^- (acid)
(f)	I_2	$S_2O_3^{2-}$
(g)	Ca^{2+}	EDTA (pH 10)

6. Derive the expression for the innate titration error of a reaction of order 2 | 2, Eq. (8.20). [*Hint:* Use reasoning analogous to the formulation of the problem for the 1 | 2 case, Eq. (8.9).] Show how the titration error becomes independent of concentration if no products are initially present.

7. Formulate a general expression for the innate titration error of a reaction of order n | 2, where n is any integer. What happens to this expression if none of the products is present when $r = 0$? (*Hint:* This is trivial; cf. Problem 6.)

8. Calculate the innate titration error titrations (a), (b), (f) and (g) in Problem 5. Assume 0.1 N solutions of substrate and titrant are used for the titration. (Assume that no products are present initially.) Refer to the Appendix for equilibrium constants.

9. Calculate the innate titration error for each of the three steps in the titration of 0.1 F H_3PO_4 with NaOH. Your results should indicate that only the first two equivalence points will be observed.

Detecting the Equivalence Point with Color Indicators

In a chemical reaction, the advantages of detecting the equivalence point with the unaided senses are too obvious to require much discussion. They are all summed up by saying "the simplest method is the best method"— a saying with which practical men from all walks of life will agree. The use of auxiliary measuring instruments is normally avoided unless it brings real improvement in accuracy or convenience.

The most common and versatile technique for allowing the equivalence point in a chemical reaction to become visible is to use a color indicator. In this chapter we shall consider the scientific aspects of color indicators: their optical and chemical properties, and the psychology of color vision.

191

THE COLOR OF SOLUTIONS

The perception of color is an intricate physiological and psychological phenomenon and bears no simple relationship to the physical phenomenon of light. Fortunately, for our purpose it is sufficient to consider only two aspects of color vision: the sensations of hue and shade. Their relationship to the absorption of light, in the case of liquid solutions, is comprehensible if not simple.

Hue

Sunlight, which is perceived as colorless or white, contains the entire visible spectrum of wavelengths in a characteristic distribution of intensities. A pure spectral color or pure hue is the sensation produced by light of a single wavelength. Spectral colors vary with wavelength according to the well-known hues of the rainbow, which may be a somewhat different perception for each individual. However, most individuals would identify hue with spectral wavelength approximately as in Table 9.1.

Table 9.1   HUE AND SPECTRAL WAVELENGTH

Spectral range (Å)	Hue	Complementary hue
4000–4250	Violet	Greenish-yellow
4250–4550	Indigo	Yellow
4550–4900	Blue	Orange
4900–5000	Turquoise	Red
5000–5550	Green	Purple or rose or magenta[a]
5550–5700	Greenish-yellow	Violet
5700–5850	Yellow	Indigo
5850–6300	Orange	Blue
6300–7200	Red	Greenish-blue

[a] This is an extra-spectral hue, distinct from violet. Sorry, there is no "purple" at the end of the rainbow!

The color of a transparent liquid is caused by the absorption of some of the light as it passes through the liquid. When the incident light is white, the hue of the transmitted light is complementary to the hue of the light that is absorbed. Complementary hues are listed also in Table 9.1. Thus, if a liquid absorbs in the spectral range corresponding to indigo, its color in white light will be perceived as yellow. Or, if a liquid absorbs in the spectral range corresponding to green, its color in white light will be perceived as purple. The sensation of "green" cannot be produced by absorption in a single spectral range; it requires absorption in at least two spectral ranges, one below 5000 Å and the other above 5550 Å.*

Shade of color

For any given hue, the color can vary from light, pastel shades to rich, dark shades, depending on the amount of white that is mixed with the saturated hue. In the case of solutions, a convenient method for producing a series of shades of nearly constant hue is to add a light-absorbing solute at a series of concentrations to a "water-white" solvent. We shall assume that the visible absorption spectrum of the solute consists of a single, fairly narrow, absorption band, so that light is absorbed only in a single well-defined spectral range. The perceived hue will remain complementary to the hue of the spectral range in which light is absorbed, and the shade will grow from light to dark as the solute concentration increases. We now wish to analyze the manner in which the color-shade varies with the solute concentration.

Let I_0 denote the incident light intensity at the center of the spectral range in which light is absorbed by the solute, and let I denote the portion of that intensity that is transmitted by the solution. The fraction of the incident light intensity that is absorbed is therefore given by $(I_0 - I)/I_0$.

To quantify the discussion of color-shade, we introduce the *percent of saturation* (percent that is absorbed) of the given hue, such that a fully absorbed hue corresponds to 100 percent saturation, while pure white (full transmission) corresponds to 0 percent saturation. For liquid solutions that absorb in a single spectral range, the quantity $100(I_0 - I)/I_0$ (as defined above) is then a direct measure of the percent of saturation of the color of the *transmitted* light. A colorless (or "water-white") solution will result in $I = I_0$ because none of the incident light is absorbed; hence, $100(I_0 - I)/I_0 = 0$ percent. Complete absorption of the incident light in the given spectral range implies that $I = 0$; hence $100(I_0 - I)/I_0 = 100$ percent. This definition is stated for later reference in Eq. (9.1).†

$$\% \text{ of saturation of the transmitted hue} = \frac{100(I_0 - I)}{I_0} \qquad (9.1)$$

---

* When we know the absorption spectrum, we can usually predict the color. However, the reverse is *not* true: When we know the color, we can *not* uniquely predict the absorption spectrum. The uncertainty results from the psychological phenomenon known as color addition. For instance, a hue that we perceive as turquoise might be due to light of a single wavelength in the range 4900–5000 Å, or it might be due to an appropriate combination of blue and green spectral colors. For further information, consult a textbook on physiology.

†In connection with Eq. (1.1) we defined $100(I/I_0)$ as percent transmission. It is thus apparent that for a given hue, percent transmission + percent saturation = 100 percent.

Equation (9.1) is a convenient starting point for deriving a relationship between the color-shade of a solution and the concentration of the light-absorbing solute. If we overlook the lack of optical perfection of a titration flask under ordinary laboratory illumination, the symbols I_0 and I have the same significance as in Eq. (1.1), which defines the absorbance A at a given wavelength, and which is repeated below.

$$\text{Eq. (1.1): } A = \log\left(\frac{I_0}{I}\right) \tag{9.2}$$

The absorbance, in turn, varies with concentration according to Beer's law.

$$\text{Beer's law [Eq. (1.2)]: } A = \epsilon l c \tag{9.3}$$

On combining Eqs. (9.1) through (9.3), we obtain an explicit relationship between color-shade and solute concentration.

$$\% \text{ saturation of the transmitted hue} = 100(1 - 10^{-\epsilon l c})\% \tag{9.4}$$

A typical plot of Eq. (9.4) is shown in Fig. 9.1. It was assumed, in constructing that figure, that $\epsilon l = 5 \times 10^4 \, M^{-1}$, which is a reasonable value for a substance that might be used as a titration endpoint indicator. Figure 9.1 has a peculiar, indeed a remarkable, shape: The hue develops

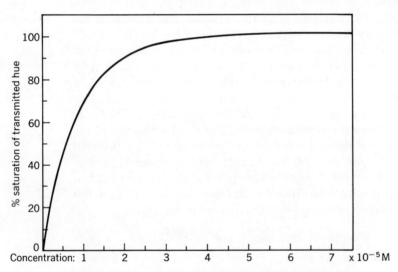

Fig. 9.1. Production of color by a light-absorbing solute. Plot of Eq. (9.4), assuming that $\epsilon l = 5 \times 10^4 \, M^{-1}$.

rapidly to within a few percent of full saturation and then levels off. Probably, under ordinary laboratory illumination, the hue first becomes evident at about 5 percent of full saturation. Changes in saturation become indistinguishable above 90 percent of full saturation. Thus the entire color change, from first perception to virtual completion, in this instance takes place within the astonishingly small concentration range of $0.05 \times 10^{-5} \, M$

to 2×10^{-5} M. Under conditions simulating those of a real titration, such a color phenomenon would appear to be almost discontinuous: If a 0.02 F solution of the light-absorbing solute were added dropwise from a buret to 50 ml of pure water, *one drop* would be sufficient to produce a practically saturated hue.

COLORED TITRANT OR SUBSTRATE AS SELF-INDICATOR

It is clear from the preceding discussion that an intensely colored titrant will act as its own endpoint indicator. Consider, for example, the titration of iron(II) with aqueous permanganate, Eq. (9.5).

$$5 \, Fe^{2+} + MnO_4^{-}(\text{purple}) + 8 \, H_3O^+ \; \rightleftharpoons$$
$$5 \, Fe^{3+} + Mn^{2+} + 12 \, H_2O \quad (9.5)$$

Before the equivalence point, the added permanganate reacts with Fe(II) and the characteristic purple hue is discharged. At or just past the equivalence point, the concentration of permanganate becomes high enough so that the solution at equilibrium is just perceptibly purple. At that point, a small further addition of permanganate will cause a sharp color change as the saturation of the purple hue rises suddenly towards 90 percent. After the sharp color change has occurred, further additions of permanganate will produce little further visual effect. The endpoint is taken as the point at which the color changes suddenly from a barely perceptible to a well developed shade of purple.

An intensely colored substrate will act similarly as its own endpoint indicator. In that case, the sudden disappearance of the characteristic hue produced by the substrate will indicate the equivalence point. A familiar example is the titration of iodine with thiosulfate in the presence of excess potassium iodide (which converts I_2 to I_3^-).

$$I_3^{-}(\text{brown}) + 2 \, S_2O_3^{2-} \; \rightleftharpoons \; S_4O_6^{2-} + 3 \, I^- \qquad (9.6)$$

The endpoint is taken at the final disappearance of the characteristic brown-yellow hue due to I_3^- (or, if starch is added, of the dark blue hue caused by the starch-iodine complex).

Molar extinction coefficients for self-indicators

A colored titrant or substrate will make a satisfactory self-indicator only if the molar extinction coefficient ϵ at the center of the visible absorption band is sufficiently large. A mathematical analysis based on Eq. (9.4) leads to the conclusion that the color-shade changes most rapidly when the concentration of the light-absorbing species is $0.434\epsilon^{-1}l^{-1}$ molar. For accurate titration, the concentration of excess titrant or unreacted substrate at the endpoint should probably be not greater than $1 \times 10^{-4}\,M$. We thus estimate a lower limit for ϵl as follows:

$$0.434\epsilon^{-1}l^{-1} \leqslant 1 \times 10^{-4}\,M$$
$$\epsilon l \geqslant 4340\,M^{-1}$$

In order for our estimate of ϵ to be a round number, let us say that the effective light path l is 4.34 cm, which is a plausible magnitude. We then conclude that *a colored titrant or substrate will make a satisfactory self-indicator if ϵ is at least 1000 $cm^{-1}M^{-1}$* at the center of its visible absorption band.

Other methods based on the unaided senses

There are of course other properties than the color of a solution that can be observed with the unaided senses. Thus at the equivalence point, a characteristic odor might appear or disappear, a gas might begin or cease to bubble out of solution, a precipitate might form or dissolve, a colloidal precipitate might coagulate, the titration mixture might become fluid or viscous, the solution might begin to foam—to name only a few of the more obvious possibilities. Such signals give valuable information, and the chemist working in the laboratory soon learns to pay attention to them. But such signals are seldom as precise or objective as those due to color indicators or instrumental methods for detecting the equivalence point. Nevertheless, if the analysis must be performed in the field—far from any well equipped laboratory—such signals may occasionally have to suffice.

Titration with soap. One perfectly respectable, though semi-quantitative, method requiring very simple means is the "hardness" titration of natural water with soap. Natural water supplies contain certain mineral impurities, mostly salts of calcium and magnesium, whose cations spoil an equivalent amount of soap by converting it into an amorphous curd without cleansing action. When feasible, the normality of such impurities is determined by titration with EDTA. However, it can also be determined by titration with a standard aqueous soap solution (see Chapter 15). After each addition of soap, the flask is stoppered and shaken vigorously. The endpoint is taken when shaking first produces an unmistakable sign of soapsuds.

STEPWISE REACTION SEQUENCE
WITH COLOR CHANGE

When substrate and titrant are both colorless, a color change can be produced by adding a solute that reacts right after the equivalence point to produce an intensely colored product. For example, in the Volhard titration of silver ion with thiocyanate in acid solution (p. 77), ferric ammonium sulfate is added to the titration flask to produce an iron(III) concentration of 0.01–0.02 F. As the solution is titrated with thiocyanate, the silver ions react first, and a white precipitate of silver thiocyanate forms. After that reaction is practically complete, further addition of thiocyanate leads to the formation of a ferric thiocyanate complex, and the solution undergoes a sudden color change from greenish-yellow to rust-orange. The two reactions, (9.7) and (9.8), constitute a stepwise reaction sequence in which the color change at the beginning of the second step serves as an endpoint signal for the first step.

$$\text{First step: } Ag^+ + SCN^- \rightleftharpoons AgSCN(s) \tag{9.7}$$

$$\text{Second step: } Fe^{3+}(\text{greenish-yellow}) + SCN^- \rightleftharpoons$$
$$FeSCN^{2+}(\text{rust-orange}) \tag{9.8}$$

Effect on innate titration error. While the addition of iron(III) in Volhard's method produces a satisfactory endpoint signal, it also has the undesirable effect of raising the innate titration error. In Volhard's method the solution contains two reactive substrates, Ag^+ and Fe^{3+}, of which only the silver ion is to be titrated. The titration therefore is a *selective* titration, and the selective titrant rule (3.38) must be applied. As a result, the effective chemical equation is (9.9), and the effective titrant is not the thiocyanate ion but the less reactive $FeSCN^{2+}$ ion.

$$FeSCN^{2+} + Ag^+ \rightleftharpoons Fe^{3+} + AgSCN(s) \tag{9.9}$$

Since the innate titration error increases whenever a reaction becomes part of a stepwise sequence, the example shows that an endpoint indicator must be chosen judiciously, taking into account not only the quality of the color change but also any adverse effect on accuracy.

Effect of indicator concentration. The increase in the innate titration error will be minimized if the indicator concentration is kept as small as possible. When we apply the methods of Chapter 8 to the selective titration of a substrate (S) in the presence of an indicator (Ind), we find that the innate titration error decreases continuously with the ratio c_{Ind}/c_S, until the limit is reached that c_{Ind} is negligibly small compared to c_S. When that limit is reached, the titration curve will be practically the same in the presence of the indicator as in its absence, and the innate titration error will be the same as that for the titration of the substrate without the indicator.

In practice, c_{Ind} is negligibly small when it becomes less than 0.25 percent of c_S. For instance, if $c_S \approx 0.01\ F$, addition of a color indicator will not affect the innate titration error if c_{Ind} is less that 0.25 percent of $0.01\ F$, or less than $2.5 \times 10^{-5}\ F$. Fortunately, there are substances that will produce distinct color changes at concentrations of the order of $10^{-5}\ F$.

When such a substance has the right kind of reactivity, it makes a fine endpoint indicator.

Let us estimate the molar extinction coefficient that such an indicator must have. As a rule of thumb, the expected color change becomes clearly visible if the absorbance that gives rise to the color is about 0.7. Applying Eq. (9.3), we write:

$$A = \epsilon l c_{\text{Ind}} = 0.7$$

By hypothesis, $c_{\text{Ind}} < 2.5 \times 10^{-5}\ F$. In order to obtain a round number for ϵ, we let $l = 2.8$ cm, which is a plausible value for a titration flask. Hence,

$$7 \times 10^{-5}\epsilon > 0.7 \quad \text{and} \quad \epsilon > 10^4\ \text{cm}^{-1}\ M^{-1}$$

Thus a color indicator will give a clearly visible color change at such low concentrations that the innate titration error is not affected, if the molar extinction coefficient at some point in the visible spectrum exceeds $10^4\ \text{cm}^{-1}\ M^{-1}$.

COLOR INDICATORS AT HIGH DILUTION

In the preceding section we considered a stepwise sequence of reactions in which the added titrant reacts first with the substrate and then with the color indicator. However, when the concentration of the indicator is very small (less than 0.25 percent of c_S), we gain additional flexibility. First of all, the reaction that indicates the equivalence point need not take place right *after* the equivalence point: It may, with equal validity, take place just before the equivalence point, or it may straddle the equivalence point. The endpoint signal need not be the sudden *appearance* of a color: It may, with equal validity, be the sudden *disappearance* of a color. The color change may involve either of the following reaction schemes:

1. The indicator reacts directly with the titrant, and this produces the desired color change.

2. The indicator reacts with the *substrate* to produce a new substance with a distinctly different color. The original indicator (with its original color) is then regenerated at the equivalence point as the concentration of free substrate becomes very small.

We shall give a practical example for each scheme.

Most color endpoint indicators are organic dyes with complex molecular structures. In the following discussion we shall show their structural formulas, but we shall also write symbolic equations in simplified notation in order that the chemical mechanisms stand out more clearly.

Ferroin in the titration of iron(II) with cerium(IV)

The use of ceric sulfate as a quantitative oxidizing agent has been described in Chapter 4. Titrations are usually done under conditions where the hydrogen ion concentration is at least $0.1\ M$. The most widely used endpoint indicator is ferroin (ferrous 1,10-phenanthroline sulfate), which reacts directly with cerium(IV). We now wish to consider the use of ferroin as endpoint indicator in the titration of iron(II).

Solutions of native ferroin in $1\ F$ H_2SO_4 are a vivid red, solutions of ferroin in the presence of cerium(IV) are pale blue or colorless. A chemical equation, showing structural formulas, for the oxidation of the indicator is given in (9.10).

$$Ce^{4+} + Fe \left[\cdots \right]_3^{2+} \;\rightleftharpoons\; Ce^{3+} + Fe \left[\cdots \right]_3^{3+} \tag{9.10}$$

Standard electrode potentials (and derived values of the equilibrium constant) for the reaction of cerium(IV) with iron(II) and with ferroin are as follows.

$$Ce^{4+} + Fe^{2+} \;\rightleftharpoons\; Ce^{3+} + Fe^{3+}; \qquad n = 1$$
$$E° = 0.77\ V; \qquad K = 10^{13} \qquad \text{in } 1\ F\ H_2SO_4 \tag{9.11}$$

$$Ce^{4+} + \text{ferroin}^{2+}(\text{red}) \;\rightleftharpoons\; Ce^{3+} + \text{ferroin}^{3+}(\text{pale blue}); \qquad n = 1$$
$$E° = 0.47\ V; \qquad K = 10^{8} \qquad \text{in } 1\ F\ H_2SO_4 \tag{9.12}$$

It is evident from the values of the equilibrium constants that both reactions have a strong tendency to go to completion, but that reaction (9.11) will be practically complete before reaction (9.12) becomes visible. The titration endpoint is taken when the red hue caused by ferroin^{2+} suddenly disappears.

Eriochrome black T in the titration of magnesium with EDTA

The use of ethylenediamine tetraacetate ion (4.36) as a titrant for magnesium(II) has been described in Chapter 4. Titration is usually done at or near pH 10. The most widely used indicator is Eriochrome Black T (Erio T). This indicator reacts with magnesium but is regenerated at the equivalence point. Solutions of the native indicator at pH 10 are blue; solutions of the indicator in the presence of magnesium are red. The reaction that produces the color change from blue to red is shown in (9.13).

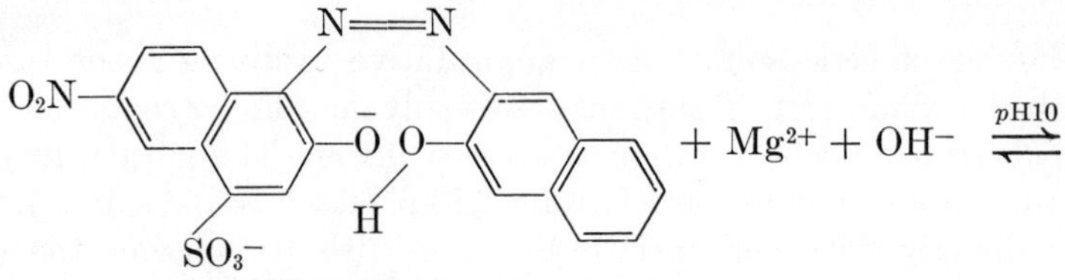

(HIn²⁻, blue)

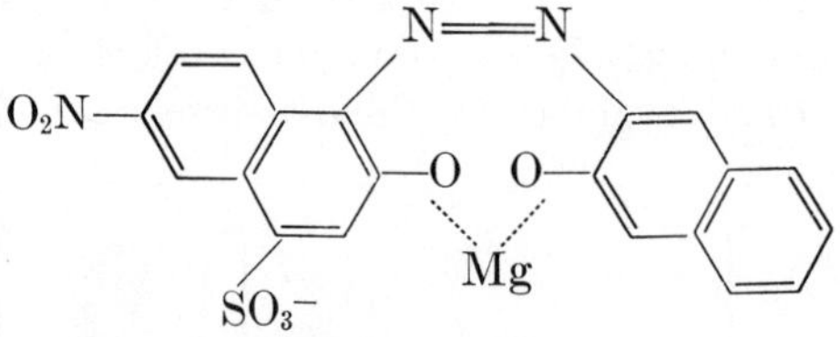

(MgIn⁻, red)

To simplify the notation, we let HIn²⁻ denote a molecule of Erio T, MgIn⁻ a molecule of its magnesium complex, and Y⁴⁻ a molecule of EDTA. At pH 10, the equilibrium in (9.13) lies far to the right. Thus, when the equilibrium mixture is titrated with EDTA, the titrant reacts almost exclusively with free magnesium ion until just before the equivalence point, when reaction (9.14) becomes competitive.

$$Y^{4-} + MgIn^-(red) + H_2O \;\rightleftharpoons\; MgY^{2-} + HIn^{2-}(blue) + OH^- \quad (9.14)$$

$$K_\psi = \frac{[MgY^{2-}][HIn^{2-}]}{[Y^{4-}][MgIn^-]} = 10^2 \quad \text{at } p\text{H } 10$$

The solution therefore remains red until just before the equivalence point, and then changes rapidly from red to magenta to blue. The endpoint is taken when the hue changes from magenta to blue.

ACID-BASE INDICATORS

Color endpoint indicators for acid-base titrations have been discussed briefly in Chapters 4 and 6. We now wish to discuss them more fully. There exist a great many dyes, covering a wide range of properties, whose color reactions are reversible, pH sensitive, and distinctive. Some of the dyes, such as litmus, come from natural sources; others are synthetic. Most of them make excellent pH indicators.

In the following discussion we shall let HIn denote the acid form of a pH indicator (of unspecified charge number z), and In its conjugate base (of unspecified charge number $z - 1$). We shall let [HIn] and [In] denote the respective molar concentrations, and refer to [In]/[HIn] as the *indicator ratio*. If the indicator is added to an aqueous solution of given pH, equilibrium will be established between HIn, In, and hydrogen ion in the solution, according to (9.15).

$$HIn^z + HOH \rightleftharpoons H_3O^+ + In^{z-1} \qquad (9.15)$$

We shall suppose that the amount of indicator is so small that its addition to the solution has no effect on the pH. Under such conditions, the indicator ratio depends on the pH and the pK_A of the indicator according to (9.16).

$$\log \frac{[In]}{[HIn]} = pH - pK_A (\text{for HIn}) \qquad (9.16)$$

We shall suppose, further, that at least one member of the pair HIn-In imparts an intense color to the solution. It then becomes possible (literally) to *see* the indicator ratio and hence to determine the pH. Let us consider how the color of the solution varies with the indicator ratio.

Indicator ratio and color

It is useful to distinguish between "one-color indicators" and "two-color indicators." In the case of one-color indicators, only one member of the pair HIn-In imparts color to the solution, and differences in color are differences in shade. In the case of two-color indicators, HIn and In both produce a characteristic color, and differences in color are largely differences in hue. The one-color indicators are sometimes preferred because their usefulness is not impaired by color blindness. The sense perception obtained from one-color indicators can be analyzed on the basis of Eq. (9.4).

For definiteness, we shall assume that In is colored while HIn is colorless, so that in applying Eq. (9.4) we must use the molar concentration [In] in place of c. To obtain [In] as a function of the indicator ratio, we let c_{HIn} denote the formal concentration of indicator (which, by hypothesis, is kept constant) and express c_{HIn} as a sum of molar concentrations as in (9.17).

$$c_{HIn} = [In] + [HIn] \qquad (9.17)$$

On dividing both sides of (9.17) by [In], and after some steps of algebra, we obtain the desired expression (9.18).

$$[In] = \frac{c_{HIn}\{[In]/[HIn]\}}{1 + \{[In]/[HIn]\}} \qquad (9.18)$$

On substituting in (9.4), we obtain (9.19), which relates the shade of the color to two physical variables, the indicator ratio [In]/[HIn] and the maximum absorbance $A = \epsilon l c_{\text{HIn}}$. (Remember that ϵ is the molar extinction coefficient for In.)

% of saturation of the transmitted hue

$$= 100 - 100 \ \text{antilog}_{10}\left(-\frac{\epsilon l c_{\text{HIn}}\{[\text{In}]/[\text{HIn}]\}}{1 + \{[\text{In}]/[\text{HIn}]\}}\right) \quad (9.19)$$

Equation (9.19) predicts that the percent of saturation of the observed hue increases from a minimum value of zero when $[\text{In}]/[\text{HIn}] = 0$ to a maximum value of $[100 - 100 \ \text{antilog}_{10}(-\epsilon l c_{\text{HIn}})]$ when $[\text{In}]/[\text{HIn}] \gg 1$. Thus the saturation will vary between 0 percent and 80 percent if $\epsilon l c_{\text{HIn}} = 0.7$, and between 0 percent and 99 percent if $\epsilon l c_{\text{HIn}} = 2$. In practice one uses the smallest amount of indicator that will allow the color to become well developed, and a choice of c_{HIn} such that $\epsilon l c_{\text{HIn}} = 0.7$ is typical. To illustrate the dependence of color on indicator ratio, we shall assume, therefore, that $\epsilon l c_{\text{HIn}}$ has a constant value of 0.7. The resulting relationship of percent of saturation versus $\log \{[\text{In}]/[\text{HIn}]\}$ is plotted in Fig. 9.2.

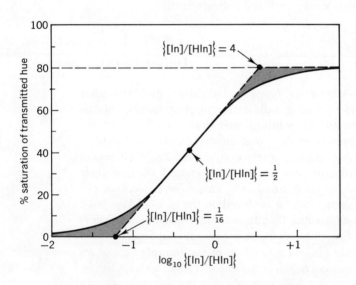

Fig. 9.2. Color change of a typical one-color pH indicator. Plot of Eq. (9.19) assuming that $\epsilon l c_{\text{HIn}} = 0.7$.

In view of Eq. (9.16), Fig. 9.2 is essentially a plot of color-shade versus pH. Note the characteristic sigmoid shape of this relationship. According to Fig. 9.2, the solution containing the indicator will be practically colorless when the indicator ratio is less than $\frac{1}{16}$, and the color will be almost fully developed when the indicator ratio is greater than 4. In the intermediate range the color sensation varies more or less linearly with pH, although the indicator seems to be most sensitive when $[\text{In}]/[\text{HIn}] \approx \frac{1}{2}$. Outside the range $\frac{1}{16} < [\text{In}]/[\text{HIn}] < 4$, the color is quite insensitive to pH. Thus the "indicator" really indicates anything by its color only in a limited range, stated in (9.20).

If In produces an intense vivid color and HIn does not, the color change takes place in the range $\frac{1}{16} < [In]/[HIn] < 4$.

If HIn produces an intense vivid color and In does not, the color change takes place in the range $\frac{1}{4} < [In]/[HIn] < 16$.

$$(9.20)$$

The pH range in which the color is sensitive to pH is called the *transition range* of the indicator. It is convenient to restate (9.20) in those terms. On taking logarithms of (9.20) and applying (9.16), we obtain (9.21), which is phrased so that it can apply also to two-color indicators.

Approximate transition range for pH *indicators*:

If In produces the more vivid color,

$$[pK_A(\text{for HIn}) - 1.2] < pH < [pK_A(\text{for HIn}) + 0.6)]$$

If HIn produces the more vivid color, (9.21)

$$[pK_A(\text{for HIn}) - 0.6] < pH < [pK_A(\text{for HIn}) + 1.2]$$

The action of two-color indicators is similar to that of one-color indicators except that the color change is primarily one of hue. In the transition range, the color is like that obtained by mixing two pigments (one matching the color produced by In, the other matching the color produced by HIn) in the same ratio as the indicator ratio $[In]/[HIn]$ of the solution. For instance, solutions of the pH indicator bromothymol blue are yellow below pH 6 and blue above pH 8. In the intervening range the color changes from yellow to green to turquoise, and finally to blue. Similarly, the color of litmus indicator changes from red to blue through intermediate hues of magenta.

In the case of two-color indicators, one color is usually sensed as more vivid and intense than the other. In that case the transition range will conform approximately to (9.21). When both colors appear equally vivid, the transition range will be spaced more symmetrically about pK_A and will be wider. Actual data for a number of common pH indicators in aqueous solution are listed in Table 9.2.

Choice of titration endpoint indicator

A pH indicator can indicate the equivalence point in an acid-base titration if its transition range begins at, includes, or terminates at the pH of the equivalence point. If the innate titration error is small, the color

Table 9.2   PROPERTIES OF SOME COMMON pH INDICATORS IN AQUEOUS SOLUTION[a]

Indicator (Common name)	pK_A (in water)	Transition range	Color change Acid	Color change Base	Charge (acid)[b]	Chemical type[c]
Crystal violet (1)	—	0.0–1.0	Yellow	Green	+3	D
Crystal violet (2)	—	1.0–2.6	Green	Violet	+2	D
Thymol blue (1)	1.65	1.2–2.8	Red	Yellow	±0	B
m-Cresol purple (1)	—	1.2–2.8	Red	Yellow	±0	B
Tropeolin 00[d]	2.0	1.3–3.0	Red	Orange	±0	C
Methyl yellow[e]	3.25	2.8–4.3	Red	Yellow	+1	C
Methyl orange	3.45	3.1–4.5	Red	Orange	±0	C
Bromophenol blue	4.1	3.0–4.6	Yellow	Blue	−1	B
Bromocresol green	4.9	3.8–5.4	Yellow	Blue	−1	B
o-Methyl red	5.0	4.4–6.3	Red	Yellow	±0	C
Chlorophenol red	6.26	4.8–6.4	Yellow	Red	−1	B
Bromothymol blue	7.30	6.0–7.6	Yellow	Blue	−1	B
Phenol red	8.00	6.4–8.2	Yellow	Red	−1	B
m-Cresol purple (2)	8.3	7.4–9.0	Yellow	Purple	−1	B
Thymol blue (2)	9.20	8.0–9.6	Yellow	Blue	−1	B
Phenolphthalein	9.3	8.0–9.8	Colorless	Pink	−1	A
Thymolphthalein	—	9.3–10.5	Colorless	Blue	−1	A
Alizarin yellow R	—	10.2–12.0	Yellow	Violet	−1	C

[a] Based primarily on data summarized by I. M. Kolthoff and C. Rosenblum, *Acid-Base Indicators* (New York: The Macmillan Company, 1937).
[b] Acid form of indicator.
[c] A—phthalein dye; B—sulfonphthalein dye; C—azo dye; D—triphenylmethane dye.
[d] Also called aniline yellow or diphenyl orange.
[e] Also called butter yellow or dimethyl yellow.

change at the equivalence point will be sharp because the pH change is sharp.

For definiteness, consider once again the titration of $0.1\ F$ acetic acid with sodium hydroxide in aqueous solution. A plot of pH versus the titrant/substrate ratio r is given in Fig. 9.3. The figure has the familiar shape of a logarithmic titration curve in which log [OH⁻] is plotted versus r. (This is because [OH⁻] = K_W/[H⁺], which implies that log [OH⁻] = $pH - pK_W$.) Thus, as the amount of hydroxide changes from "not quite enough" to a stoichiometric "excess," there is an abrupt increase in pH, centered at the pH 8.87 of the equivalence point under these conditions. The figure also shows transition ranges for two common pH indicators. For phenolphthalein, the transition range from colorless to purple is entirely within the steeply ascending portion of the titration curve, and it includes the equivalence point. As sodium hydroxide is added to the titration flask, the solution will remain colorless until $r = 0.999$. The next drop of titrant then produces the sudden appearance of a pink color, and the color change to an intense pink (approaching red) is complete when $r = 1.001$. The entire color signal therefore involves only 0.2 percent of the equivalent amount of base and includes the equivalence point, making a very satisfactory display of the equivalence point.

For bromothymol blue, the transition range from yellow to blue begins at $r \approx 0.9$, where the titration curve is still fairly flat, and ends at $r = 0.999$, where the titration curve is already very steep. This two-color indicator acts differently from phenolphthalein, giving us advance warning as the titration approaches the equivalence point. The color will remain yellow until $r \approx 0.9$ and then will change, first slowly to greenish-yellow, then more quickly to green, then quickly to turquoise, and finally, suddenly, to blue. The sudden appearance of blue, which occurs just before the equivalence point, is taken as the titration endpoint. The advance warning in the form of color changes before the endpoint keeps us from overshooting the endpoint, but the multiplicity of hues may be objectionable. It also happens that the colors produced by bromothymol blue in water are not particularly vivid.

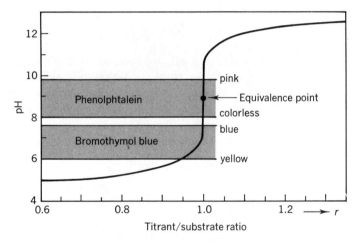

Fig. 9.3. Selection of color endpoint indicator for the titration of 0.1 F acetic acid with sodium hydroxide in aqueous solution. The shaded areas show the transition ranges for the pH indicators, phenolphthalein and bromothymol blue.

Finding the pH at the equivalence point. It is clear from the preceding example that when the pH at the equivalence point is known, it becomes a simple matter to choose an appropriate endpoint indicator: We simply select an indicator with a suitable transition range. There are two methods for finding the pH at the equivalence point: by experiment and by calculation.

In the experimental method, one prepares a synthetic sample that replicates the titration mixture at the equivalence point and measures its pH. It is useful to remember here that titration mixtures at their equivalence point need not be prepared from substrate and titrant: They can also be prepared from the pure products, and often with greater accuracy. For example, to replicate the first equivalence point in the stepwise titration of

phosphoric acid with sodium hydroxide, it may be convenient to prepare a
solution of pure NaH_2PO_4 in water.

$$H_3PO_4 + (Na^+) + OH^- \rightleftharpoons (Na^+) + H_2PO_4^- + HOH \quad (9.22)$$

The concentration of NaH_2PO_4 in the replicate should of course be similar
to that in a real titration at the equivalence point.

To find the pH at the equivalence point by calculation, we use the
methods described in Chapter 4, or the more general methods of Chapter 7.
Of course, the effective chemical equation and the required values of the
equilibrium constants must be known. For instance, in the titration of
phosphoric acid to the first equivalence point, the effective chemical equation is (9.23) and the equilibrium constant is given in (9.24).

$$H_3PO_4 + HPO_4^{2-} \rightleftharpoons 2\,H_2PO_4^- \quad (9.23)$$

$$K = \frac{[H_2PO_4^-]^2}{[H_3PO_4][HPO_4^{2-}]} = \frac{K_{A1}(\text{for } H_3PO_4)}{K_{A2}(\text{for } H_3PO_4)} \quad (9.24)$$

To calculate the pH, we need to know both K_{A1} and K_{A2}.

Formulas for the pH at the equivalence point in common cases are
stated in Eqs. (9.25) through (9.30). These equations apply to acid-base
titrations not only in water (where $K_{IP} = K_W$), but also in nonaqueous
solvents of high dielectric constant if we define pH = $-\log [SH^+]$, where
SH^+ denotes the lyonium ion.

Strong acid + strong base:

$$pH = \tfrac{1}{2}pK_{IP} \quad (9.25)$$

Weak acid (pK_A; c_S) + strong base:

$$pH = \tfrac{1}{2}(pK_{IP} + pK_A + \log c_S) \quad (9.26)$$

Weak base (pK_B; c_S) + strong acid:

$$pH = \tfrac{1}{2}(pK_{IP} - pK_B - \log c_S) \quad (9.27)$$

Weak acid (pK_A) + weak base (pK_B):

$$pH = \tfrac{1}{2}(pK_{IP} + pK_A - pK_B) \quad (9.28)$$

First equivalence point in titration of dibasic acid with strong base:

$$pH = \tfrac{1}{2}(pK_{A1} + pK_{A2}) \quad (9.29)$$

Selective titration with strong base of a weak acid $HX(pK_{A,HX}; c_{HX})$ in the
presence of a much weaker acid $HY(pK_{A,HY}; c_{HY})$:

$$pH = \tfrac{1}{2}(pK_{A,HX} + pK_{A,HY} + \log c_{HX} - \log c_{HY}) \quad (9.30)$$

Practical test of endpoint indicator. Having found the pH at the
equivalence point and chosen a color indicator with a suitable transition
range, the final step is to test that indicator. A practical test is desirable,

not only to provide final proof of the accuracy of the chosen indicator, but also to make sure that you can see the color change easily.

The test is made as follows: Prepare a synthetic sample that simulates the titration mixture at the equivalence point. Add indicator. Then add one drop of titrant or one drop of a typical solution of the original substrate. The addition of a small amount of either titrant or substrate should produce a distinct color change. If the pH at the equivalence point depends on the substrate concentration, test the indicator at both ends of the expected concentration range.

The use of color indicators becomes unsatisfactory when the pH at the equivalence point depends on two or more independent concentrations, while at the same time the slope of the titration curve is not very steep. This situation is common in the selective titration of mixtures of acids, and it arises when the innate titration error is well above 0.25 percent. Consider, for example, the titration of acetic acid in the presence of phenol, which was discussed in Chapter 8. The innate error in that titration is fairly large, being 0.62 percent when $c_{HAc} = 0.05\ F$ and $c_{HOPh} = 0.1\ F$, so that the equivalence point must be located with care. If a color indicator were used, the color corresponding to the pH at the equivalence point would have to be matched precisely. But that would be impractical in a real situation because the pH at the equivalence point depends on two variables, c_{HAc} and c_{HOPh}, neither of which is known accurately until after the analysis has been completed. We are therefore faced with a vicious cycle, being unable to titrate accurately with a color indicator unless we know the pH at the equivalence point, and being unable to predict the pH at the equivalence point unless we can analyze the solution. The vicious cycle is broken if we use potentiometric titration; the equivalence point then is simply the point of maximum slope of the experimental titration curve, and the pH need not be predicted in advance.

Molecular structure of typical pH-indicators

The indicators listed in Table 9.2 are organic dyes with complex molecular structures. They can be classified chemically as phthalein dyes, sulfonphthalein dyes, azo dyes, and triphenylmethane dyes. The light absorption is associated with the presence in such molecules of one or more electron-deficient atoms that are bonded through an extensive system of conjugated double bonds to one or more atoms with unshared electron pairs. The addition of a proton to an unshared electron pair produces the color change.

Phthalein dyes owe their tongue-twisting name to phthalic anhydride, an important chemical first derived from na*phthal*ene, and which is one of the reactants from which all phthalein dyes are formed. Phenolphthalein, whose molecular structure is shown below, is a typical member of the family.

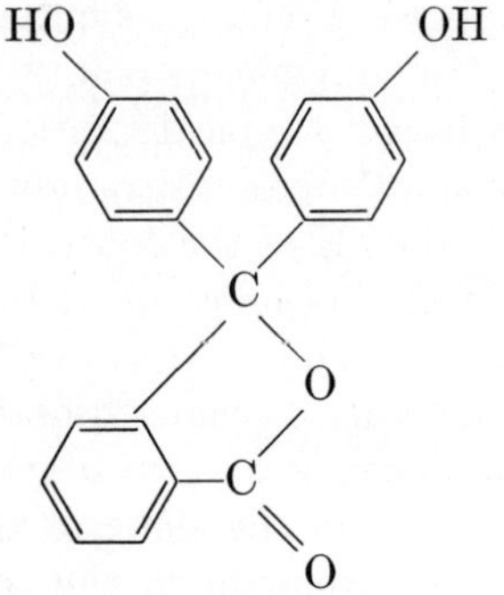

Phenolphthalein (colorless crystals, m.p. 261°C)

In aqueous solution phenolphthalein reacts with water to produce an acidic hydrate; the color change in the pH range 8.0–9.6 is due to the reaction shown in (9.31).

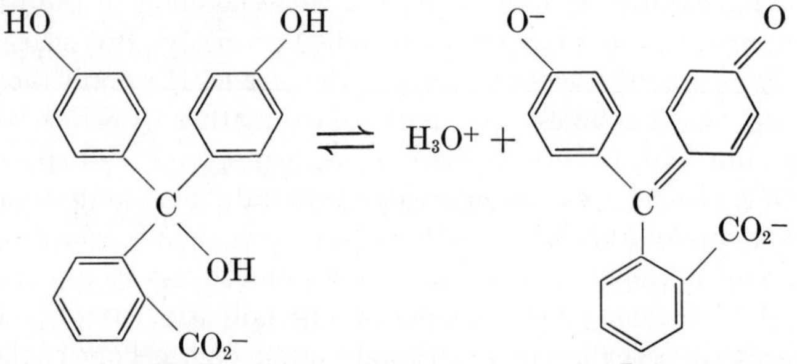

(9.31)

In strong alkali the purple color fades slowly, owing to the formation of the colorless compound shown in (9.32), whose molecular structure is similar to that of HIn^-.

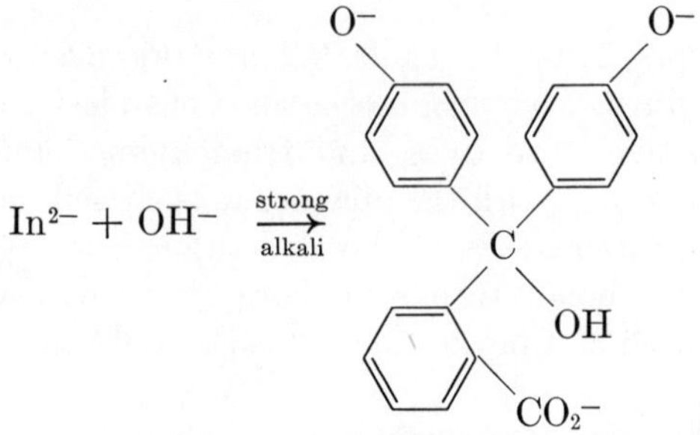

$$In^{2-} + OH^- \xrightarrow[\text{alkali}]{\text{strong}}$$

(9.32)

"Bleached" form of phenolphthalein (colorless)

The structure and reactions of sulfonphthalein dyes are similar to those of phthalein dyes. The structurally simplest sulfonphthalein dye is phenol red. Below pH 1, phenol red is an equilibrium mixture of two isomers, shown in (9.33). (Note that the phthalein and sulfonphthalein dyes produce color in solution only if the central carbon atom has a trigonal structure.)

Phenol red below pH 1

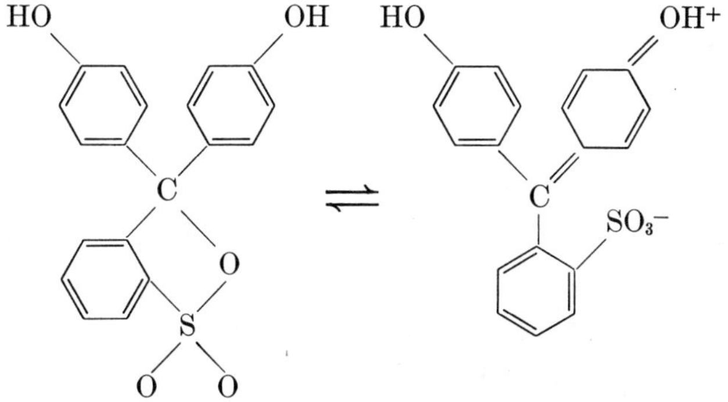

(9.33)

(Colorless, predominant at low dielectric constant)   (Red, predominant at high dielectric constant)

Although both isomers in (9.33) have an overall charge number of zero, the red isomer has an unusual charge distribution, with unit plus and minus charges separated by several atomic diameters. Such a molecule is called a *zwitterion* and behaves in some respects like a pair of separate ions. For instance, the formation of the zwitterion at equilibrium in (9.33) is *opposed* by a *decrease* of the dielectric constant, other things being equal. Thus solutions of phenol red in slightly acidified anhydrous acetic acid at a dielectric constant of 6.3 are quite colorless.

In aqueous solution, phenol red shows two color changes, one between pH 1 and 3, and the other between pH 6.4 and 8.0. The corresponding transformations of the indicator are shown in (9.34).

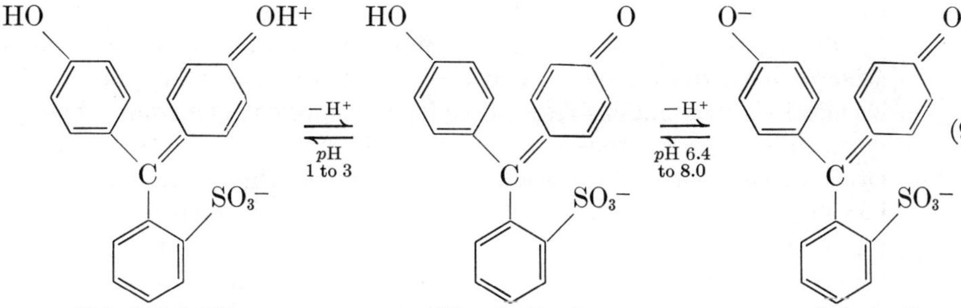

(9.34)

$H_2In(\pm)$ (red) HIn^- (yellow) In^{2-} (red)

Azo dyes are characterized by the presence of one or more —N=N— groups in the molecule. The molecular formulas of two typical azo dyes, methyl yellow (butter yellow) and methyl orange, are shown in (9.35).

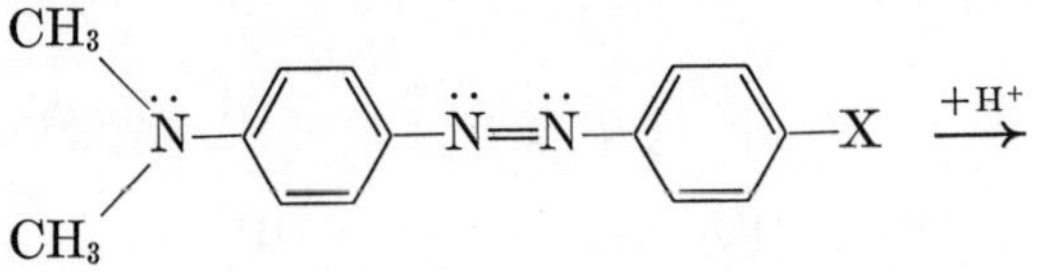

In

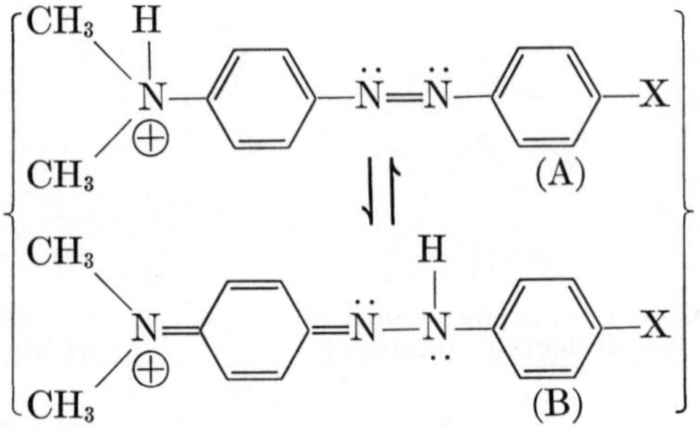

(9.35)

HIn⁺

Methyl yellow (X = H):
(In), yellow; (HIn⁺), red orange.
Methyl orange (X = SO₃⁻):
(In⁻), orange; (HIn±), red.

The conjugate acid (HIn⁺) of these dyes is an equilibrium mixture of two isomers; isomer (B) causes the color to change to red. Methyl yellow is carcinogenic and should be handled with care. The conjugate acid of methyl orange provides another example of a zwitterion. Note that the sulfonate group (X = SO₃⁻) bears a negative charge.

Triphenylmethane dyes consist of molecules in which a central carbon atom is bonded to three aromatic rings. Crystal violet is typical. The chemical transformations responsible for the color changes from yellow to green to violet in aqueous solution below *p*H 3 are shown in (9.36).
Other well-known triphenylmethane dyes are malachite green and rosaniline. The triphenylmethane dyes are no longer used much for dyeing of textiles because the colors are not "fast," but they continue to be used for staining biological specimens and bacteria. The color of triphenylmethane dyes fades at alkaline *p*H; hydroxide ions combine with the dye molecules under those conditions in a reaction that is analogous to (9.32).

(Yellow)

$$\xrightarrow[\text{pH 0 to 1.0}]{-\text{H}^+}$$

(Green)

$$\xrightarrow[\text{pH 1.0 to 2.6}]{-\text{H}^+}$$

(Violet)

$$(9.36)$$

Acid-base indicators in nonaqueous solvents

The same dyes that serve as pH indicators in aqueous solution will also serve as acid-base indicators in nonaqueous solvents. The color change associated with a given acid-base reaction of the indicator remains approximately the same in all hydroxylic solvents. For example, bromothymol blue undergoes the same change, from yellow to blue, in methanol that it does in water. However, the transition in methanol occurs at concentrations of hydrogen ion $(CH_3OH_2^+)$ between 10^{-11} and $10^{-13}\,M$, while that in water occurs between pH 6.0 and 7.6. This marked solvent effect on the transition range for the color can be explained in terms of the solvent effect on the acid-dissociation constant of the indicator.

pK_A values for a series of indicators in water, methanol, and ethanol are listed in Table 9.3. The solvent effects on pK_A vary from small to large, depending on the chemical type and charge type of the indicator, and are usually consistent with rules stated in Chapters 5 and 6. Exceptions occur when the indicator acid or its conjugate base is a zwitterion. For example, the acid form of methyl orange [see (9.35), $X = SO_3^-$] is a zwitterion. Yet methyl orange shows practically the same solvent effect on pK_A as does methyl yellow [see (9.35), $X = H$], where the acid form is a cation.

In hydroxylic solvents of high dielectric constant such as methanol or ethanol, the behavior of acid-base indicators is much like that in water. The transition range of the indicator can be predicted from its pK_A according to (9.21) if pH is now defined by $pH = -\log[SH^+]$, that is, the hydrogen ion is identified with the lyonium ion of the solvent. The selection of a titration endpoint indicator then follows precisely the same rules as for aqueous titrations, and the calculation of the hydrogen ion concentration at the equivalence point is analogous. However, the equilibrium constants used in the calculation must be appropriate for the chosen solvent.

In hydroxylic solvents of low dielectric constant such as acetic acid, the proton-transfer reactions of a color indicator are complicated by the tendency of ionic solutes to associate to ion pairs, as described in Chapter 5. It turns out, however, that ionic association in hydroxylic solvents has little or no effect on the absorption of light by the associating solutes. The transition range of an acid-base indicator in such a solvent is therefore given by (9.37), where $\Sigma\,[In]$ and $\Sigma\,[HIn]$ denote the *total* concentration of molecular species with the characteristic light absorption of In and HIn, respectively, regardless of the state of association.

Approximate transition range of acid-base indicators in hydroxylic solvents of low dielectric constant:

$$\frac{1}{16} < \frac{\Sigma\,[In]}{\Sigma\,[HIn]} < 4 \qquad \text{if In produces the more vivid color}$$

$$\frac{1}{4} < \frac{\Sigma\,[In]}{\Sigma\,[HIn]} < 16 \qquad \text{if HIn produces the more vivid color}$$

(9.37)

pK_A VALUES FOR SOME ACID-BASE   **Table**
INDICATORS IN WATER AND ALCOHOL[a]   **9.3**

Indicator (Common name)	pK_A in Water	Meth- anol	Eth- anol	Color change Acid	Base	Charge (acid)[b]	Chemical type[c]
Tropeolin 00	2.0	2.2	2.3	Red	Orange	±0	C
Methyl yellow	3.25	3.4	3.55	Red	Yellow	+1	C
Methyl orange	3.45	3.8	3.4	Red	Orange	±0	C
p-Methyl red	2.3	4.1	3.55	Red	Yellow	+1	C
o-Methyl red	5.0	9.2	10.45	Red	Yellow	±0	C
Neutral red	7.4	8.2	8.2	Red	Yellow	+1	C
Thymolbenzein (1)	—	3.5	3.3	Red	Yellow	+1	D
Thymolbenzein (2)	—	13.15	13.9	Yellow	Magenta	0	D
Phenolphthalein	9.3	—	15.3	Colorless	Pink	−1	A
Thymol blue (1)	1.65	4.7	5.35	Red	Yellow	±0	B
Thymol blue (2)	9.2	14.0	15.2	Yellow	Blue	−1	B
Bromophenol blue	4.1	8.9	9.5	Yellow	Blue	−1	B
Bromocresol green	4.9	9.8	10.65	Yellow	Blue	−1	B
Bromocresol purple	6.4	11.3	12.05	Yellow	Purple	−1	B
Bromophenol red	6.4	11.3	—	Yellow	Red	−1	B
Bromothymol blue	7.3	12.4	13.2	Yellow	Blue	−1	B
Phenol red	8.0	12.8	13.55	Yellow	Red	−1	B
Cresol red	8.4	13.2	—	Yellow	Red	−1	B
meta-Cresol purple	8.6	13.5	—	Yellow	Purple	−1	B

[a] Based primarily on data reported by L. S. Guss and I. M. Kolthoff, *J. Am. Chem. Soc.*, **62**, 249 (1940). See also Table 9.2.
[b] Acid form of indicator.
[c] A—phthalein dye; B—sulfonphthalein dye; C—azo dye; D—triphenylmethane dye.

Because of the greater molecular complexity of solutes at low dielectric constant, the selection of a titration endpoint indicator in practice is more likely to be the result of trial and error than of deliberate planning.

In aprotic solvents, and especially in aprotic solvents of low dielectric constant, the behavior of color indicators can become exceedingly complex, because hydrogen bonding and ionic association tend to change the absorption of light by the affected form of the indicator so as to change the *hue* of the solution. Thus, simple "one-color" or "two-color" indicators often become "many-color" or "variable-color" indicators in aprotic solvents, so that visual endpoint detection becomes inconvenient or even impractical.

PROBLEMS

1. Analyze the statement "the color of this glass of tea is pale orange" in terms of what you know about the origins of color and the physiology of color perception. Predict the effect of doubling the diameter of the glass (using the same tea) on the shade and hue of the observed color.

2. On comparing Eqs. (9.2) and (9.3), we see that

$$\log \left(\frac{I}{I_0}\right) = -\epsilon l c \quad \text{or} \quad \frac{I}{I_0} = 10^{-\epsilon l c}$$

Show how these relationships may be derived from the assumptions that the decrease in light intensity of a certain hue $(-dI)$ through an infinitesimally small layer of solution is directly proportional to (a) the actual value of the intensity (I) and (b) the number of absorbing particles dN in the layer. [*Hint:* Write down the assumptions in the form of a mathematical equation, separate the variables, and (using the calculus) integrate over the *entire* light path in the solution.]

3. Consider a solution for which the percent saturation of a particular hue is 50 percent. What will be the effect of doubling the molar concentration of the absorbing species on (a) the absorbance, (b) the percent transmission, and (c) the percent saturation of this hue.

4. A particular titrant has a molar extinction coefficient $\epsilon \approx 600$ cm^{-1} M^{-1} in the center of its visible absorption band. How might experimental conditions be arranged so that, contrary to the statement in the text that ϵ should be at least 1000 cm^{-1} M^{-1}, this titrant may be used as a self-indicator? (*Hint:* Examine the assumptions made in the text.)

5. Briefly discuss the chemistry of the following.
  (a) Self-indicators
  (b) Redox indicators
  (c) Indicator reactions involving complex formation
  (d) Acid-base indicators
Give an example of each. How does the presence of the indicator influence the accuracy of a titration?

6. Derive an equation for the percent saturation of transmitted hue for a solution where the HIn form of the indicator (only) absorbs visible light. Plot $\log_{10}$ {[In]/[HIn]} versus percent saturation (cf. Fig. 9.2) for this situation assuming that $\epsilon l c_{HIn} = 0.7$ (ϵ = extinction coefficient for HIn). Your results should verify the second statement in (9.20).

7. Verify the validity of Eqs. (9.25) to (9.30), noting any approximations in their derivation. You may assume, of course, that equilibrium constants and formal concentrations are such as to permit accurate titration.

8. Calculate the pH at the equivalence point(s) for the following aqueous titrations. Use data from Table A.1, as needed, and allow for dilution owing to the mixing of the reactants. What endpoint indicator(s) would you use for each titration?

Substrate	Titrant
(a) 0.1113 F HCl	0.04256 F NaOH
(b) 0.1300 F NH$_3$	0.1100 F HNO$_3$
(c) 0.1000 F Na$_2$CO$_3$	0.1500 F HCl
(d) 0.1300 F HN$_3$	0.1300 NH$_3$

9. Calculate the equivalence point pH for the selective titration of 0.1000 F acetic acid in the presence of 0.0500 F glycine (in the substrate solution) with 0.1000 F NaOH. What endpoint indicator would you use?

10. Calculate the equivalence point pH for the titration in (100 percent) methanol of 0.1000 F anilinium ion with 0.1150 F NaOCH$_3$. What indicator would you select for the titration?

11. Why are the pK_A values for some indicators (such as thymol blue and phenol red) markedly different in water and in ethanol, while other indicators (such as tropeolin 00 and methyl yellow) have pK_A values that are about the same in both solvents?

12. Predict the transition range(s) for the following indicators in methanol and in ethanol: (a) tropeolin 00, (b) *o*-methyl red, (c) neutral red, (d) bromophenol blue, (e) bromocresol purple, and (f) phenolphthalein (ethanol only). Indicate, in each case, whether you are choosing HIn or In as the species with the more vivid color.

chapter **10**

Instrumental Analysis of Liquid Solutions

If there were any doubt about the power of modern technology, the phenomenal growth and impact of instrumental methods in analytical chemistry would surely dispel it. By probing unknown substrates with sophisticated instrumentation, by measuring their physical, spectroscopic, and electrochemical properties with high precision, it becomes possible to push quantitative analysis to levels of specificity, sensitivity, and efficiency that are entirely beyond the reach of chemical methods alone. Equally important, it becomes possible to let routine tasks of chemical analysis be done by automatic machinery.

In this chapter we shall give a brief introduction to the instrumental analysis of liquid solutions, describing how instruments are used to measure concentrations and determine stoichiometry. Explanations of specific instruments will be given in later chapters, as part of the laboratory procedures, or by your instructor.

CHEMISTS AND "BLACK BOXES"

Instrumental analysis differs from the kind of analytical chemistry discussed in earlier chapters because the chemist is much more dependent on experts from other fields. The difference is one of degree rather than of kind, but you can sense it right away. Modern analytical instruments are so sophisticated, in terms of physics and electronics, that their inner workings are not entirely understood by most persons with only a college degree in chemistry. Many laboratories therefore employ an electronics engineer to service the instruments and keep them in working order, while the chemists make the actual measurements and interpret the results. For this reason, your instructor is likely to stress those matters that will enable you to do the chemist's part in this teamwork. The chemist must understand the physical principles of the measurement, know how the instrument operates by understanding the function of every knob, switch, push button, and meter, check that the instrument is in good working order by doing control experiments on known samples, and—most important of all—translate the instrument readings into the language of chemistry and deduce the desired information.

If the instrument is simple and rugged, your instructor may let you inspect its inside and trace the mechanical and electrical connections. However, if the instrument is finely tuned and expensive, he may justly object to your opening it up, and you may end up with the feeling that you are merely manipulating the outside of a mysterious "black box." That feeling is upsetting because a scientist is expected to work in full mastery of his experiment. How can you be the master of a "black box" whose inner workings you do not know? The answer is, of course, that you are part of a team that includes a member who *does* understand the "black box," and your own control experiments will warn you if something is wrong.*

BULK SOLUTION PROPERTIES

Instrumental analysis can be applied to matter of all kinds—animal, vegetable, and mineral; solid, liquid, gaseous, and colloidal; matter on

---

* For an opposing point of view, we quote the following remark by James Clerk Maxwell, made about 1875, which still rings true: "The student who uses home-made apparatus which is always going wrong often learns more than one who has the use of carefully adjusted instruments to which he is apt to trust and which he dare not take to pieces."

this planet and matter in interstellar space. However, we shall limit our discussion to the instrumental analysis of liquid solutions. It is convenient to divide this subject into two parts: the measurement of physical properties that are characteristic of the entire solution, and the measurement of the emf of electrochemical cells.

Properties that are characteristic of the entire solution are called *bulk solution properties*. Familiar examples are the density, refractive index, boiling point, freezing point, conductivity for heat, conductivity for electricity, absorbance of light (or radiant energy) at specific wavelengths, dielectric constant, and viscosity.

In the limit of infinite dilution, all bulk solution properties must of course approach the values characteristic of the pure solvent. Moreover, as solute is added to the pure solvent, the rate of change of the bulk solution property must be finite. These conditions are stated in Eqs. (10.1) and (10.2), where $\mathcal{P}$ denotes the bulk solution property, $\mathcal{P}_0$ the corresponding property of the solvent, c_S the solute concentration, and a_S a finite constant that is characteristic of the solute.

$$\lim_{c_s \to 0} \mathcal{P} = \mathcal{P}_0 \tag{10.1}$$

$$\left(\frac{d\mathcal{P}}{dc_S}\right)_{c_s=0} = a_S \; (a_S \neq \pm\infty) \tag{10.2}$$

It follows from (10.1) and (10.2) that at sufficient dilution, any bulk solution property becomes a linear function of the solute concentration.

$$\mathcal{P} = \mathcal{P}_0 + a_S c_S \qquad \text{if } c_S \text{ is small} \tag{10.3}$$

Although in general $\mathcal{P}$ is a linear function of c_S only when c_S is small, chemists are clever at expressing their results in such a way that the linear relationship will hold up to moderately high concentrations. For example, the quantity measured with a spectrophotometer is actually the intensity ratio I_0/I. However, we usually take the logarithm and work with the absorbance $A = \log (I_0/I)$. We do this because the absorbance remains a linear function of c_S up to a much higher concentration than does I_0/I, a fact that is stated expressly by Beer's law, Eq. (1.2). For another example, the quantity measured with an electrical conductivity bridge is the ohmic resistance (R) of the solution. However, we usually calculate the conductance $(1/R)$, which remains a nearly linear function up to a much higher concentration. In both examples, the variable which gives the better linear relationship can be derived from theory.

When the solution contains several solute components, each at a low concentration, Eq. (10.3) may be generalized as shown in (10.4).

$$\mathcal{P} = \mathcal{P}_0 + \sum_i a_i c_i \tag{10.4}$$

The summation in (10.4) extends over all solute components.

SOLUTE CONCENTRATIONS FROM BULK
SOLUTION PROPERTIES

It is clear from Eqs. (10.3) and (10.4) that bulk solution properties can be used to measure solute concentrations. Indeed, Eq. (10.4) is a general version of Eq. (7.53), and bulk solution properties therefore lead to linear titration curves. We shall return to this subject later. At this point we wish to consider the use of bulk solution properties for obtaining solute concentrations without titration.

Interpolation on a calibration curve

If the solution is known to contain only a single solute, we may determine the concentration of that solute as follows.

1. We choose a convenient bulk solution property for measurement and make sure that the measuring instrument is in good working order.

2. We measure a series of known solutions and construct a calibration curve showing $\mathcal{P}$ as a function of c_S. The calibration curve need not be a straight line, but must become a straight line at low concentrations.

3. We measure $\mathcal{P}$ for the unknown solution and determine c_S by interpolation on the calibration curve.

For example, Fig. 10.1 shows a plot of density versus molar concentration of pyridinium perchlorate in water at 25°C. Figure 10.2 shows a similar plot of refractive index versus molar concentration. To determine an unknown concentration of pyridinium perchlorate in this concentration range, we measure either the density or the refractive index of the solution and read the corresponding concentration off the figure. An example is shown in Fig. 10.1.

This method is simple in concept, but it has its pitfalls. Because most solution properties vary with the temperature, the calibration curve and the unknown concentration must be determined *at the same temperature*. Moreover, because we are measuring an unknown, we are never quite certain that the solution really contains only the one solute for which the calibration curve is designed, and it may be necessary to measure more than one solution property.

Temperature control. Most analytical instruments are designed with some provision for controlling the sample temperature. A common method is to pump water from a nearby constant-temperature bath through hollow channels in a brass block that surrounds the sample. Brass is a good material for this purpose: It can be machined to fit snugly around the

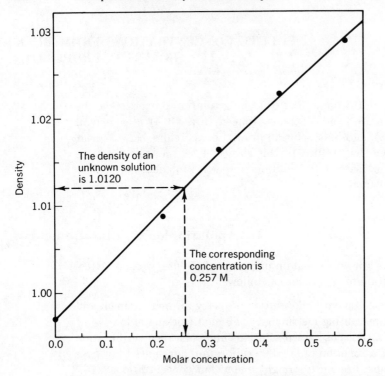

Fig. 10.1. Plot of density versus molar concentration for pyridinium perchlorate in water at 25°C. (Courtesy of Bruce Barnett.)

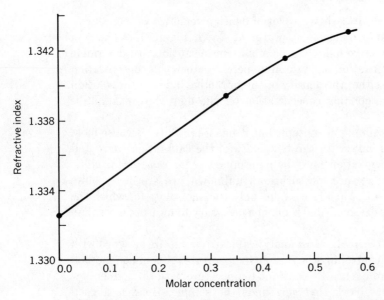

Fig. 10.2. Plot of refractive index versus molar concentration for pyridinium perchlorate in water at 25°C. (Courtesy of Bruce Barnett.)

sample tube, and it has a high thermal conductivity and heat capacity so that the entire brass block will be at a uniform temperature.

The required accuracy of the temperature control varies greatly with the nature of the measurement. Mathematical analysis (details of which are beyond the scope of this book*) leads to Eq. (10.5).

$$\delta T \approx \left| \frac{a_S}{(d\mathcal{P}_0/dT) + c_S(da_S/dT)} \right| \delta c_S \qquad (10.5)$$

where

δT = maximum allowable variation in sample temperature

δc_S = maximum allowable error in c_S

da_S/dT = change in a_S [Eq. (10.3)] per degree

$d\mathcal{P}_0/dT$ = change in $\mathcal{P}_0$ per degree, for the pure solvent

In practice, Eq. (10.5) often simplifies as follows.

1. When the measured property is the density, refractive index, dielectric constant or viscosity of dilute solutions, $|c_S(da_S/dT)| \ll |d\mathcal{P}_0/dT|$; hence

$$\delta T \approx \left| \frac{a_S}{d\mathcal{P}_0/dT} \right| \delta c_S \qquad (10.6)$$

2. When the measured property is the absorbance or (for solutions of electrolytes) the electrical conductivity, $|d\mathcal{P}_0/dT| \ll |c_S(da_S/dT)|$; hence

$$\delta T \approx \left| \frac{a_S}{c_S(da_S/dT)} \right| \delta c_S \qquad (10.7)$$

As an actual example, let us estimate the required temperature control in the measurement of the refractive index for the data shown in Fig. 10.2. In this case, $\mathcal{P}$ is the refractive index (n) and c_S is the molar concentration of pyridinium perchlorate. We shall use Eq. (10.6), with $d\mathcal{P}_0/dT$ equal to dn_0/dT. For pure water, $dn_0/dT = -1 \times 10^{-4}$ per degree C at 25°. To find a_S, we examine the initial linear portion in Fig. 10.2, which is reproduced by the equation $n = 1.3325 + 0.0217\, c_S$; hence $a_S = 0.0217$.

On substituting the preceding values in Eq. (10.6), we obtain the following result:

$$\delta T \approx \frac{0.0217}{1 \times 10^{-4}} \delta c_S \approx 200 \delta c_S$$

Thus, if the maximum allowable error in c_S were $\pm 0.0001\ M$, we should have to control the temperature to $\pm 0.02°C$. That is to say, the temperatures at which the known and unknown samples are measured may not differ by more than 0.02°C.

In practice, temperature control to $\pm 0.1°C$ is fairly easy to obtain. However, if the control must be better than that, it is advisable to measure

---

* The mathematical analysis begins with the equation

$$\left(\frac{\partial T}{\partial c_S} \right)_{\mathcal{P}} = -\frac{(\partial \mathcal{P}/\partial c_S)_T}{(\partial \mathcal{P}/\partial T)_{c_S}}$$

the known and the unknown sample in rapid succession, or side-by-side.

Specificity. A good analytical method should confirm the identity of the substrate while measuring its concentration. In a chemical method we can choose a reaction that will proceed in a certain characteristic manner only for a small group of substances, including the substrate. Similarly, in a method based on bulk solution properties, we can choose a property that is highly sensitive only to a small group of solutes, including the substrate. Let us consider some examples.

The electrical conductivity will be high, compared to that of the pure solvent, if free ions are present in solution. The electrical conductivity is therefore used to measure the concentration of electrolytes. Nonelectrolytes at low concentrations rarely interfere.

The viscosity of a solution will be high, compared to that of the pure solvent, if solutes of high molecular weight are present.* The viscosity is often used for the analysis of polymer solutions and for the determination of high molecular weights.

The plane of polarization of plane-polarized light passing through a solution in an optically inactive solvent is rotated only if optically active solutes are present. The optical rotation is often used to characterize, or measure the concentration of, such solutes.

In solvents of low dielectric constant, a marked increase in the dielectric constant signifies the presence in solution of highly polar solutes, such as ion pairs.

The most reliably specific bulk solution property is the absorption spectrum. Spectrophotometers are available that measure absorbance with quantitative precision, from about 2000 Å in the ultraviolet to well into the near-infrared. In such a wide spectral range there are bound to be at least a few spectral regions in which the substrate absorbs strongly and possible impurities absorb only weakly. If the absorbance is measured in such spectral regions, and also at a few wavelengths where the impurities would absorb strongly, we have an almost ideal method of analysis: one which is precise, specific, able to detect impurities, and (since all measurements are made with one instrument) convenient. It turns out, also, that precise control of the temperature is not required: For 0.25 percent accuracy in the measured absorbance, temperature control to $\pm 1°$ is usually quite adequate.

Linear titration curves

When a bulk solution property is measured throughout a titration, the resulting plot of $\mathcal{P}$ versus the titrant/substrate ratio r is a linear titration

---

* Strong molecular interactions result in high viscosity for concentrated solutions of some substances with low formula weight, e.g., NaOH in water or H_3PO_4 in water.

curve. Examples have been given in Figs. 7.1 and 7.2. The equivalence point is the point of maximum curvature.

For measuring solute concentrations, linear titration curves are sometimes preferred to interpolation on a calibration curve. Both methods make use of bulk solution properties. However, a linear titration curve displays the entire course of a specific chemical reaction and thus gives a great deal of additional information. It can show if the reactant is other than the expected substrate, or if the unknown sample is a mixture of reactants. Inert impurities do not interfere. The necessary measurements can be made in rapid succession so that the temperature control need not be elaborate. In most cases it is quite satisfactory to work at room temperature, especially if the titration apparatus can be set up in a draft-free corner of the laboratory, away from radiators and windows. One can often get excellent results with rather inexpensive instrumentation because the measuring instrument need not be absolutely accurate—it need only be precise enough to locate the equivalence point. If the curvature of the titration curve is severe, the endpoint signal will be sharp and easy to detect. A favorable example, the titration of uranium(IV) with ceric

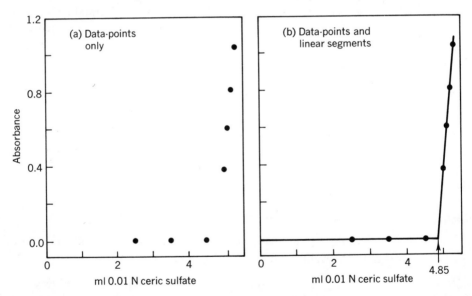

Fig. 10.3. Photometric titration of about 5 mg of uranium(IV) with 0.01 N ceric sulfate. The absorbance is measured at 3600 Å. Based on data reported by C. E. Bricker and P. B. Sweetser, *Anal. Chem.*, **25**, 764 (1953).

sulfate, is shown in Fig. 10.3. In this case the measured property, the absorbance at 3600 Å, remains constant at zero until the equivalence point, where it suddenly rises sharply. The sudden increase is due to the accumulation of excess Ce(IV), which absorbs strongly at 3600 Å. Solutions of uranium(IV) and of the reaction products, Ce(III) and UO_2^{2+}, have an absorbance of zero at this wavelength.

Locating the equivalence point. In Fig. 10.3, the equivalence point is taken as the point of intersection of two straight-line segments, one fitting the data before the sudden change in slope, the other fitting the data after the sudden change in slope. We now wish to show that under ideal conditions [constant temperature, negligible dilution, and strict adherence to Eq. (10.4)], this procedure is theoretically correct.

As shown earlier [Eq. (7.54)], the linear titration curve, under ideal conditions, takes the form of Eq. (10.8), where the subscripts S, T, and P denote the substrate, titrant, and product, respectively.

$$\mathcal{P} = \mathcal{P}_0 + a_S c_S + (a_S + a_T - a_P)x + (a_P - a_S)c_S r \qquad (10.8)$$

A typical plot of $\mathcal{P}$ versus r according to (10.8) was given in Fig. 7.6. Note that the titration curve consists of three distinct regions: (1) a straight line until shortly before the equivalence point; (2) a curve that includes the equivalence point; (3) another straight line beginning slightly past the equivalence point.

The mathematical condition for obtaining the straight line (1) is that the molar concentration of unreacted titrant x be small compared to both c_S and rc_S. In that case the term proportional to x in (10.8) is negligible and we obtain (10.9), which represents a linear variation of $\mathcal{P}$ with r.

Before the equivalence point, when $x \ll rc_S$:

$$\mathcal{P} = \mathcal{P}_0 + a_S c_S + (a_P - a_S)c_S r \qquad (10.9)$$

The mathematical condition for obtaining the straight line (3) is that $x = (r - 1)c_S$. [See Eq. (7.56).] On making appropriate substitution in (10.8), we obtain the linear equation (10.10).

Past the equivalence point, when $x = (r - 1)c_S$:

$$\mathcal{P} = \mathcal{P}_0 + (a_P - a_T)c_S + a_T c_S r \qquad (10.10)$$

To find the intersection of the straight lines (10.9) and (10.10), we equate the right-hand sides and solve for r.

$$\mathcal{P}_0 + a_S c_S + (a_P - a_S)c_S r = \mathcal{P}_0 + (a_P - a_T)c_S + a_T c_S r$$

$$\therefore (a_S - a_P + a_T)c_S = (a_S - a_P + a_T)c_S r$$

$$\therefore r = 1$$

The solution is $r = 1$; that is, the intersection is at the equivalence point.

Actual linear titration curves tend not to be quite ideal because the temperature may be drifting, dilution may not be negligible, and there may be small deviations from Eq. (10.4). As a result, actual titration curves tend to consist of two almost, but not quite, linear segments, which are connected by a short, sharply curved segment. To find the equivalence point we use portions of the titration curve near the equivalence point, using the following procedure.

Plot the data and find (approximately) the point of maximum curvature. Then decide where the approximately linear regions (1) and (3) begin. This will be a subjective decision, but in most cases the sharply curved segment (which includes the equivalence point) is quite short.

Using a transparent ruler, draw a "best straight line" through the three or four data points nearest the curved segment in the linear region before the equivalence point. Then draw a similar "best straight line" through the three or four data points just past the curved segment.

Note the volume of titrant at the intersection of the two "best straight lines." It is your titration endpoint. Calculate the substrate concentration.

This method is laborious (an entire titration curve must be determined and plotted) and somewhat subjective, but as a pragmatic procedure it works remarkably well. In favorable cases it will give quite reasonable results even when the innate titration error, as estimated by the engineering formulas in Chapter 8, is discouragingly high. For example, Fig. 10.4

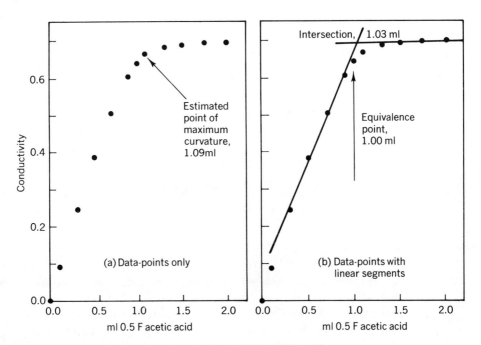

Fig. 10.4. Conductometric titration of 50.0 ml of 0.0100 F NH_3 with 0.500 F acetic acid in methanol. The conductivity is in kilohm^{-1} cm^{-1}. Data of E. Grunwald.

shows the conductometric titration curve for the titration of ammonia with acetic acid in methanol.

$$NH_3 + HAc \underset{\text{methanol}}{\rightleftharpoons} NH_4^+ + Ac^- \tag{10.11}$$

$$K \approx 50$$

Under the experimental conditions, $K \approx 50$ and the innate titration error, according to (8.22), is 25 percent! Nevertheless, Fig. 10.4 shows that the titration endpoint falls within a few percent of the equivalence point at 1.00 ml. Of course, there are better methods for determining the concentration of ammonia in methanol, but the example shows that linear titration methods are especially valuable for the solution of difficult analytical problems.

STOICHIOMETRY FROM BULK SOLUTION PROPERTIES

If the formal concentrations of substrate and titrant are known, the linear titration curve will establish the stoichiometry of the reaction. One carries out the titration and locates the equivalence point in the manner described before, and then evaluates the formula weight ratio of the reactants. Linear titration curves are especially suitable for this purpose because (as noted in Fig. 10.4) the equivalence point can be found with fair accuracy even if the equilibrium constant is less than huge. In cases of complex stoichiometry, one tries to keep the reaction conditions close to ideal by careful control of the temperature and by minimizing any dilution.

For example, Fig. 10.5 shows the variation of the dielectric constant during the acid-base titration of picric acid (HPic) with triethylamine (B) in benzene.

Picric acid   **Triethylamine**

The reaction clearly proceeds with one-to-one stoichiometry. Since picric acid neither associates nor dissociates appreciably at 0.002 F concentration in benzene, and since the marked increase in the dielectric constant suggests that the product is an ion pair, we represent the reaction by (10.12).

$$HPic + B \underset{\text{benzene}}{\rightleftharpoons} BH^+ \cdot Pic^- \tag{10.12}$$

It is interesting to contrast Eq. (10.12) with Eqs. (5.25) and (5.26), which

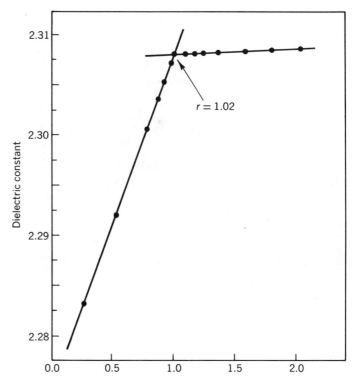

Fig. 10.5. Dielectric constant during the titration of 0.002 F picric acid (HPic) with triethylamine (B) in benzene. $r = c_B/c_{HPic}$. The temperature and c_{HPic} remain constant. Based on data of R. Megargle, G. L. Jones, Jr. and D. Rosenthal, *Anal. Chem.*, **41**, 1214 (1969).

represent the far more complicated reactions of *acetic* acid with triethylamine in a solvent of low dielectric constant.

Stepwise reaction sequences

If the reaction between substrate and titrant takes place in a series of more or less discrete steps, the linear titration curve is likely to show it. If the reaction steps are well resolved and consecutive, the titration curve is likely to show sharp curvature at the end of each step. If the steps are only approximately consecutive and poorly resolved, the titration curve may nevertheless give a clear indication of the stepwise stoichiometry.

For example, Fig. 10.6 shows an oscillometric titration curve (essentially, a plot of the dielectric constant) for the reaction of *secondary-*butyllithium with acetone in a solvent consisting of hexane and benzene.

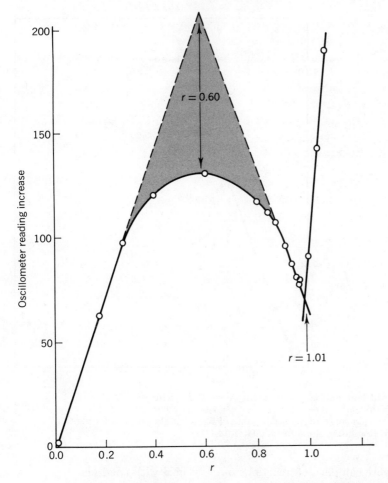

Fig. 10.6. Oscillometer reading during the titration of 0.3 *F sec*-butyl-lithium with 0.9985 *F* acetone in a mixed solvent consisting of hexane and benzene. The oscillometer is a high-frequency alternating current instrument that measures changes of the dielectric constant within a narrow range. The "oscillometer reading increase" is in arbitrary units. The actual dielectric constant is in the neighborhood of 2.2. Based on data by S. C. Watson and J. F. Eastham, *Anal. Chem.*, **39**, 171 (1967).

In terms of simplest molecular formulas, the expected reaction is stated in Eq. (10.13).

$$\underset{\textit{sec-}\textbf{Butyllithium}}{\underset{H_3C}{\overset{H}{CH_3CH_2\overset{|}{\underset{|}{C}}{:}^- Li^+}}} + \underset{\textbf{Acetone}}{\underset{CH_3}{\overset{CH_3}{\overset{|}{\underset{|}{C}}{=}O}}} \rightleftharpoons \underset{H_3C\ \ CH_3}{\overset{H\ \ CH_3}{CH_3CH_2\overset{|}{\underset{|}{C}}{-}\overset{|}{\underset{|}{C}}{-}O^- Li^+}} \qquad (10.13)$$

The titration curve does in fact show a sharp "break" at the expected one-to-one formula weight ratio. However, before that point is reached, there is a curved region, in which the slope changes gradually, such that the curvature reaches a maximum near $r = 0.60$. In interpreting the titration curve, we assume that Eq. (10.13) is basically correct but that the reaction takes place in steps: In the first portion of the titration curve, acetone reacts with butyllithium to form a series of molecular complexes such that (on the average) one formula weight of substrate reacts with 0.6 formula weight of acetone. These complexes appear to be quite polar, for the dielectric constant rises sharply.

EQUILIBRIUM CONSTANTS FROM SPECTRAL ABSORPTION DATA

We have seen that bulk solution properties may be used to measure concentrations, to detect the existence of chemical reactions, and to determine their stoichiometry. We now wish to describe how such data may be used to calculate equilibrium constants.

Rather than talk in generalities, we shall limit ourselves to a single important property, the absorption spectrum, and to data for one specific reaction, which we shall treat in detail. Our example will bring out the general strategy and certain specific tactics but is hardly more than an introduction to what is in fact a vast subject. One of the fascinations in the measurement of equilibrium constants from bulk solution properties is that each chemical reaction presents its own unique problems. Past experience may suggest likely solutions, but each problem deserves to be analyzed in its own right.

Spectral curves when there is chemical reaction

Our example will be the reaction of *p*-nitrophenol with triethylamine in benzene at 25°C.

OH

CH_3CH_2—N—CH_2CH_3

CH_2CH_3

p-Nitrophenol (HA)   **Triethylamine (B)**

NO$_2$

p-Nitrophenol resembles picric acid (p. 226) in its molecular structure, but is a weaker acid. We have seen that picric acid reacts with triethylamine with one-to-one stoichiometry. We now wish to show on the basis of ultraviolet spectral data that p-nitrophenol reacts similarly with triethylamine in benzene solution, and to evaluate the equilibrium constant.

To simplify the notation in the following, we shall let HA denote p-nitrophenol and B denote triethylamine. Solutions of HA in benzene show a characteristic absorption spectrum between 2800 Å and 4000 Å, shown as curve 0 in Fig. 10.7. Dilute ($ca.$ 10^{-4} F) solutions of HA obey Beer's law ($A = \epsilon_{HA} l c_{HA}$), and curve 0 is a plot of the formal extinction coefficient (ϵ_{HA}) versus wavelength.

Although solutions of B in benzene show only negligible absorbance between 2800 Å and 4000 Å, when B is added to HA the absorption spectrum undergoes marked changes. This is shown in Fig. 10.7 for a series of

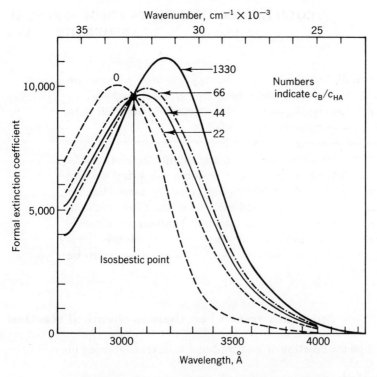

Fig. 10.7. Spectral absorption curves of p-nitrophenol (1.00×10^{-4} F) with various concentrations of triethylamine in benzene at 25°C. The number shown next to each curve denotes c_B/c_{HA}. From M. M. Davis, *Acid-Base Behavior in Aprotic Organic Solvents* (Washington, D.C.: U.S. Government Printing Office, 1968) National Bureau of Standards Monograph 105, Fig. 25.

solutions in which the *formal* concentration c_{HA} remains constant at 10^{-4} F while c_B increases progressively. The number shown next to each curve (22, 44, 66, 1330) denotes the respective value of c_B/c_{HA}. For reasons that will become apparent, it is again convenient to plot the formal extinction coefficient, defined by $\epsilon_{\text{formal}} = A/lc_{\text{HA}}$, where A is the measured absorbance of the solution containing both HA and B. Since solutions of B alone do not absorb in this spectral range, we may say in a purely formal way that the addition of B to solutions of HA causes deviations from Beer's law.

Beer's law on a molar basis

Marked deviations from Beer's law in the absence of chemical reaction are virtually unknown. Therefore, when marked deviations do occur, we can be reasonably sure that a chemical reaction is taking place. In the present case, since the deviations occur when B is added to HA (but not when HA is the only solute), we may conclude that the reaction involves the interaction of B with HA.

In interpreting absorbance data when there is a chemical reaction, we assume that Beer's law will apply precisely if we use *molar* concentrations rather than *formal* concentrations. The absorbance of a solution is therefore given by Eq. (10.14), where J represents the jth molecular species and the summation extends over all molecular species present in the solution.

$$A = l \Sigma \, \epsilon_J [J] \qquad (10.14)$$

When there is a chemical reaction, the equation must include an explicit term for each reactant and each product. If we leave out any of these terms, and if we use formal instead of molar concentrations (as we did in Fig. 10.7), we must not be surprised if there appear to be deviations.

Nature of reaction. The isosbestic point

Having concluded that B reacts with HA, the next logical question is: What is the nature of the reaction? Is it a single reaction with well-defined stoichiometry, or is it a stepwise series of reactions?

Again, Fig. 10.7 indicates the answer. Note that all spectral curves intersect *precisely* at a single point, close to 3050 Å in the figure. A point of precise coincidence of the formal extinction coefficients in an otherwise changing series of spectral curves is called an *isosbestic point*. It turns out that an isosbestic point is likely to be observed only if the underlying chemical reaction is simply a single reaction. The footnote on p. 233 will explain why this is so.

Having concluded that B reacts with HA in a single reaction, the final question is: What is the stoichiometry? The answer to that question can be given only after a careful mathematical analysis of the spectral curves.

We begin the analysis by *assuming* a stoichiometry—any plausible stoichiometry will do. We then analyze the spectral curves and obtain an equilibrium "constant" from each curve. If these equilibrium "constants" are indeed constant, showing only random fluctuations consistent with the experimental errors, then both the assumed stoichiometry and the equilibrium constants may be accepted as correct. However, if the equilibrium "constants" vary systematically with the concentrations of the reactants, we must repeat the analysis for a different assumed stoichiometry until an acceptable result is obtained.*

Absorbance and molar composition of solute component

We shall assume that the unknown reaction is (10.15) and thus assume a one-to-one stoichiometry.

$$HA + B \; \rightleftharpoons \; BH^+A^- \qquad (10.15)$$

This guess is plausible because picric acid reacts in this manner [Eq. (10.12)]. The reaction product may be either an ion pair or a hydrogen-bonded complex of one-to-one composition. On this basis, the equations of mass balance are (10.16) and (10.17), and K is given by (10.18).

$$c_{HA} = [HA] + [BHA] \qquad (10.16)$$

$$c_B = [B] + [BHA] \qquad (10.17)$$

$$K = \frac{[BHA]}{([B][HA])} \qquad (10.18)$$

The absorbance, according to (10.14), is given by (10.19).

$$A = l(\epsilon_{HA}[HA] + \epsilon_{BHA}[BHA] + \epsilon_B[B]) \qquad (10.19)$$

However, solutions of B alone have negligible absorbance. Hence, $\epsilon_B = 0$ and (10.19) simplifies to (10.20).†

$$A = l(\epsilon_{HA}[HA] + \epsilon_{BHA}[BHA]) \qquad (10.20)$$

Since $[HA] + [BHA] = c_{HA}$, Eq. (10.20) states that the absorbance is a function of the molar composition of the HA component.

To analyze the spectral curves in Fig. 10.7, we need an expression involving ϵ_{formal}. Such an expression is readily derived if we rewrite (10.20) in terms of the degree of conversion α of the HA component to BH^+A^-. α

---

* This trial-and-error procedure is closely analogous to the procedure, in chemical kinetics, of testing various rate laws to find the (hopefully unique) rate law that fits the data.

† When ϵ_B is not equal to zero, we introduce a "corrected absorbance" $A' = (A - \epsilon_B c_B)$ and continue with the analysis as described in the text, using A' in place of A. It can be shown [from (10.17) and (10.19), and rearranging the result] that $A' = \epsilon_{HA}[HA] + (\epsilon_{BHA} - \epsilon_B)[BHA]$, which is of the same mathematical form as (10.20).

is defined in (10.21); the expression for $(1 - \alpha)$ follows directly from (10.21) and (10.16).

$$\alpha = \frac{[BHA]}{c_{HA}} \tag{10.21}$$

$$(1 - \alpha) = \frac{[HA]}{c_{HA}} \tag{10.22}$$

On introducing these expressions in (10.20) and rearranging, we obtain (10.23).

$$\frac{A}{lc_{HA}} = \epsilon_{HA}(1 - \alpha) + \epsilon_{BHA}(\alpha)$$
$$= \epsilon_{HA} + (\epsilon_{BHA} - \epsilon_{HA})\alpha \tag{10.23}$$

By definition, $A/lc_{HA} = \epsilon_{formal}$. The desired relationship between ϵ_{formal} and composition of the HA component is therefore (10.24).

$$\epsilon_{formal} = \epsilon_{HA} + (\epsilon_{BHA} - \epsilon_{HA})\alpha \tag{10.24}$$

Equation (10.24) tells us that ϵ_{formal} varies linearly with α. If we measure ϵ_{formal} and know ϵ_{HA} and ϵ_{BHA}, we can deduce the actual molar composition of the equilibrium mixture.*

Analysis of data

To deduce α, we chose three wavelengths at which the spectral curves in Fig. 10.7 are well separated: 3200 Å, 3400 Å and 3500 Å. Each wavelength will allow us to deduce an independent set of α, and by comparing three independent sets we can confirm the accuracy of the analysis. The values of ϵ_{formal} obtained by reading the spectral curves at the chosen wavelengths are listed in Table 10.1.

According to Eq. (10.24), to compute α we must know both ϵ_{HA} and ϵ_{BHA}. Of course, ϵ_{HA} is simply the extinction coefficient in the absence of B and is given directly by curve 0. On the other hand, ϵ_{BHA} is the extinction coefficient when the substrate is completely converted to BHA. Since the

---

* The condition for an isosbestic point is that $\epsilon_{BHA} = \epsilon_{HA}$, so that $\epsilon_{formal} = \epsilon_{HA}$ for all values of α. If we construct a graph showing the variation of ϵ_{HA} with wavelength, and also (on the same scale) the variation of ϵ_{BHA} with wavelength, the isosbestic point will be the point at which the two curves cross. The argument that an isosbestic point indicates a single reaction rather than a series of reactions is as follows: Because the spectral curve for each substance is unique, it is unlikely that more than two spectral curves will cross at precisely the same point. An isosbestic point therefore indicates the presence of two molar species (two spectral curves) in varying proportion. Since one of the two species must be the original substrate, we find (by simple counting) that there is one product, and hence a single reaction.

Table 10.1   CALCULATION OF α FROM FORMAL EXTINCTION COEFFICIENTS. ORIGINAL SPECTRAL CURVES ARE SHOWN IN FIG. 10.7

$\dfrac{c_B}{c_{HA}}$	3200 Å		3400 Å		3500 Å	
	ϵ_{formal}	α	ϵ_{formal}	α	ϵ_{formal}	α
$0(\epsilon_{HA})$	5540	0.000	980	0.000	590	0.000
22	7730	0.382	3470	0.404	2220	0.410
44	8800	0.568	4650	0.596	2880	0.575
66	9310	0.657	5240	0.692	3300	0.681
1330	11140	0.976	7030	0.982	4500	0.982
$\infty\,(\epsilon_{BHA})$	11280	1.000	7140	1.000	4570	1.000

equilibrium in (10.15) will shift progressively to the right as c_B increases, we may conclude that ϵ_{formal} will approach ϵ_{BHA} as c_B approaches infinity.

$$\epsilon_{BHA} = \lim_{c_B \to \infty} \epsilon_{formal} \qquad (10.25)$$

Although Eq. (10.25) is mathematically correct, a practical person likes to deal with finite limits; infinity is a concept he leaves to philosophers. Fortunately, as $c_B \to \infty$, $1/c_B \to 0$, and (10.25) is therefore equivalent to (10.26).

$$\epsilon_{BHA} = \lim_{1/c_B \to 0} \epsilon_{formal} \qquad (10.26)$$

According to (10.26), to obtain ϵ_{BHA} we must plot ϵ_{formal} versus $1/c_B$ and extrapolate the resulting curve to $1/c_B = 0$. This is done for the present data in Fig. 10.8. The extrapolation is short and appears to introduce little error. The results are included in Table 10.1.*

Given ϵ_{HA} and ϵ_{BHA}, it then becomes a simple matter to solve Eq. (10.24) to obtain α. The results are included in Table 10.1. The values obtained for any given values of c_B/c_{HA} at three wavelengths show only small random fluctuations, and the mean value of α is established with a mean deviation of about ± 0.01.

Having established α, we next calculate molar concentrations as shown in Table 10.2, and then apply Eq. (10.18) to obtain the equilibrium "constant." If the results obtained for K turn out to be constant, then the stoichiometry assumed in Eq. (10.15) is justified. As shown in the final column of Table 10.2, the results obtained for K are indeed quite constant,

---

* Fortunately, many reactions have equilibrium constants that are large enough to allow the *direct* determination of all pertinent extinction coefficients. For example, the dissociation constant for a weak acid, HA, can sometimes be determined by measuring the absorbance of solutions of HA under the following conditions: (a) in the presence of a strong acid, so that $\alpha = 0$ and $\epsilon_{formal} = \epsilon_{HA}$; (b) at high pH, so that HA is fully converted to A^- ($\alpha = 1$) and $\epsilon_{formal} = \epsilon_{A^-}$; (c) at a known intermediate pH where both HA and A^- contribute significantly to the measured absorbance.

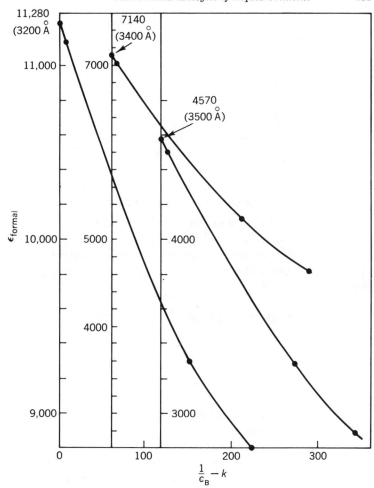

Fig. 10.8. Evaluation of ϵ_{BHA} by extrapolation to $1/c_B = 0$. On the abscissa, $k = 0$ for 3200 Å, $k = 60$ for 3400 Å, and $k = 120$ for 3500 Å.

CALCULATION OF EQUILIBRIUM CONSTANT
FOR REACTION (10.15) FROM DATA IN
TABLE 10.1. $c_{HA} = 1.00 \times 10^{-4}$ F

Table 10.2

$\dfrac{c_B}{c_{HA}}$	α (mean)	[BHA] ($= \alpha c_{HA}$)	[HA] ($= [1-\alpha]c_{HA}$)	[B] ($= c_B - $ [BHA])	K Eq. (10.18)
22	0.399	0.399×10^{-4}	0.601×10^{-4}	21.6×10^{-4}	307
44	0.580	0.580×10^{-4}	0.420×10^{-4}	43.4×10^{-4}	318
66	0.677	0.677×10^{-4}	0.323×10^{-4}	65.3×10^{-4}	321
1330	0.980	0.980×10^{-4}	0.020×10^{-4}	1329×10^{-4}	$(370 \pm 80)^a$

(Mean $\pm$ m.d.) 315 ± 6

[a] This value is very inaccurate because $(1 - \alpha)$ is so small. It was not used in computing the mean.

with a mean deviation of ± 2 percent as c_B/c_{HA} varies from 22 to 66. Indeed, even the (less accurate) value obtained when $c_B/c_{HA} = 1330$ is consistent with the other results.

Concluding remarks. The calculation of the equilibrium constant concludes our example. Let us appraise what we have done.

We have given a detailed description of a specific chemical system. We have

> detected the existence of chemical reaction,
> identified the reactants,
> found that the reaction is simple rather than complex,
> determined the stoichiometry, and
> measured the equilibrium constant.

Most notable of all, we have been able to do all this without the traditional operations of chemistry, without having to isolate or purify any reaction product or characterize any equivalent weight, but simply by measuring the ultraviolet absorption spectrum for a series of solutions.

ELECTROCHEMICAL CELLS

An electrochemical cell (also called a *voltaic cell* or *galvanic cell*) is a device for harnessing a redox reaction in such a way that an emf is developed in an external circuit. Its essential features are shown in Fig. 10.9.

The cell is arranged so that oxidation and reduction take place in separate compartments, called *half-cells*, which are separated by a barrier that is solid enough to prevent bulk mixing, yet porous enough to permit the passage of ions from one compartment into the other. Each compartment contains an electrode that is capable of conducting electrons into and out of the compartment, and which provides a catalytically active surface at which the desired redox half-reaction can take place rapidly.

The following terminology* should be memorized:

---

* The terms electrode, anode, and cathode are due to Michael Faraday. This is how Faraday introduced them: "In place of the term pole, I propose using that of *electrode* (ἤλεκτρον, and ὁδός, *a way*), and I mean thereby that substance, or rather surface, whether of air, water, metal, or any other body, which bounds the extent of the decomposing matter in the direction of the electric current." . . . "If the magnetism of the earth is due to electric currents passing round it, the latter must be in a constant direction, which, according to present usage of speech, would be from east to west, or, which will strengthen this help to the memory, that in which the sun appears to move. If in any case of electrodecomposition we consider the decomposing body as placed so that the current passing through it shall be in the same direction, and parallel to that supposed to exist in the earth, then the surfaces at which the electricity is passing into and out of the substance would have an invariable reference, and exhibit constantly the same relations of powers. Upon this notion we purpose calling that towards the east the anode

The electrode at which oxidation occurs is called the *anode*.

The electrode at which reduction occurs is called the *cathode*.

The points at which the cell is connected to an external circuit are called *terminals* and are labeled $+$ and $-$. When the cell is connected to an external resistance or *load* (that is, when the switch S in Fig. 10.9 is closed), the electron current leaves the cell at the $-$ or negative terminal, and flows into the cell at the $+$ or positive terminal. In this mode of operation, the negative terminal is connected to the anode and the positive terminal is connected to the cathode.

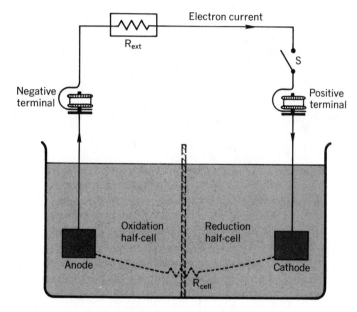

Fig. 10.9. Schematic diagram of an electrochemical cell connected to an external load. R_{cell} represents the internal resistance of the cell. R_{ext} represents the resistance of the external circuit, including the load. S is an "on-off" switch.

An electrochemical cell, by providing a flow of electrons through an external circuit, can do work on an attached load. Commercial voltaic cells and batteries,* such as the MnO_2-zinc dry cell, the nickel-cadmium cell, the lead-acid storage battery, or the H_2-O_2 fuel cell, have all the features of the cell shown in Fig. 10.9.

(ἄνω, *upwards*, and ὁδός, *a way*, the way which the sun rises), and that towards the west the *cathode* (κατά, *downwards*, and ὁδός, *a way*, the way which the sun sets)." M. Faraday, *Experimental Researches in Electricity*, Series VII, **1**, 197–198 (1834).

* A battery is two or more voltaic cells hooked together. For an elementary review of electrochemical cells, consult your general chemistry textbook or E. J. Lyons, Jr. *Introduction to Electrochemistry* (Boston: D. C. Heath and Company, 1967).

Measurement of cell emf

The emf developed between the negative and positive terminals of a voltaic cell becomes independent of cell geometry (which affects the internal cell resistance) only if the emf is measured under conditions where no current is flowing through the cell. Two common methods of measurement are indicated in Fig. 10.10.

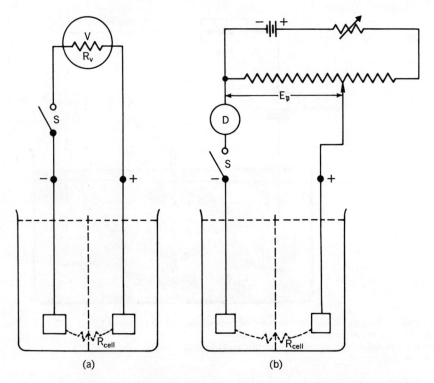

(a)   (b)

Fig. 10.10. Two methods for measuring the emf of voltaic cells. (a) The cell is connected to a voltmeter V whose electrical resistance (R_V) must be very much greater than the internal resistance of the cell. (b) The cell is connected to a potentiometer which supplies a variable, known voltage E_P in opposition to the cell emf. When the detector D shows that no current is flowing through it and the cell, the cell emf and E_P are equal.

In method (a), which is more convenient but less exact, the cell is connected to a voltmeter of high internal resistance and the cell emf is read directly on the voltage scale. In this method it is essential that the internal resistance R_V of the voltmeter be much greater than that of the cell.

In method (b), which is exact and has been developed to very high precision, the cell emf is measured with a potentiometer. The electric circuit is wired so that the cell emf (E_{cell}) is placed in opposition (minus to minus and plus to plus) to a carefully calibrated voltage E_P developed by the potentiometer. The magnitude of E_P can be varied by moving a contactor along a slide wire; see Fig. 10.10(b). When the switch S is closed, if E_P is different from E_{cell}, current will flow along a path consisting of the detector D in series with the cell, and the detector will show a characteristic deflection. However, when E_P and E_{cell} are equal, no current flows and the detector remains at its null point. In practice, one varies E_P until the deflection of the detector goes to zero. At that point, $E_P = E_{cell}$.

The use of a potentiometer with a sensitive detector also provides an automatic check on the reversibility of the cell reaction. If the cell is reversible, then even a tiny change in E_P, from one side of the null point to the other, will cause a reversal in the direction of current flow.

Equations for cell emf

We shall assume that you are familiar with Nernst's equation for the emf of an electrochemical cell such as that shown in Fig. 10.9. The emf for the entire cell can be represented as the difference of a cathode half-cell emf and an anode half-cell emf, plus a correction term δE:

$$E = E(\text{cathode}) - E(\text{anode}) + \delta E \tag{10.27}$$

The half-cell emf's in (10.27) are represented by exact analogs of Nernst's equation. Using standard reduction potentials and the American convention for the sign of electrode potentials as described in Chapter 2, they are represented by (10.28) and (10.29), in which the mass-action quotient Q is given by (10.30).

$$E(\text{cathode}) = \left(E_{red}^{\circ} - \frac{RT}{n\mathsf{F}} \ln Q \right)_{\text{cathode}} \tag{10.28}$$

$$E(\text{anode}) = \left(E_{red}^{\circ} - \frac{RT}{n\mathsf{F}} \ln Q \right)_{\text{anode}} \tag{10.29}$$

$$Q = \frac{\text{mass-action expression for reduced state in half-cell}}{\text{mass-action expression for oxidized state in half-cell}} \tag{10.30}$$

The correction term δE accounts for the emf developed at the junction between the half-cell compartments (we must expect a difference in electrical potential whenever there is a boundary between unlike materials), and for possible deviations from Eqs. (10.28) and (10.29). For cells of appropriate design, δE will be small or even zero.

Depending on the expression for Q, some half-cells are sensitive to the concentration of a specific species of ions in solution, while others are sensitive to a specific redox couple in solution. Let us consider some examples. (Assume $\delta E = 0$.)

1. A pure element in equilibrium with its ions in solution.

Example: $Ag^+ + e^- \rightleftharpoons Ag(s)$

$$E^\circ_{red} = +0.7991 \text{ V at } 25°C; n = 1; Q = \frac{1}{[Ag^+]}$$

$$\therefore E = 0.7991 - \frac{RT}{F} \ln \frac{1}{[Ag^+]}$$

$$= 0.7991 + 0.05915 \log [Ag^+] \text{ V at } 25°C$$

The potential of this half-cell depends on the silver ion concentration.

2. A highly insoluble salt in equilibrium with a pure element and ions in solution.

Example: $Hg_2Cl_2(s) + 2 e^- \rightleftharpoons 2 Hg(l) + 2 Cl^-$

$$E^\circ_{red} = +0.2679 \text{ V at } 25°C; n = 2; Q = [Cl^-]^2$$

$$E = 0.2679 - \frac{RT}{2F} \ln [Cl^-]^2$$

$$= 0.2679 - 0.05915 \log [Cl^-] \text{ V at } 25°C$$

This half-cell potential depends on the chloride ion concentration.

3. Two soluble oxidation states of the same element.

Example: $Fe^{3+} + e^- \rightleftharpoons Fe^{2+}$

$$E^\circ_{red} = +0.771 \text{ V at } 25°C; n = 1; Q = \frac{[Fe^{2+}]}{[Fe^{3+}]}$$

$$E = 0.771 - \frac{RT}{F} \ln \frac{[Fe^{2+}]}{[Fe^{3+}]}$$

$$= 0.771 - 0.05915 \log \frac{[Fe^{2+}]}{[Fe^{3+}]} \text{ V at } 25°C$$

The emf of this half-cell is determined by the ratio of iron(II) concentration to iron(III) concentration.

4. Two soluble oxidation states of the same element in an equilibrium involving other solute species.

Example: $H_3AsO_4 + 2 H^+ + 2 e^- \rightleftharpoons H_3AsO_3 + H_2O$

$$E^\circ_{red} = +0.56 \text{ V at } 25°C; n = 2; Q = \frac{[H_3AsO_3]}{[H_3AsO_4][H^+]^2}$$

$$E = 0.56 - \frac{RT}{2F} \ln \frac{[H_3AsO_3]}{[H_3AsO_4][H^+]^2}$$

$$= 0.56 - 0.02958 \log \frac{[H_3AsO_3]}{[H_3AsO_4][H^+]^2} \text{ V at } 25°C$$

In this case, the emf depends not only on the arsenic(III)-arsenic(V) redox couple but also on the *p*H.

SOLUTE CONCENTRATIONS FROM EMF MEASUREMENTS ON ELECTROCHEMICAL CELLS

In the normal measuring arrangement, one half-cell is filled with the unknown "solution X" and equipped with an electrode capable of sensing the unknown substrate. The other half-cell, called the *reference* half-cell, is merely a convenient dummy of fixed composition, designed in such a way that the correction term δE in (10.27) is either zero or very small or nearly constant. The conventional way of representing such a cell is shown in (10.31).

$$\text{Sensing electrode} \left| \text{Solution X} \right| \text{Reference solution} \left| \text{Reference electrode} \right. \tag{10.31}$$

In this convention, the sensing electrode is on the left, the reference half-cell is on the right, and each vertical line denotes a phase boundary across which the ion current in the cell must flow. The emf of cell (10.31) is given by (10.32).

$$\mathsf{E} = \mathsf{E}(\text{reference half-cell}) - \mathsf{E}(\text{sensing half-cell}) + \delta\mathsf{E} \tag{10.32}$$

Neglecting δE, the cell emf indicates how much the emf of the sensing half-cell *differs* from that of the reference half-cell. The difference (**E**) may be positive or negative. When **E** is positive, the sensing electrode becomes the negative terminal of the cell; when **E** is negative, the sensing electrode becomes the positive terminal of the cell.

Because the sensing half-cell is of special interest, we shall henceforth use the symbols E_{red}° and Q (without additional subscript or label) to denote these quantities for the sensing half-cell. The half-cell emf is then given by (10.33).

$$\mathsf{E}(\text{sensing half-cell}) = \mathsf{E}_{red}^{\circ} - \frac{RT}{n\mathsf{F}} \ln Q \tag{10.33}$$

On substituting in (10.32) and rearranging, we obtain (10.34).

$$\mathsf{E} = \frac{RT}{n\mathsf{F}} \ln Q - \mathsf{E}_{red}^{\circ} + \mathsf{E}(\text{reference half-cell}) + \delta\mathsf{E} \tag{10.34}$$

Assuming that δE is virtually constant or zero, if the same sensing electrode and reference half-cell are used throughout a series of measurements, the last three terms on the right in (10.34) will be individually and collectively constant. On introducing **E*** to denote the collective constant, we obtain (10.35) for the emf of cell (10.31).

$$\mathsf{E} = \mathsf{E}^* + \frac{RT}{n\mathsf{F}} \ln Q \tag{10.35}$$

E* is a characteristic "constant" for the given measuring cell and is called the *cell constant*. If we measure E and know E*, Eq. (10.35) will enable us to find the value of Q for the unknown "solution X." Knowing the value of Q in turn will enable us to find the substrate concentration.

Sensing half-cells

A half-cell can sense the concentration of a particular substrate only if the mass-action quotient Q contains the concentration of that substrate. Moreover, the concentrations of other species that appear in the expression for Q must be either known, or known to be constant.

Specific-ion sensing half-cells. The most direct method of sensing the concentration of an unknown substrate is to choose a half-reaction such that Q (or $1/Q$) is simply equal to the unknown concentration. Half-cells of this type are specific-ion sensing half-cells. Some common examples are shown in Table 10.3.*

Table
10.3

A FEW PRACTICAL HALF-CELLS FOR
SENSING SPECIFIC IONS

Half-cell	Ion detected
$Au(s) \mid Au^{3+}$	Gold(III)
$Pd(s) \mid Pd^{2+}$	Palladium(II)
$Ag(s) \mid Ag^{+}$	Silver ion
$Hg(l) \mid Hg_2^{2+}$	Mercurous ion
$Cu(s) \mid Cu^{2+}$	Copper(II)
$Pb(s) \mid Pb^{2+}$	Lead(II)
$Ni(s) \mid Ni^{2+}$	Nickel(II)
$Tl(s) \mid Tl^{+}$	Thallous ion
$Cd(s) \mid Cd^{2+}$	Cadmium(II)
$Zn(s) \mid Zn^{2+}$	Zinc(II)
$Ag\text{-}AgI(s) \mid I^{-}$	Iodide ion ($> 10^{-8}\ M$)
$Ag\text{-}AgBr(s) \mid Br^{-}$	Bromide ion ($> 10^{-6}\ M$)
$Ag\text{-}AgCl(s) \mid Cl^{-}$	Chloride ion ($> 10^{-5}\ M$)
$Hg(l)\text{-}Hg_2Cl_2(s) \mid Cl^{-}$	Chloride ion ($> 10^{-6}\ M$)

Many of the examples in Table 10.3 involve a metal in equilibrium with its ions in solution. The electrode may be a foil or wire of the pure metal, a thin deposit of the metal on a piece of platinum, or (in the case of active metals such as zinc) a dilute amalgam, that is, a dilute solid solution of mercury in the metal.

Other examples in Table 10.3 involve a metal, an anion, and the corresponding highly insoluble salt. Such electrodes measure the concentration

---

* The familiar glass electrode, which senses specifically the concentration of hydrogen ions, is actually an ion-selective membrane and will be discussed later.

of the anion. To facilitate the half-cell reaction, the salt is placed in direct contact with the metal. For instance, the Ag-AgCl electrode is prepared by electrolytic deposition of a thin coating of AgCl on a clean silver anode. The $Hg(l)$-$Hg_2Cl_2(s)$ electrode is prepared by allowing finely ground calomel to float on pure mercury.

The range of anion concentrations that can be detected by electrodes of this type is limited by the solubility product constant of the highly insoluble salt. For example, let the unknown "solution X" be silver-free and be probed for chloride ion with the Ag-AgCl electrode. After insertion of the electrode, "solution X" will of course become saturated with silver chloride. Let $[Cl^-]_0$ denote the chloride concentration before insertion of the electrode, and let s denote the solubility of silver chloride in "solution X." Then, in the solution in which the actual measurement is made, $[Ag^+] = s$, $[Cl^-] = ([Cl^-]_0 + s)$, and $K_{SP} = s([Cl^-]_0 + s)$. If $[Cl^-]_0 \gg s$, $[Cl^-]$ will be practically equal to $[Cl^-]_0$ and the cell measures essentially the desired concentration. However, when $[Cl^-]_0 \ll s$, $[Cl^-] \approx s$, and $K_{SP} \approx s^2$. Hence, when $[Cl^-]_0 \ll \sqrt{K_{SP}}$, the cell measures $\sqrt{K_{SP}}$ rather than $[Cl^-]_0$.

The preparation of good electrodes that come into equilibrium rapidly with their respective solutions is as much an art as it is a science. The electrode must be active without having excessive surface area or surface porosity (on a microscopic scale). Otherwise portions of substrate from the preceding measurement will not be removed completely when the electrode is rinsed "clean," and will introduce error or sluggish electrode response into the next measurement.

Redox-couple sensing half-cells. A few practical half-cells (and the corresponding mass-action quotients) for specific redox couples are listed in Table 10.4. Such half-cells are valuable for potentiometric

A FEW PRACTICAL HALF-CELLS FOR SENSING REDOX COUPLES		**Table 10.4**
Half-cell	***n***	***Q***
Pt-$H_2(g) \mid H^+$	1	$P_{H_2}^{1/2}/[H^+]$
Pt-$Cl_2(g) \mid Cl^-$	1	$[Cl^-]/P_{Cl_2}^{1/2}$
$Pt \mid I_3^-,\ I^-$	2	$[I^-]^3/[I_3^-]$
$Pt \mid Tl^{3+},\ Tl^+$	2	$[Tl^+]/[Tl^{3+}]$
$Pt \mid V^{3+},\ V^{2+}$	1	$[V^{2+}]/[V^{3+}]$
$Pt \mid Ti^{4+},\ Ti^{3+}$	1	$[Ti^{3+}]/[Ti^{4+}]$
$Pt \mid Fe^{3+},\ Fe^{2+}$	1	$[Fe^{2+}]/[Fe^{3+}]$
$Pt \mid Ce^{4+},\ Ce^{3+}$	1	$[Ce^{3+}]/[Ce^{4+}]$
$Pt \mid Co^{3+},\ Co^{2+}$	1	$[Co^{2+}]/[Co^{3+}]$
$Pt \mid Fe(CN)_6^{3-},\ Fe(CN)_6^{4-}$	1	$[Fe(CN)_6^{3-}]/[Fe(CN)_6^{4-}]$
$Pt \mid VO^{2+},\ V^{3+},\ H^+$	1	$[V^{3+}]/[VO^{2+}][H^+]^2$
$Pt \mid UO_2^{2+},\ U^{4+},\ H^+$	2	$[U^{4+}]/[UO_2^{2+}][H^+]^4$
$Hg(l) \mid HgY^{2-},\ Y^{4-}$	2	$[Y^{4-}]/[HgY^{2-}]$
(Y^{4-} = EDTA anion)		

titration and for measuring equilibrium constants of redox reactions. However, because the mass-action quotients involve the concentrations of at least two different substances, such half-cells are less useful for sensing the concentration of a specific substrate. There are some notable exceptions, of course. The half-cell Pt-H_2(g) | H^+ is of extraordinary importance, being the primary standard for the measurement both of *p*H and of standard electrode potentials. In practice, pure hydrogen gas is bubbled over the electrode in contact with the solution at atmospheric pressure, so that variations in P_{H_2} naturally remain small.

The most common electrode material for redox-couple sensing half-cells is the noble metal, platinum. A clean platinum surface will adsorb a wide variety of ions and nonelectrolytes and will act as catalyst for a great many redox processes. However, because the range of substances that can be adsorbed is so wide, a platinum electrode is easily spoiled or "poisoned." The offending substance or "poison" acts like the proverbial dog in the manger, occupying sites and thus preventing the use of the electrode's surface by the redox couple, in many cases without being oxidized or reduced itself. A poisoned platinum surface can usually be reactivated by brief immersion in aqua regia (a three-to-one mixture of concentrated hydrochloric and nitric acids).

Ion-selective membranes. The glass electrode

An alternative approach to the measurement of ionic concentrations is to use a cell in which the unknown "solution X" is placed on one side of an ion-selective membrane. Ideally, an ion-selective membrane is a partition between two liquids that will allow one, and only one, kind of ions to pass. If there is a difference in the concentrations of passable ions on opposite sides of the membrane, a characteristic emf will develop.

Membrane emf. To present the ideas, Eq. (10.36) shows a typical cell in which a known "solution A" and an unknown "solution X" are separated by an ion-selective membrane that is permeable only to ions of species M^+.

$$\text{AgCl-Ag} \left| \begin{array}{c} \text{Solution A} \\ [M^+]_A, [Cl^-]_A \end{array} \right| \begin{array}{c} \text{Solution X} \\ [M^+]_X \end{array} \left| \begin{array}{c} \text{Reference} \\ \text{solution} \end{array} \right| \begin{array}{c} \text{Reference} \\ \text{electrode} \end{array} \quad (10.36)$$

$$\uparrow$$
$$\text{Ion } (M^+)\text{-selective membrane}$$

The emf of this cell is given by an equation containing two half-cell emf's and δE, as before, plus the membrane emf.

$$E = E(\text{reference half-cell}) - E_{red}^{\circ}(\text{Ag-AgCl}) + \frac{RT}{F} \ln [Cl^-]_A$$

$$+ \delta E + E(\text{membrane}) \quad (10.37)$$

For univalent cations M^+, the membrane emf is given by (10.38); for ions N^z of general charge type z, it is given by (10.39).

$$E(\text{membrane}) = -\frac{RT}{F} \ln \frac{[M^+]_X}{[M^+]_A} \tag{10.38}$$

$$E(\text{membrane}) = -\frac{RT}{zF} \ln \frac{[N^z]_X}{[N^z]_A} \tag{10.39}$$

Note that z and $E(\text{membrane})$ may be positive or negative.

Ion-selective solid membranes. Solid materials that can be fabricated into practical ion-selective membranes are listed in Table 10.5. At

ION-SELECTIVE SOLID MEMBRANES   **Table 10.5**

Membrane material	Ions determined	Ions that may interfere
Glass	H^+	Na^+, at high pH
LaF_3	F^-, La^{3+}	OH^-
Ag_2S	S^{2-}, Ag^+	Hg^{2+}
CuS in Ag_2S[a]	Cu^{2+}	Hg^{2+}, Ag^+
PbS in Ag_2S[a]	Pb^{2+}	Hg^{2+}, Ag^+, Cu^{2+}
CdS in Ag_2S[a]	Cd^{2+}	Hg^{2+}, Ag^+, Cu^{2+}
AgI in Ag_2S[a]	I^-	S^{2-}, CN^-
AgI in Ag_2S[a]	CN^-	S^{2-}, I^-
AgBr in Ag_2S[a]	Br^-	I^-, CN^-, S^{2-}, NH_3
AgSCN in Ag_2S[a]	SCN^-	Br^-, I^-, CN^-, S^{2-}, NH_3
AgCl in Ag_2S[a]	Cl^-	Br^-, I^-, CN^-, S^{2-}, NH_3

[a] The second salt is dispersed in a very fine form in solid silver sulfide.

this writing the list is quite brief, but research is active in this field, and we can look forward to important developments.

An ionic solid or glass will function as an ion-selective membrane if the entire current is carried by just one kind of ions. For example, electric current in solid silver sulfide is almost entirely a current of silver ions, which can move without great hindrance through interstices in the crystal lattice. Solid silver sulfide therefore functions as a specific membrane for silver ions. However, because liquid solutions in contact with solid silver sulfide become saturated and conform to a solubility product, $K_{SP} = [Ag^+]^2[S^{2-}]$, a silver sulfide membrane will also determine sulfide ions.

In the glass used to fabricate a glass electrode, the charge carriers are protons which "jump" without great hindrance between oxide oxygen atoms that are present in the glass. As a result, the glass functions as a specific membrane for hydrogen ions. To show how an ion-selective membrane is used to measure concentrations, let us consider the glass electrode in greater detail.

Glass electrode. The elements of a typical cell with a glass electrode are shown below.

$$\mathrm{Ag\text{-}AgCl} \left| \begin{array}{c} \text{Solution A} \\ [\mathrm{H^+}]_A,\ [\mathrm{Cl^-}]_A \end{array} \right| \left. \begin{array}{c} \text{Solution X} \\ [\mathrm{H^+}]_X \end{array} \right| \left. \begin{array}{c} \text{Saturated} \\ \text{KCl solution} \end{array} \right| \mathrm{Hg_2Cl_2(s)\text{-}Hg(l)}$$

$$\underbrace{\qquad\qquad\qquad\qquad}_{\text{Glass electrode}} \quad \underset{\underset{\text{Glass membrane}}{\uparrow}}{} \quad \underbrace{\qquad\qquad\qquad\qquad\qquad}_{\text{Reference half cell}}$$

In real life, the glass electrode and the reference half cell are likely to be assembled in separate containers, shaped like test tubes, which can be dipped into the unknown "solution X" to establish a current path. See Fig. 10.11.

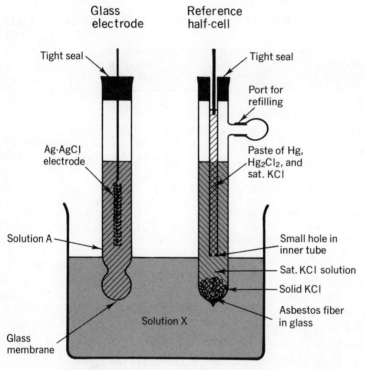

Fig. 10.11. A typical pH-measuring cell.

To derive a relationship between the cell emf and the pH of "solution X", we begin with Eq. (10.37) and express the emf developed across the glass membrane according to (10.38), writing [H⁺] in place of [M⁺]. The result is (10.40).

$$\mathsf{E} = \frac{RT}{\mathsf{F}} \ln [\mathrm{Cl^-}]_A - \mathsf{E}^\circ_{\mathrm{red}}(\mathrm{Ag\text{-}AgCl})$$

$$- \frac{RT}{\mathsf{F}} \ln \frac{[\mathrm{H^+}]_X}{[\mathrm{H^+}]_A} + \mathsf{E}(\text{reference half cell}) + \delta\mathsf{E} \quad (10.40)$$

On applying Eq. (10.40), we note that "solution A" is hermetically sealed inside the glass electrode. (See Fig. 10.11.) Although the manufacturer keeps the composition of "solution A" a secret, we may assume that stable materials are used. Hence $[Cl^-]_A$ and $[H^+]_A$ are characteristic constants for the given glass electrode. Assuming as before that δE is virtually constant, and denoting all constant terms collectively by a single "cell constant" E^*, we obtain (10.41), or its equivalent form (10.42), at 25°C.

$$E = E^* - \frac{RT}{F} \ln [H^+]_X \qquad (10.41)$$

$$E = E^* + 0.05915\ pH_X \text{ V at } 25°C \qquad (10.42)$$

Both equations show that the cell emf is a simple linear function of the unknown pH in "solution X."

Ion-selective liquid membranes. If "solution A" and "solution X" in (10.36) are aqueous, the ion-selective membrane may be made of a porous water-repellent solid disk (e.g., a cellulose acetate filter disk) that is saturated with a suitable organic liquid. To pass across the membrane, an ion must enter, migrate through, and leave the organic liquid. Normally, this process will be quite slow. However, by dissolving specific electrolytic catalysts in the organic liquid, the passage can be greatly facilitated for specific ions, and a nearly ion-selective liquid membrane results. Ions whose concentration can be measured selectively in this way include Ca^{2+}, Pb^{2+}, NO_3^-, ClO_4^-, Cl^-, and BF_4^-.

Complete cells

Before an electrochemical cell is used to measure unknown concentrations, the "cell constant" E^* is determined by measurements with known solutions. Since E^* includes the emf developed by the reference half-cell, the latter need not be a half-cell whose properties are known precisely in advance: Any convenient, stable, and reversible half-cell will do.

In order for the measured concentrations to be accurate, the reference half-cell must be linked to the sensing half-cell in such a way that the correction term δE in Eq. (10.27) remains approximately constant in all measurements. To accomplish this, we may use the theory of liquid-liquid junction potentials as a trusted guide,* but final proof must come from control experiments.

Experience with aqueous solutions indicates that the following aqueous reference half-cells bring about small or nearly constant values of δE:

$$| \text{ Saturated KCl } | \text{ AgCl-Ag(s)} \qquad (10.43)$$

---

* See, for example, H. Rossotti, *Chemical Applications of Potentiometry* (Princeton, N.J.: D. Van Nostrand Co., Inc., 1969).

$$| \text{ Saturated KCl } | \text{ Hg}_2\text{Cl}_2(\text{s})\text{-Hg}(\text{l}) \tag{10.44}$$

$$\begin{array}{|l|c|l|} \text{Salt bridge of} & & \\ \text{concentrated } (> 1 \ F) & 0.1 \ F \ \text{KCl} & \text{AgCl-Ag}(\text{s}) \\ \text{NH}_4\text{NO}_3 \text{ or NaClO}_4 & & \end{array} \tag{10.45}$$

The junction between "solution X" and the reference solution need not be larger than a pinhole. The junction is often established with the aid of an asbestos fiber, as shown in Fig. 10.11. (Using a glass-blowing torch, a small hole is blown into the glass wall of a reference half-cell, an asbestos fiber is threaded through, and the glass is reheated until it collapses onto the fiber.) The asbestos fiber acts as a wick which brings the liquids into contact.

The reference half-cells shown in (10.43) and (10.44) are normally preferred because of their greater simplicity. The half-cell (10.44) is most common in pH measurements; it is the *saturated calomel electrode*, or S.C.E. for short.* However, when mutual contamination of "solution X" and "saturated KCl" must be absolutely avoided, a salt bridge is used, as in (10.45). The bridge solution may be set in an agar gel to slow down its mixing with the half-cell solutions.

In practice, the great virtue of electrochemical cells is that they allow us to measure exceedingly low concentrations. For instance, by using a cell with a Pt-H$_2$(g)| electrode, hydrogen ion concentrations as low as 10^{-13} M can be measured with ease, and comparably dramatic claims can be made for many other electrodes. The reason for this almost unique capability† is that the emf measures the *logarithm* of the concentration. As a result, a given error in emf will cause precisely the same *percentage* error at low concentrations as at high concentrations. [For example, an error of 1 millivolt (mV) in the emf of a pH-measuring cell will cause a 4 percent error in the hydrogen ion concentration, regardless of the magnitude of that concentration.]

Because electrochemical cells can measure exceedingly low concentrations, they are often used to measure equilibrium constants. Thus, much of our quantitative information concerning solubility-product constants of highly insoluble salts, association constants for the formation of stable complex ions, and acid- and base-dissociation constants comes from concentration measurements with electrochemical cells. Moreover, the sensing electrode and reference half-cell can be assembled in the form of miniature probes so that measurements can be made in cramped spaces or on small samples.

The disadvantage of electrochemical cells for measuring unknown concentrations is that the results, though reliable, are seldom highly

---

* In common usage, the term "electrode" is sometimes used loosely to denote an entire half-cell.

† Certain radioactive tracer and activation techniques can also determine exceedingly low concentrations.

accurate. The cell emf can be measured readily with an accuracy of ± 0.1 mV which, in the absence of other errors, would define concentrations to better than 1 percent. However, there *are* other sources of error, the most troublesome of which is the possible inaccuracy of the cell constant E^*. We have seen that E^* is not a genuine constant, but incorporates a variable correction term δE whose exact value depends on the electrolytic composition of "solution X." For best results, the electrolytic makeup of the known solutions used in the measurement of E^* should be a close match to that of the unknown solution. If the match is indeed close, the measured concentration may be accurate to 1 percent or better; if not, the real accuracy may be no better than 10 percent. Similarly, in the measurement of pH it is easy to mistake good precision for high accuracy. Do not be misled if the pH of your standard buffer is stated on the label to the thousandth of a pH unit. Even with the best of precision, you will be fortunate if you can really guarantee the pH of your unknown solutions to the nearest hundredth of a unit.

POTENTIOMETRIC TITRATION

In a potentiometric titration, the unknown substrate is titrated directly in the "solution X" compartment of an electrochemical cell and the emf is monitored throughout the titration. The sensing electrode must be able to sense the molar concentration of titrant or substrate at least in the vicinity of the equivalence point. The plot of emf versus volume of titrant under such conditions is in effect a logarithmic titration curve, and the equivalence point is the point of maximum slope. Actual examples of potentiometric titration curves have been shown in Figs. 7.3 and 7.4.

Because we normally use the potentiometric data only to find the equivalence point, there is considerable flexibility in the choice of electrodes. A sensing electrode will be suitable if it can sense any of the following.

1. The molar concentration x of titrant.
2. The molar concentration [S] of substrate.
3. The molar concentration [M] of any species that remains in equilibrium with the titrant or substrate so that log [M] varies linearly with log x or with log [S], at least near the equivalence point.*
4. A redox electrode, if one member of the redox couple is either the titrant, the substrate, or a solute M as described in (3).

Before considering the detection of the equivalence point, we shall present two theorems.

---

* For example, if the titrant is hydroxide ion, a pH-sensing electrode will be suitable.

Theorem: With any of the electrodes described above, the cell emf becomes a linear function of log x in the region of the titration curve immediately surrounding the equivalence point.

The explanation is simply that in the region immediately surrounding the equivalence point, the concentrations of the reaction products (and of most other solutes) change relatively little and may be regarded as constant. The only concentrations that change a great deal—by orders of magnitude—are those of the titrant, the substrate, and any solute M that exists in equilibrium with titrant or substrate as described in (3) above.

The second theorem concerns redox titrations that are carried out in redox-sensing half-cells. The electrode then dips into a solution that contains *two* redox couples. For example, in the titration of Fe(II) with Ce(IV) according to Eq. (10.46), the redox couples are $Fe^{3+} + e^- \rightleftharpoons Fe^{2+}$ and $Ce^{4+} + e^- \rightleftharpoons Ce^{3+}$.*

$$Fe^{2+} + Ce^{4+} \rightleftharpoons Fe^{3+} + Ce^{3+} \tag{10.46}$$

Theorem: In a potentiometric titration, it will be sufficient if the sensing electrode responds rapidly and reversibly to only one of the redox couples, provided that the other redox couple does not poison the electrode.

The proof is simply that if the parts of a system are in equilibrium, the whole of the system is in equilibrium. Let the two redox couples be "couple S" and "couple T." In solution, couple S is in equilibrium with couple T. Therefore, when couple S comes into equilibrium with the electrode, couple T is also, necessarily, in equilibrium with the electrode, even though the direct reaction of couple T at the electrode might be slow.

There is a useful corollary. Since both redox couples exist in equilibrium with the same electrode, both redox couples define the *identical* half-cell emf. As a result, there will always be two alternate ways of stating the half-cell emf.

For definiteness, let us consider again the titration of Fe(II) with Ce(IV), and let the half-cell emf be sensed with a platinum electrode. If we regard the titration half-cell as a Pt | Fe^{3+}, Fe^{2+} half-cell, we write Eq. (10.47). If, with equal validity, we regard it as a Pt | Ce^{4+}, Ce^{3+} half-cell, we write Eq. (10.48).

$$E(\text{half-cell}) = E^\circ_{red}(Fe^{3+}, Fe^{++}) - \frac{RT}{F} \ln \frac{[Fe^{2+}]}{[Fe^{3+}]} \tag{10.47}$$

$$E(\text{half-cell}) = E^\circ_{red}(Ce^{4+}, Ce^{3+}) - \frac{RT}{F} \ln \frac{[Ce^{3+}]}{[Ce^{4+}]} \tag{10.48}$$

---

* Following our usual practice, we represent both redox couples in the form of half-reactions for reduction, even though the spontaneous complete reaction is such that iron(II) becomes oxidized.

Since the two equations express the identical emf, they must be equal to each other.

$$E[\text{half-cell (10.47)}] = E[\text{half-cell (10.48)}] \qquad (10.49)$$

This approach, of writing two alternative equations for the half-cell emf, becomes especially useful at the equivalence point. On adding (10.47) and (10.48), we obtain (10.50).

$$2E(\text{half-cell}) = E^\circ_{red}(Fe^{3+}, Fe^{2+}) + E^\circ_{red}(Ce^{4+}, Ce^{3+})$$

$$- \frac{RT}{F} \ln \frac{[Fe^{2+}][Ce^{3+}]}{[Fe^{3+}][Ce^{4+}]} \qquad (10.50)$$

At the equivalence point $(r = 1)$, $[Ce^{3+}] = [Fe^{3+}]$ and $[Fe^{2+}] = [Ce^{4+}]$. [See Eq. (10.46).] As a result, Eq. (10.50) reduces simply to (10.51).

At the equivalence point $(r = 1)$:

$$E(\text{half-cell}) = \tfrac{1}{2}[E^\circ_{red}(Fe^{3+}, Fe^{2+}) + E^\circ_{red}(Ce^{4+}, Ce^{3+})] \qquad (10.51)$$

Locating the equivalence point

There are two methods for finding the equivalence point in a potentiometric titration.

1. In the predetermined-endpoint method one adds titrant until the emf (or the pH) reaches the known value characteristic of the equivalence point.

2. In the maximum-slope method one adds titrant and measures the emf after each addition, until one has enough data to locate the point of maximum slope which indicates the equivalence point.

Predetermined-endpoint method. In this method, the procedure for finding the emf at the equivalence point is analogous to the one described in Chapter 9 for selecting and testing color endpoint indicators. One prepares a solution that matches the titration mixture at the equivalence point as nearly as possible, places it in the titration cell and measures the emf, which will be that of the endpoint. One then adds a drop of titrant and notes the change in emf to determine the precision with which the endpoint must be detected. The solution of the unknown substrate is then titrated to this predetermined emf. The emf may be expressed in any convenient units—for example, in volts, millivolts, or pH units. The "cell constant" E^* need not be calculated expressly but must remain constant throughout the series of experiments.

If the redox potentials and/or equilibrium constants are available, an alternative procedure is to *calculate* (rather than measure) the emf at the equivalence point, using appropriate equations given in this chapter and in Chapters 8 and 9. In this case the cell must be calibrated with known solutions and the "cell constant" evaluated expressly.

Maximum-slope method. In this method, the emf may be expressed in any convenient units, and the "cell constant" of the titration cell need not be known expressly. The equivalence point is that point on the plot (see Figs. 7.3 and 7.4) of emf versus titrant volume V at which the slope, $d\mathsf{E}/dV$, goes through a maximum. However, it would be time consuming if one had to make a plot of emf versus V for each titration. Fortunately, numerical methods are available that are not only faster but also more objective.

For definiteness, consider the potentiometric titration data in Table 10.6, which were taken from the research notebook of Dr. Alice Y. Ku at Brandeis University. In this case the substrate is 0.0024 F HCl, which is titrated with 0.01145 N NaOH in the presence of the weak acid 0.50 F NH_4Cl in aqueous solution. The emf is expressed in pH units and gives directly the pH of the titration mixture. The titrant is delivered from a microburet with a precision of about 0.001 ml. It is added in fairly large increments when the pH is changing slowly, and in small increments when the pH is changing rapidly. After each addition of titrant, the pH and the total volume of added titrant are recorded.

In analyzing the results listed in Table 10.6, we shall write in general terms (E in place of pH; V_a, V_b, V_e in place of specific volumes) because the method of analysis may be generally applied. The first step in finding the point of maximum slope is to calculate the average slope, $\Delta\mathsf{E}/\Delta V$, for each interval in the region where the emf is changing rapidly. The data usually follow a pattern such as that shown in Fig. 10.12, with one large slope (S_{max}) adjacent to two unequal medium slopes (S_i and S_f), and with more distant slopes becoming progressively smaller. The equivalence point (where the volume of titrant is V_e) lies somewhere in the interval of maximum slope, that is, between V_a and V_b. In Fig. 10.12, S_f is greater than S_i, and the equivalence point therefore will lie closer to V_b than to V_a. To find V_e more precisely, we use Eq. (10.52).

$$V_e = V_a + (V_b - V_a)\frac{S_f}{S_f + S_i} \tag{10.52}$$

Reflection will show that Eq. (10.52) is plausible also if S_f is smaller than S_i, or if the two are equal.*

The use of Eq. (10.52) is illustrated in Table 10.6. In Trial A, the

---

* Equation (10.52) can be given a theoretical justification. Since S_i and S_f are significantly smaller than S_{max}, we may assume that Eq. (7.55) is a valid approximation in the region $V \leqslant V_a$, and that Eq. (7.56) is valid when $V \geqslant V_b$. It then follows from Eq. (8.7) that the *derivative* $d\mathsf{E}/dV$ is inversely proportional to the distance $|V - V_e|$ from the equivalence point. As a result, Eq. (10.52) is exact if S_i is a valid measure of $d\mathsf{E}/dV$ at V_a, and if S_b is a valid measure of $d\mathsf{E}/dV$ at V_b.

POTENTIOMETRIC TITRATION OF 5.00 ML
OF 0.0024 N HCl IN THE PRESENCE OF
0.50 F NH₄Cl IN WATER

**Table
10.6**

Volume of 0.01145 N NaOH	E and $\Delta E/\Delta V$, as measured by	
	pH	$\Delta pH/\Delta V$

Trial A

0.000	3.49	
0.598	3.86	
0.800	4.13	
		$0.25/0.104 = 2.4$
0.904	4.38	
		$0.44/0.094 = 4.7 \ (S_i)$
0.998 (V_a)	4.82	
		$0.86/0.095 = 9.0 \ (S_{max})$
1.093 (V_b)	5.68	
		$0.45/0.107 = 4.2 \ (S_f)$
1.200	6.13	
		$0.22/0.100 = 2.2$
1.300	6.35	
1.400	6.52	

$$V_e = 0.998 + 0.095\,\frac{4.2}{4.2 + 4.7} = 1.045 \text{ ml}$$

Trial B

0.000	—	
0.608	3.90	
		$0.23/0.190 = 1.2$
0.798	4.13	
		$0.66/0.205 = 3.2 \ (S_i)$
1.003 (V_a)	4.79	
		$0.49/0.055 = 8.9 \ (S_{max})$
1.058 (V_b)	5.28	
		$0.63/0.090 = 7.0 \ (S_f)$
1.148	5.91	
		$0.34/0.104 = 3.3$
1.252	6.25	
		$0.20/0.106 = 1.9$
1.358	6.45	

$$V_e = 1.003 + 0.055\,\frac{7.0}{7.0 + 3.2} = 1.041 \text{ ml}$$

titrant is added in nearly equal increments of 0.1 ml and S_i is slightly greater than S_f. V_e is found to be 1.045 ml. In Trial B, the titrant is added in unequal increments of 0.205 ml, 0.055 ml, and 0.090 ml, respectively, in the region of interest, and S_f is much greater than S_i. V_e is found to be 1.041 ml. The example shows that the increments of titrant need not be

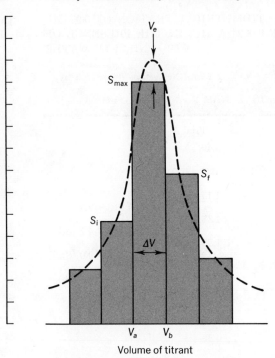

Volume of titrant

equal in the region where the slope is changing rapidly. However, for best results they should be approximately equal.

The size of $(V_b - V_a)$ has an important effect on the accuracy of the potentiometric titration. With typical emf data of millivolt precision (0.01–0.02 pH unit), the error in V_e will be about 10 percent of $(V_b - V_a)$. If we aim at 0.1 percent titration accuracy, therefore, $(V_b - V_a)$ should be about 1 percent of V_e; it will not be worthwhile to make it smaller. When the innate titration error [as estimated by means of Eq. (8.5)] becomes greater than 0.2 percent, a good "rule of thumb" is to let $(V_b - V_a)$ amount to five to ten times the innate titration error.

For the titration described in Table 10.6, the innate titration error is calculated to be 1.3 percent.* Accordingly, $(V_b - V_a)$ is 9 percent of V_e in Trial A and 5 percent of V_e in Trial B. The discrepancy of the two results is 0.4 percent, and subsequent experiments have shown that the mean result is accurate to 0.5 percent.

Although Fig. 10.12 represents most cases, occasionally one obtains a titration curve like that in Fig. 10.13, with two adjacent and nearly equal large slopes surrounded on both sides by a series of progressively smaller

---

* The effective chemical equation for this titration is $H_3O^+ + NH_3 \rightleftharpoons NH_4^+ + H_2O$. $K = 1.7 \times 10^9$. The calculation is based on Eq. (8.26).

slopes. In that case, we may calculate V_e with somewhat greater accuracy by means of Eq. (10.53), where the symbols have the significance shown in Fig. 10.13.

$$V_e = V_b + (V_c - V_a) \frac{S_2 - S_1}{4(S_2 + S_1)} \tag{10.53}$$

Note that V_e will be greater or smaller than V_b depending on whether S_2 is greater or smaller than S_1.

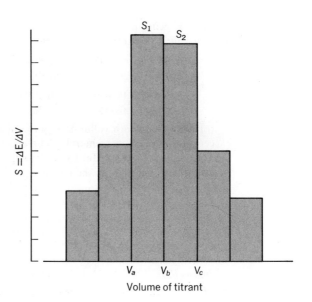

Fig. 10.13. Histogram showing a less common variation of $\Delta E / \Delta V$ near the equivalence point. In this case, V_e is calculated from Eq. (10.53).

REFERENCES

A. J. de Bethune and N. A. Swendeman Loud, *Standard Aqueous Electrode Potentials and Temperature Coefficients at 25°C*, Clifford A. Hampel, Skokie, Ill., 1964.

F. Daniels, J. W. Williams, R. A. Alberty, P. Bender, C. D. Cornwell, and J. E. Harriman, *Experimental Physical Chemistry*, 7th Ed., McGraw-Hill Book Company, New York, 1969.

R. A. Durst, Ed., *Ion-Selective Electrodes*, U.S. Government Printing Office, Catalog No. C 13.10:314, Washington, D.C., 1969.

G. W. Ewing, *Instrumental Methods of Chemical Analysis*, 3rd Ed., McGraw-Hill Book Company, New York, 1969.

J. B. Headridge, *Photometric Titrations*, Pergamon Press, Oxford, England, 1961.

E. H. Lyons, Jr., *Introduction to Electrochemistry*, D. C. Heath & Company, Boston, 1967.

H. Rossotti, *Chemical Applications of Potentiometry*, D. Van Nostrand Co., Inc., Princeton, N.J., 1969.

H. H. Willard, L. L. Merritt, Jr., and J. A. Dean, *Instrumental Methods of Analysis*, 4th Ed., D. Van Nostrand Co., Inc., Princeton, N.J., 1965.

See also the instruction manuals for particular instruments. They often contain a great deal of useful information.

PROBLEMS

1. The following results were obtained for the dielectric constant of solutions of LiCl in anhydrous acetic acid at 25°C:

[LiCl]:	0.0000	0.00104	0.00220	0.00300
$\mathcal{E}$:	6.271	6.295	6.320	6.332

[LiCl]:	0.00392	0.00470	0.00590	0.00711
$\mathcal{E}$:	6.354	6.366	6.380	6.392

(a) Draw a smooth calibration curve of dielectric constant versus molar concentration of lithium chloride. The experimental error in $\mathcal{E}$ is about ± 0.004. The plot of $\mathcal{E}$ versus [LiCl] will be slightly curved. Find the slope a_S in Eq. (10.3) from the data at the four lowest concentrations.

(b) The dielectric constant of an unknown LiCl solution was found to be 6.341 at 25°C. Find the LiCl concentration by interpolation on the smooth calibration curve and estimate the error of the result.

(c) How accurately must the temperature be controlled if the concentration is to be measured to $\pm 1 \times 10^{-5}$ M? Apply Eq. (10.6). Use a_S as obtained in (a) and calculate $d\mathcal{E}_0/dT$ from the following data: The dielectric constant of acetic acid is 6.22 at 20°C and 6.31 at 30°C.

2. What possible sources of error arise when one tries to determine solute concentration from the measurement of a single bulk solution property? How may these errors be eliminated (or minimized)? Use a specific example and outline an experiment which would be expected to yield a solute's concentration through measurement of bulk properties.

3. 5.00 ml of an unknown HCl solution was diluted with about 100 ml of pure water, and the HCl was titrated conductometrically with 0.1150 N NaOH. The following data were obtained. The scale readings are in arbitrary units but the instrument is calibrated so that a zero reading corresponds to no conductance.

Ml NaOH added	Scale reading
0	828
1.10	745
2.80	650
5.00	500
7.10	345
10.00	300
12.75	465
16.05	640
18.80	730

(a) Construct the titration curve, locate the equivalence point, and calculate the HCl concentration in the original 5.00 ml sample. (Volume changes during the titration may be neglected.)

(b) Should the conductance be close to zero at the equivalence point? Explain your answer.

4. An unknown weak organic acid (HA) is to be identified by measuring its pK_A in water. The acid is only slightly soluble in water. After a bit of experimentation it was found that pK_A can be determined spectrophotometrically by measuring the absorbance (A) at 2890 Å. The following data were obtained: Separate 15.0 ml portions of a clear saturated aqueous solution of HA were diluted to 50.0 ml with

    (i) 0.1 N HCl; A (at 2890 Å) = 1.30

    (ii) 0.1 N NaOH; A (at 2890 Å) = 0.059

    (iii) acetate buffer; A (at 2890 Å) = 0.680; the pH of the buffered solution is 4.21.

(a) From this information, determine the pK_A of the weak organic acid. You may assume that the solubility of HA in water is well below 0.1 F. Explain why it is not necessary to know c_{HA}.

(b) Identify the acid by comparison with the entries in Table A.1. How would you confirm this identification?

5. Dilute solutions of pyridine (Py) in acetic acid are partly ionized. The reaction that takes place is: $Py + HAc \rightleftharpoons PyH^+Ac^-$. The composition of the equilibrium mixture can be determined spectrophotometrically by measuring extinction coefficients (ϵ) at 2600 A. The following results were obtained at 30°C:

    For un-ionized pyridine: $\epsilon_{Py} = 1720$ $(M^{-1}cm^{-1})$

    For PyH^+Ac^-: $\epsilon_{PyH^+Ac^-} = 4040$ $(M^{-1}cm^{-1})$

    For the equilibrium mixture: $\epsilon_{formal} = 3230$ $(M^{-1}cm^{-1})$

Calculate the base ionization constant, $K_i = [PyH^+Ac^-]/[Py]$.

6. Describe a suitable electrochemical cell for each of the following potentiometric titrations in aqueous solution. In each case, specify the species being detected by the sensing electrode, whether a salt bridge is needed, and any other pertinent information. Then write an equation relating the measured emf to the concentration of substrate or titrant.

Substrate	Titrant
(a) $HClO_4$	NaOH
(b) $AgNO_3$	NaCl
(c) $Fe(CN)_6{}^{4-}$	Ce^{4+}
(d) I^-	$Na_2S_2O_3$

7. Find the solubility and pK_A of benzoic acid in water at 25°C from the following potentiometric titration data:

    Volume of clear saturated (at 25°) HBz solution taken for analysis: 50.00 ml

    Normality of NaOH: 0.0610 N

    Potentiometric procedure: The reaction was monitored by measuring the pH with a glass electrode versus a saturated calomel electrode. The pH scale of the pH meter was standardized by comparison with buffers of known pH.

V^a	pH	V	pH	V	pH
0	2.99	18.92	4.91	22.86	9.22
1.33	3.23	19.92	5.02	22.92	9.61
3.56	3.55	20.70	5.19	23.05	10.08
5.14	3.70	21.50	5.44	23.27	10.34
7.36	3.90	21.92	5.65	24.80	11.00
10.50	4.12	22.22	5.89	26.65	11.28
12.23	4.25	22.42	6.19	31.27	11.60
14.40	4.42	22.74	7.79		
17.20	4.66	22.83	8.73		

^{a} V = ml of titrant

Proceed as follows:

(a) Plot the titration data and estimate the point of maximum slope, which is the equivalence point.

(b) Calculate $\Delta p\text{H}/\Delta V$ in the region near the equivalence point. Then apply Eq. (10.52) or Eq. (10.53) to find V_e precisely. Calculate the solubility of benzoic acid in formula weights per liter of saturated solution at 25°C.

(c) Determine the pK_A of benzoic acid from the value of pH at the equivalence point. Obtain the latter by interpolation, if necessary. In this calculation, allow for dilution with titrant and use $pK_W = 14.00$.

(d) Determine pK_A of benzoic acid from the pH at the midway point of the titration.

(e) Compare the accuracy of the two different methods for finding pK_A used in (c) and (d). Which is more reliable? Why?

8. Suppose you were given a laboratory equipped with instruments and glassware of every description but no chemicals except distilled water, a bottle of reagent grade cadmium chloride, and a cadmium chloride solution of unknown concentration. Describe in detail at least three experiments that you could perform to determine the unknown $CdCl_2$ concentration. For each experiment state the criteria which must be met for a successful result. What effect would small amounts of other solutes have on each determination? How could you get reliable results even if such impurities did exist?

9. The following two sets of data were obtained in the potentiometric titration of two different acid substrates with 0.1 N NaOH in water. V denotes the volume of titrant added to the titration flask. Using Eq. (10.52) or Eq. (10.53), find V_e.

V^a	pH	V	pH
4.050	5.376	3.550	4.431
4.100	5.700	3.600	4.912
4.150	6.510	3.625	7.126
4.200	7.336	3.650	10.684
4.250	7.798	3.675	11.036

^{a} V = ml of titrant

10. The cell: AgCl-Ag(s) ¦ solution X | glass electrode was used to measure the concentration of a series of aqueous HCl solutions at 25°C.

(a) Show that the cell emf is given by

$\mathsf{E} = \mathsf{E}^* + 0.05915 \log [\text{H}^+] + 0.05915 \log [\text{Cl}^-]$.

(b) Find E^* from the following result: For 0.01273 F HCl, $\mathsf{E} = 0.2072$ V.

(c) Using the value of **E*** obtained in (b), find the HCl concentration of two HCl solutions for which **E** is 0.2252 V and 0.1911 V, respectively. Compare these concentrations with values of 0.01823 F and 0.00913 F, respectively, obtained by acid-base titration.

(d) A buffered solution of pH $\approx$ 9.2 was prepared which contained the following solutes in water: $Na_2B_4O_7 \cdot 10 \, H_2O$ (borax), 0.01038 F; NaCl, 0.01926 F. The emf measured for this buffer was -0.2095 V. Find the pH.

PART TWO

Laboratory Experiments and Procedures

11 chapter

S.O.P. (Standard Operating Procedure)

In this chapter we wish to introduce some of the apparatus and techniques which are widely employed in analytical chemistry.

Since every laboratory has its own method of operation (you are, no doubt, already familiar with yours), we will not dwell on such subjects as laboratory organization, storeroom procedures, safety regulations, etc. These aspects of standard operating procedure (S.O.P.) are, however, extremely important and you should be certain you understand how they pertain to you.

Neatness counts

To be effective as an analytical chemist, a student must discipline himself to be a meticulous worker. The work area must be kept neat and all glassware should be scrupulously cleaned *before* it is put away. If you think your work habits are not yet up to par, arrange for some reminders. For example, make up a few gummed labels saying "NEATNESS COUNTS" and place them strategically around your work area. This may seem like a fatuous idea, but it could prove very rewarding if taken seriously.

Notebook and reports

Although every instructor has his own ideas as to the keeping of a laboratory notebook, there are a few universal guidelines for keeping records and reporting results. In keeping a laboratory notebook, it is imperative that you record *everything* you do and *any* observations you make which might be of value in the interpretation of results or the tracing of errors. Thus, all raw data from weighing, pipetting, and titration should be entered as part of a permanent record rather than on scrap paper where it may be lost or miscopied before the final results are calculated. If you make a mistake or discard a reading, draw a single line through the entry in the notebook and, unless the reason is obvious, write a short note explaining the change. Never erase or otherwise obliterate an entry. You may find a need for an item that previously appeared worthless or even bogus.

Reporting of analytical results is normally done by handing a small card to the instructor. Information to be reported includes individual determinations, an average value, and the deviation from the mean stated in absolute and relative terms. For most experiments the relative error should be well under 0.5 percent, and thus all reported values should contain four significant figures. For a review of the mathematics needed for analytical calculations, see Chapter 2. Remember that a slide rule gives results with only three significant figures. For more precise calculations you should use either a desk calculator or logarithms. A table of logarithms is provided in Table A.6.

LABORATORY APPARATUS

The analytical balance

The single pan analytical balance is the most common weighing device in today's quantitative analysis laboratory. This balance, which is illustrated in Fig. 11.1, normally has a sensitivity of 0.1 milligram (mg) or 0.0001 gram. Thus, even with a built-in uncertainty of ±0.1 mg in reading the scale, weighings (actually, determinations of mass) of as little as

0.1000 g can be made with a relative error of one part per thousand or less. In the single pan balance, weights are *removed* from above the pan to keep a constant mass. In the double pan system of weighing, weights are *added* to one pan to counteract the weight of the sample on the other side.

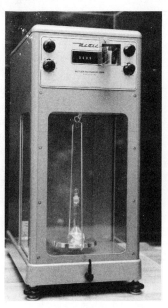

Fig. 11.1. Single pan analytical balance.

Although several different models are available, operation of all standard single pan balances is basically the same.

1. Clean the pan by brushing lightly with a camelhair brush.

2. Close the doors, set the pan release to the "full release" or "fine weigh" position, and zero the balance. This operation is not normally necessary since most weighings are done by difference. However, zeroing the balance is a good practice and also helps spot trouble.

3. Set the pan release to the "arrest" position and place the object to be weighed on the pan using tweezers or crucible tongs (not fingers!). Be sure that the sample is at room temperature. Close the doors.

4. Set the pan release to the "load" or "coarse weigh" position and adjust the weight knobs until an increase of the smallest increment (usually 0.1 g) will cause a full deflection on the illuminated scale.

5. Set the pan release to the "fine weigh" position and read the scale when it comes to rest. Do not lean on the table or otherwise disturb the balance.

6. Record the weight in your notebook and arrest the pan. Before removing the sample from the balance, set all knobs to their zero position.

7. Check the balance to make sure the pan is fully arrested, weights are

set to zero, and doors are closed. Do not leave the balance area before cleaning up all spills, removing paper, etc.

When using the balance for the first time, it is a good idea to practice operating the controls and reading the scale by weighing a cork or other small item. Remember—the analytical balance is a delicate instrument and must be treated gently.

Desiccators and drying ovens

A drying oven is a device for removing water (or other volatile substance) from a sample. A desiccator helps keep the sample dry. For purposes of driving off excess moisture from a solid, the drying oven should normally be kept at 110°C. At this temperature, most powdered or granular samples will come to constant weight (a test for dryness) within an hour or so. Samples left in the oven should be labeled and protected by placing the sample (in its container) inside a beaker and partially covering with a watch glass.

Fig. 11.2. Dessicator.

The desiccator (Fig. 11.2) is a glass, plastic, or metal container with an airtight cover containing a drying agent (called a desiccant) and a platform for supporting small beakers, crucibles, or weighing bottles containing the sample. Some common desiccants include phosphorous pentoxide, the perchlorates of barium and magnesium, concentrated sulfuric acid, silica gel [containing some cobalt(II) chloride which changes from blue to pink as the desiccant becomes exhausted], and the oxide, sulfate, and chloride of calcium. Your desiccator will most likely contain $CaCl_2$.

Weighing bottles and crucibles

These items (Fig. 11.3) are used to collect and/or hold a solid sample which is to be weighed. Thus, they must all be handled only with a strip of paper (for tall weighing bottles), tweezers, or tongs. Since most experiments are done in triplicate, you should have three each of porcelain crucibles and filter crucibles. These should be numbered (including crucible lids) since each must be weighed more than once.

Common weighing bottles

Crucible with lid

Sintered glass   Gooch   Sintered porcelain

Filter crucibles

Fig. 11.3. Weighing bottles and crucibles.

Weighing bottles. These are normally used for weighing out solid samples to be used in preparing standard solutions or as unknowns. The most convenient method of using a weighing bottle (and this takes practice) when several samples are required is to start with a bottle containing enough sample for *all* experiments, weigh it, dump the appropriate amount of sample into a beaker, reweigh the bottle, dump out a second sample, reweigh, and so forth. This procedure is valid for most analyses since the sample size (especially for unknowns) is relatively unimportant as long as you have weighed the sample carefully. By using this "method of successive dumpings," or *difference weighing*, three samples can be weighed out with only four weighings rather than the six determinations required if the samples were prepared individually. To avoid touching the weighing bottle with your hands, wrap a strip of paper around it and lift by the tab ends. For single weighings of substances which are not affected by brief exposure to air, a piece of weighing paper or a specially designed plastic cup may be used instead of a weighing bottle. Remember, however, that if you are *adding* material to the paper, cup, or weighing bottle, subsequent transfer to a beaker or volumetric flask must be quantitative.

Porcelain crucibles with covers. These are used for high-temperature experiments such as the decomposition of carbonate salts or ignition of metallic hydroxides. You have probably been introduced to the use of this item in a previous course. For analytical work, the crucible must be cleaned thoroughly (in hot nitric acid when all else fails) and dried to constant weight (at the same temperature at which it will later be used) before introduction of the sample. Heating or ignition of the sample should be carried out until the crucible again comes to constant weight. The crucible should be allowed to reach room temperature in a desiccator before each weighing, as a warm object in a balance sets up convection currents which can cause large errors.

Filter crucibles. The most common and conveniently used of these have a sintered bottom fused to the walls of the container. Construction of the crucible can be glass or, if high temperatures are necessary, porcelain. A *Gooch crucible* has holes in the bottom that must be covered with an asbestos mat before each use. For quantitative collection of a precipitate, the (weighed) filter crucible is used with a suction apparatus such as the one pictured in Fig. 11.4. The ballast bottle is used to prevent water from

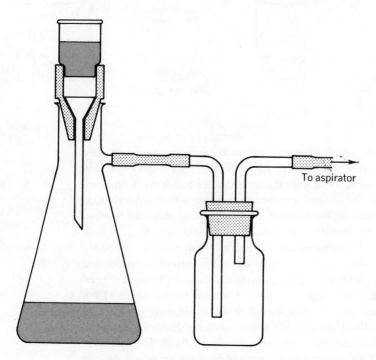

To aspirator

Fig. 11.4. Suction filtration apparatus.

the aspirator from flowing back into the filter flask when the water is turned off. After the precipitate has been collected, the crucible is dried to constant weight. Filter crucibles are not suitable for use with gelatinous precipitates (e.g., iron hydroxide) because these tend to clog the sintered disk. In these cases, it is best to use "ashless" filter paper in a glass funnel. The filter paper and precipitate are placed in a solid porcelain crucible and ignited to burn away the paper and drive off water.

Volumetric glassware: burets, pipets, and volumetric flasks

Good, calibrated, volumetric glassware (Fig. 11.5) is indispensable for most of the work of the analytical chemist. These items must be treated with the utmost respect, kept scrupulously clean, and used properly. Otherwise, analytical precision is impossible. You will probably be supplied with "Class A" (a designation of tolerance set by the National Bureau of Standards) glassware which is accurate enough to be used without further calibration. However, it is a good exercise in proper technique, as well as a check on the apparatus, for you to calibrate at least one buret and one pipet with water before the first volumetric experiment. Your instructor can supply you with procedural details.

Volumetric flasks. Volumetric flasks are available in an assortment of sizes from 1 ml to several liters capacity. The most common sizes in analytical chemistry are of 50, 100, 250 and 500 ml capacity. Volumetric flasks are calibrated "to contain" so many milliliters of solution at 20°C. They are usually marked "TC" meaning "to contain," while pipets and burets are inscribed with "TD," meaning "to deliver." To prepare a solution of known composition in a volumetric flask, weigh out the solid material (or measure the appropriate amount of liquid with a buret or pipet), transfer quantitatively to the flask containing some solvent, dissolve or otherwise mix the solute by swirling, and fill carefully to the mark with solvent. To effect quantitative addition of a solid, the solute should be added through a small funnel which (along with the weighing bottle) should be rinsed with solvent and the rinsings added to the solution. Remember that glassware and solutions both change volume with changing temperature. Therefore, if you are preparing a solution that has a significant heat of solution or dilution, the flask should be brought to the correct temperature before the last few ml of solvent are added.

The Cassia flask (Fig. 11.5c) is especially convenient to use when a standard solution of a particular concentration is desired. The most common size Cassia flask has graduations for every 0.1 ml from 100 to 110 ml. It is possible, for example, to prepare a 0.01000 F solution by weighing out anywhere between 0.01000 and 0.01100 formula weights of solute, using the amount weighed to *calculate* the needed volume, and preparing that volume of solution. This method is far superior to the more tedious (but

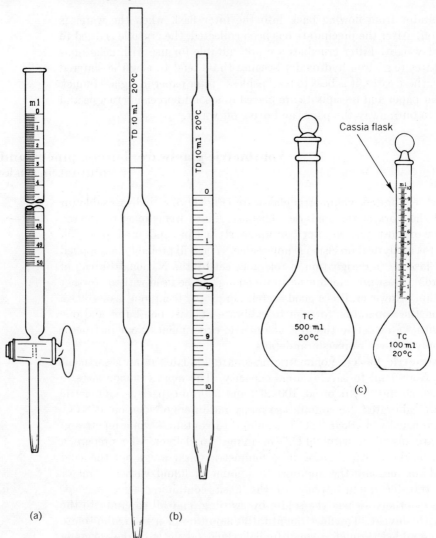

Fig. 11.5. Volumetric glassware. (a) Buret, (b) Pipets, and (c) Volumetric flasks.

common) practice of trying to weigh out exactly 0.01000 formula weight of solute and diluting to 100 ml in a standard flask. Remember, however, that it is only rarely necessary to prepare a solution which contains precisely the amount of solute indicated. As long as you *know* how much solute you have, a solution of approximately the indicated concentration is to be preferred to the risk of contamination involved in trying to weigh out an "exact" amount of the solute.

Pipets. The *volumetric* pipet (the word is a diminutive of "pipe tube") is the type most commonly employed in the analytical laboratory. It is calibrated *to deliver* a fixed volume of liquid with very high precision. Graduated or *Mohr pipets*, while convenient to use, are subject to large errors and are employed only rarely for quantitative deliveries. Pipets come in a variety of sizes from less than 1 ml to 200 ml. Since one drop of water measures about 0.05 ml, it is easy to see that incorrect use of a pipet, especially in the smaller sizes, can lead to enormous errors. To assure accurate and reproducible results, the pipet must be kept so clean that it drains completely with no droplets left behind. The amount of time that it takes for complete delivery is stamped on the outside of the pipet.

In use, the pipet is first rinsed with small portions of the solution to be delivered. The liquid is then brought up *above* the calibration mark. At this point, the tip of the pipet is wiped with a clean tissue to remove excess liquid, after which the liquid level is allowed to fall until the bottom of the meniscus (curved surface of the solution) is sitting exactly on the mark. Touch the tip to the surface of a beaker to drain off any clinging liquid and you are ready to deliver the solution. During delivery, the pipet should be held vertically, with the tip in contact with the vessel receiving the solution. It should be held this way until drained completely, then removed carefully so as not to pick up any liquid from the container. For filling the pipet, use of a rubber suction bulb is safer and (unless you can't find it) usually more convenient than sucking by mouth. Figure 11.6 demonstrates the correct use of a rubber bulb in pipetting. Note that after the pipet is filled above the mark, the bulb is *pushed* off by the thumb from below rather than *pulled* off from above. This minimizes the chance of sucking liquid up into the bulb or pushing it down below the mark. Before your first titration, it is a good idea to practice using the pipet (calibration is an excellent exercise) with water until you feel comfortable with it.

Burets. A buret is a graduated tube with a stopcock and tip at the bottom allowing variable amounts of solution (titrant) to be delivered into the reaction vessel. The buret most often used in the elementary analytical laboratory has a 50 ml capacity and is graduated in 0.1 ml increments. With practice, you should be able to estimate a reading to ±0.01 ml. In order to minimize the effects of errors in reading the buret, a titration experiment should be designed so that at least 50 percent of the buret's capacity is required for each run. The experiments in this text assume that a 50 ml buret is to be used. If you are using the 10 ml size (another common capacity), the volume of substrate added to the reaction vessel should be reduced accordingly.

As with pipets, your buret should be so clean that it drains evenly. Before use, rinse the buret with small portions of the titrant solution. Then fill to above the "zero" point and drain slowly to just below zero. Be sure that no air bubbles are trapped inside the tip.

In a titration the initial point is read on the buret, then the correct amount of titrant is added (it's not that simple, of course), a final reading is made, and the volume of titrant delivered is determined. Because the liquid along the walls of the buret needs time to "catch up" to the flow, you should wait a few seconds before taking a reading. To help in reading a buret, take advantage of the fact that the graduations extend around the cylinder at each milliliter marking. Use these etchings as guides to help eliminate parallax and try a white file card ruled with a black line behind the buret to aid in locating the meniscus to the nearest 0.01 ml.

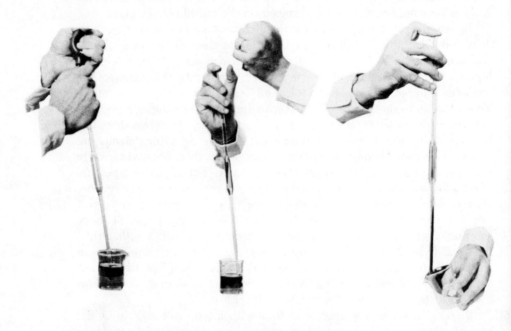

Fig. 11.6. Use of pipet bulb. (a) Squeeze bulb and place over the very end of the pipet. Use your thumb to block the bulb from going lower than $\sim\frac{1}{8}$ in. (b) Draw the liquid above the mark. Then flick the bulb off with your thumb and cover pipet end with your forefinger. (Some people feel more comfortable using the thumb on the pipet end.) Wipe any clinging liquid from the pipet tip with a lintless tissue, drain to the mark. A final touch of the tip to the inside of a glass beaker will remove any clinging drop. (c) Hold the pipet vertically and allow the liquid to drain into the receiving vessel. Keep the tip in contact with the glass for the appropriate time to allow proper drainage. Photo by Bruce Buteau.

The "correct" posture for using a buret is with one hand approaching the stopcock from behind for best control while the other hand swirls the titration flask (Fig. 11.7). An alternate method of mixing the solution is with a magnetic stirring bar. This method is preferred when a beaker is used, as in a pH titration. Since small amounts of added solvent will not affect the location of the equivalence point, the buret tip can be squirted with your wash bottle in order to add the last little bit of titrant to the vessel. In this way it is often possible to add titrant to within 0.01 ml of the actual equivalence point.

Fig. 11.7. Buret posture.

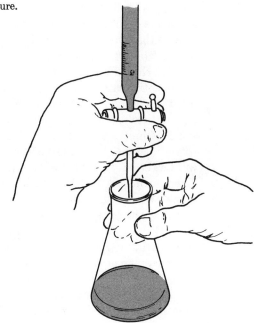

A note about cleaning glassware. Once a piece of equipment is clean, it should stay clean. A piece of apparatus that is used only for solutions might have to be scrubbed before the first use, but after it has been put in good shape, you should only have to rinse with solvent (three small portions will be enough) to keep it that way. For initial cleaning, a commercial soap, detergent, or laboratory cleaner usually suffices. Always wash the outside first! (Why?) For really stubborn marks, especially in a flask or pipet that can't be reached by a bottle brush, a bit of potassium dichromate cleaning solution will usually do the job. However, labware to which any cleaner has been added should be rinsed *thoroughly* with distilled water before use. When dichromate cleaning solution or another strong acid has been used, the final set of rinsings should be preceded by a rinsing with dilute ammonia. This assures that the acid, which tends to cling tenaciously to glass surfaces, is completely removed.

PREPARATION AND DISSOLUTION OF THE SAMPLE

The necessary steps preceding analysis are never trivial, and sometimes are more difficult than the analysis itself. Obviously one must have a working knowledge of the major components and possible contaminants, and some qualitative analysis may be necessary. It is always prudent to assume that the sample might be nonuniform. In many cases this assumption can be acted on early. If the sample is a homogeneous liquid, simple swirling should be sufficient; if it is a fine powder, slow mechanical rotation of the entire sample for several hours should assure uniformity.

If the sample consists of coarse particles, it becomes necessary to reduce the particle size by crushing or grinding, because the probability that a small sample, selected at random for analysis, is representative of the whole increases with the number of particles. If the sample is small (as will be the case for classroom unknowns), grinding with mortar and pestle is most convenient. The grinding tools must be made of a hard material, such as agate, which does not chip; porcelain ware is not recommended for samples with abrasive particles, such as mineral ores. If the sample is large—say, a day's ore production in a copper mine—obtaining a truly representative small portion for analysis would be a major task to challenge the ingenuity of the chemical engineer and statistician.

If the sample needs to be dried prior to analysis, placing it in a ventilated drying oven at 110°C for one or two hours is often sufficient. However, if the water is hard to remove, placing the sample in a vacuum oven and pumping continuously for several hours will be more effective. Heat-sensitive substances must, of course, be dried at room temperature.

After the sample has been prepared and dried, it must be brought into solution. Here again, the degree of difficulty will vary widely. Many substances will dissolve readily in water. Organic substances that are insoluble in water often dissolve in an organic solvent and may be titrated in that medium. Acidic substances may dissolve in dilute alkali; basic substances and metals may dissolve in dilute acid.

For many ores and metals, a stronger solvent action than simple acid-base reaction with, or oxidation by, hydrogen ion is required. Nitric acid is an effective solvent when stronger oxidizing action is needed. The nitrate ion is reduced largely to NO in dilute nitric acid, and largely to NO_2 in the concentrated acid. Hydrochloric acid is an effective solvent for substances that form chloride complexes. The chloride ion also acts as a mild reducing agent $(2 Cl^- \rightarrow Cl_2 + 2 e^-)$. Aqua regia, a 3:1 mixture of concentrated hydrochloric and nitric acid, acts both as an oxidizing and as a complex-forming acid. It is therefore a very potent solvent and will dissolve even the noble metals, gold and platinum.

Whenever chemical agents are needed to bring a sample into solution, it is a good rule to use the smallest amount of the mildest reagent that will do the job, not only for safety's sake, but also to minimize the introduction of substances that might interfere with the subsequent analysis. Even so, it is often necessary to remove such chemical agents—by boiling, evaporation, precipitation, or extraction—before proceeding with the actual analysis. We shall offer further suggestions on the preparation of samples for analysis as the need arises.

A FINAL NOTE BEFORE YOU PROCEED

The subsequent chapters consist mainly of "procedures" (abbreviation: P.) for a variety of experiments in quantitative analysis. To carry them out effectively, be as fully prepared as possible *before* you come into the laboratory. The chemistry involved in the procedures is normally discussed in Chapters 1–10, so look in the index to find the relevant sections and study them.* Then read the day's procedure and make a list of the required steps, paying special attention to calculations that may be required for the preparation of samples or solutions or for determining the final result.

The description of apparatus and techniques in this chapter is only a beginning. There is much more to be learned for good analytical work, but this is best taught in the laboratory, by the instructor and by practice. Operations such as the quantitative transfer of a solid or solution from one container to another, the handling of filter paper, the "splitting of drops" near the endpoints of titrations, the effective (and sparing) use of the wash bottle, evaporation of liquids without spattering, and many others, are best mastered "in the field." Do not hesitate to ask for help from your instructor. His assistance on a seemingly trivial matter can often mean the difference between a successful analysis and a failure.

* For additional information, consult the references listed at the end of Chapters 4, 6, and 10, or references suggested by your instructor.

chapter **12**

Gravimetric Analysis

Gravimetric methods of chemical analysis have been invaluable in laying the groundwork for much of our modern chemical knowledge. Elucidation of the laws of stoichiometric chemical combination, discovery of the composition of natural substances, and the determination of atomic and molecular weights were all made possible by various gravimetric procedures, often carried out with astonishing precision.

However, because many gravimetric procedures rely on intricate and time-consuming operations before the sample can be converted into a form suitable for weighing, alternate and more convenient methods of analysis are usually preferred.

Only a few of the more common gravimetric methods are presented here as experiments. These are of two types: decomposition of a solid and precipitation of an ion from solution. The practices of drying, accurate weighing, and sample handling, which are of prime importance in these experiments, are essential preludes to analytical procedures of all types.

DECOMPOSITION OF A SOLID

P.12.1 Analysis of sodium bicarbonate

The purity of a sodium bicarbonate sample (commercial baking soda, for instance) is to be determined by taking advantage of the decomposition reaction

$$2\,NaHCO_3(s) \; \rightleftharpoons \; Na_2CO_3(s) + CO_2(g) + H_2O(g) \qquad (12.1)$$

The decomposition temperature of sodium bicarbonate is 270°C (Table 1.1). Thus, it is possible to dry the sample at 110°C and then to effect the decomposition with the relatively cool (600°C) flame of a Bunsen burner.

For analysis in triplicate, put about 2 g (weighed crudely) in a weighing bottle and dry in the oven at 110°C for about an hour or until constant weight is reached. (*Note:* Remember never to weigh a hot object—let it cool!) Using the "method of successive dumpings," divide the sample into three porcelain crucibles (with lids) which have been (1) cleaned, (2) fired over a Bunsen flame for at least five minutes, (3) allowed to cool in a desiccator* and, (4) weighed. Be certain that each crucible is numbered to prevent mixups. After the crucibles plus lids plus samples have been weighed, you are ready to begin heating.

Each of the three samples should be placed on a ring support and porcelain triangle with the crucible lids slightly ajar to permit evolved gases to escape. This arrangement is shown in Fig. 12.1. If possible, arrange to do all three determinations simultaneously, but be sure you can keep an eye on them all. Begin heating gently and continue heating for about fifteen minutes. Since the decomposition occurs at a much lower temperature than the burner produces, it is wise to keep the flame at a safe distance. Do not allow the crucible to get red hot. If heating is excessive, there is a chance of spattering. Obviously, the experiment becomes worthless if any sample is lost.

After heating, let the crucibles cool and weigh them. Determine the weight loss for each sample and use the stoichiometry indicated by Eq. (12.1) to calculate the percent of $NaHCO_3$ in the original sample. What substances would have to be absent from the sample for this to be a legitimate analysis?

---

* Before putting a hot crucible into the desiccator, it should be allowed to cool for a few minutes in the air. Otherwise a partial vacuum is created inside the desiccator and the lid may be difficult to remove.

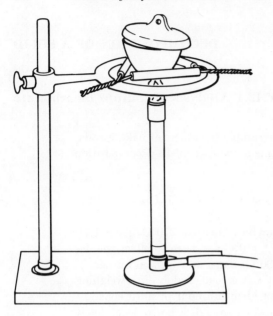

Fig. 12.1. Crucible heated on a Bunsen flame.

PRECIPITATION OF AN ION FROM SOLUTION

P.12.2 Gravimetric chloride analysis

The amount of chloride in a soluble solid is to be determined by precipitation of insoluble AgCl.

Preparation of crucibles and samples. Prepare three filter crucibles (be sure they are labeled) and dry them at 110°C to constant weight. If you are using Gooch crucibles (Fig. 11.3), be sure that the holes are completely covered with asbestos but that the mat is not so thick as to cause filtration to be sluggish. Your instructor can give you some help here. Dry the chloride sample in a weighing bottle for one to two hours at 110°C and let it cool in the desiccator.

For each of three determinations, accurately weigh out 0.4 to 0.5 g of the chloride sample into a 250 ml beaker, dissolve in water and dilute to ~150 ml. Make the solution slightly acidic by adding about 10 drops (0.5 ml) of concentrated HNO_3. Acid conditions are necessary to prevent precipitation of AgOH or of silver salts with anions such as PO_4^{3-}, CO_3^{2-}, $C_2O_4^{2-}$, etc.

Formation and digestion of the precipitate. Silver ion is to be added as a 0.5 F $AgNO_3$ solution. Before coming into the laboratory, calculate how much of this solution would be needed to precipitate all the chloride from a 0.5 g sample if your unknown were 100 percent NaCl. Now repeat the calculation for your particular sample weights. This will serve as a guide for the maximum volume of $AgNO_3$ solution required for each determination. Do not take more than this amount from the reagent bottle.*

Add the silver nitrate solution slowly and with constant stirring of the sample. By the time half of the precalculated amount of $AgNO_3$ has been added, the precipitate of silver chloride should be coagulated sufficiently to begin settling in the beaker. At this point, let the solution sit until the supernatant liquid becomes clear and test for completeness of precipitation with a few drops of silver nitrate. If AgCl still forms, add another 2–3 ml of $AgNO_3$ (while stirring), wait, and test again. Repeat this procedure until no more AgCl precipitates. Of course, you should not need more than the calculated maximum of silver solution.

The white AgCl precipitate decomposes on exposure to light [Eq. (12.2)].

$$AgCl \xrightarrow{\text{light}} Ag^\circ + Cl_2 \qquad (12.2)$$

This often results in a purple tinge to the precipitate and may cause error in the analysis if exposure to intense light is prolonged.† However, if normal care is exercised, this error will be negligible.

When precipitation is complete, heat the solution nearly to boiling and stir. Allow the precipitate to digest in this manner for about ten minutes or until complete coagulation leaves the supernatant liquid free from turbidity. Check again for completeness of precipitation with a few drops of $AgNO_3$ solution.

This is a good stopping point, so cover the beaker with a watch glass and store overnight (or until the next laboratory period) in a dark place. If you suspect the presence of HCl fumes in the air, a larger beaker, inverted over the 250 ml beaker, may be used for added protection.

Filtration, drying, and weighing. Set up the filtration apparatus shown in Fig. 11.4 with a tared (weighed) filter crucible. Carefully decant most of the supernatant liquid through the crucible. Add about 25 ml of dilute ($\sim 0.01\ F$) HNO_3 to the precipitate, stir, decant (into the crucible), and repeat with 25 ml more. Nitric acid is used as a wash-liquid because it prevents the precipitate from becoming colloidal. It is easily volatilized during the subsequent drying procedure.

Now, with the aid of a rubber policeman, transfer the remaining

---

* Once removed, no chemical should ever be returned to the laboratory reagent bottle. Any solid or solution that is left over is wasted. Thus, it becomes important in terms of cost, conservation of natural resources, and problems of waste disposal to be frugal in the use of reagents.

† After the precipitate has been washed and collected, escaping chlorine tends to make the weight of the precipitate too low. However, while Ag^+ is in excess, the reaction

$$3\ Cl_2 + 3\ H_2O + Ag^+ \longrightarrow AgCl + ClO_3^- + 6\ H^+$$

can take place, resulting in an increase in the weight of the precipitate.

precipitate to the filter crucible, being careful not to lose any solid. Rinse the last bit of precipitate from the beaker into the crucible with small portions of the HNO₃ wash solution. Wash the precipitate three more times (about 5 ml each time) or until the last traces of silver ion are gone. You may test for completeness of washing by putting a test tube containing a few drops of dilute NaCl or HCl solution into the filter flask (suspend the tube from a wire for easy handling) to collect liquid from just the last washing.

Place the filter crucible in a small beaker, cover with a fluted watch glass,* and dry in the oven at 110°C (150–160°C may be used for quicker drying) until constant weight is reached.

Calculate the percent chloride in the original sample for each determination. The results should agree to within a few tenths of one percent.

PRECIPITATION OF BARIUM SULFATE

The two procedures which follow both involve analysis by precipitation of barium sulfate and are quite similar.

$$\text{Ba}^{2+} + \text{SO}_4^{2-} \;\rightleftharpoons\; \text{BaSO}_4(\text{s}) \qquad (12.3)$$

The analysis of barium is less common than the sulfate analysis, but since it is somewhat easier to perform and less subject to error, we offer it first. Errors may arise from the tendency of barium sulfate to precipitate from solution in the form of finely divided crystals that cannot be filtered easily, from the relative stability of its supersaturated solutions, from the coprecipitation of barium hydroxide unless the solution is acid, and from possible reaction with carbon from the filter paper during ignition.† As a result, it is especially important to practice proper technique and pay close attention to details of procedure.

P.12.3 Analysis of barium as BaSO₄

Your sample will probably contain a hydrated barium salt, so do not dry it unless your instructor advises otherwise.

Preparation. Prepare three porcelain crucibles (with lids) as in P.12.1. Since the product, BaSO₄, is to be ignited over a flame, glass filter crucibles cannot be used. If you have sintered porcelain or Gooch filter crucibles,

---

* If you don't have a fluted watch glass, three bent pieces of glass rod or metal placed on the beaker lip will raise a regular watch glass enough to allow vapors to escape.

† See I. M. Kolthoff and E. B. Sandell, *Textbook of Quantitative Inorganic Chemical Analysis*, 3rd Ed., pp. 325 ff. (New York: The Macmillan Company, 1952), for a detailed discussion of these errors.

put them inside the ordinary crucibles and heat each one to constant weight over the full heat of a Bunsen flame. Weigh each pair of crucibles together with the cover on the larger one.

If filter crucibles are not available, the precipitate of $BaSO_4$ may be collected on a fine-porosity "ashless" filter paper (such as Whatman No. 42), which in turn sits on a glass filter funnel. After the entire precipitate has been collected, the filter paper plus precipitate is transferred to a sintered porcelain crucible, and the crucible with its contents is ignited over the full heat of a Bunsen flame. Under these conditions the "ashless" filter paper burns off, leaving practically no solid residue, and the precipitate of $BaSO_4$ becomes thoroughly dry.

Accurately weigh out three portions of about 0.5 g of the unknown sample. Dissolve each portion in 75–100 ml of water to which one or two drops of 6 N HCl has been added. (If any solid residue remains, it should be filtered off before you proceed. After filtration, wash the filter paper with three 10 ml portions of water containing a little HCl, and add the washings to the solution of the sample.) Place these solutions in a safe place. Later, they will be added quantitatively to acid solutions of sulfate.

To prepare sulfate solutions for the precipitation, place about 300 ml of water and 1.0 ml of concentrated (18 F) H_2SO_4 into a 600 ml beaker (one for each analysis), bring almost to boiling, and keep the solution hot. It is best to bring the solution up to temperature over a flame and then transfer it to the steam bath or hot plate.

Formation and digestion of the precipitate. Very slowly and with stirring (this is important!) add the barium solution to the large beaker containing the acidic sulfate solution. Use a stirring rod to direct the liquid, being careful not to lose any. Wash the small beaker several times with water from your wash bottle to make the transfer quantitative, then wash all precipitate down the walls of the large beaker and off the stirring rod. The entire precipitate of barium sulfate is now in the large beaker.

$BaSO_4$ precipitates in a fine enough form that it would pass through filter paper if filtration were attempted at this point. In addition, coprecipitation of other salts is lessened if the precipitate is allowed to digest. Therefore, cover the beaker with a watch glass and keep the solution hot for at least an hour. The precipitate will become more coarse-grained. After digesting, the sample may be allowed to cool if the analysis must be interrupted, but it should be reheated before filtering. Add water to replace any water lost by evaporation.

Filtration and washing. Although the solution above the precipitate is to be discarded, it is prudent to use a *clean* filter flask or beaker for collecting the filtrate. If any solid passes through the filter, that portion of the solution must be filtered again.*

---

* Mixing the solution with a bit of ashless filter paper pulp helps when some precipitate refuses to stay in the filter.

Therefore, begin filtering by decanting the hot, clear supernatant liquid into the filter and discard the (clear) filtrate. Then wash the precipitate in the beaker twice with small portions of a (cold) solution containing *one drop* of concentrated H_2SO_4 in 500 ml of water and transfer the $BaSO_4$ quantitatively to the filter. Wash several times more with the H_2SO_4 solution until the filtrate gives no AgCl precipitate when tested with a few drops of silver nitrate solution.* Use a small test tube as in P.12.2 to collect filtrate for the test.

Drying by ignition and weighing. When the precipitate has been collected and washed, it is to be ignited over a flame to constant weight. If a (tared) filter crucible was used, place it in a larger, ordinary crucible with the lid slightly ajar (Fig. 12.1), gradually bring the heat up until the crucible glows red, and ignite for 15–20 minutes. If ashless filter paper was used, fold over the top of the paper to make a neat package and place it inside a tared crucible with the lid ajar. Lay the crucible at a 45° angle on the porcelain triangle (this improves air circulation) and heat rather gently at first to burn off the filter paper. The paper will first char. If it starts to burn, use the lid of the crucible to put the fire out. If heated too strongly, the carbon in the filter paper can reduce barium sulfate to the sulfide.† Thus, it is important to control the temperature so that the carbon just barely glows red and is oxidized slowly. When all the carbon is gone,‡ set the crucible in the upright position and heat to dull red for another 10–15 minutes. If any black deposit remains on the crucible lid, ignite the lid inside up to remove it. Cool the crucible in the desiccator and weigh to determine the weight of the $BaSO_4$ precipitate. It is advisable, to insure constant weight, that a second ignition be performed.

Calculate the percentage of barium in your unknown.

P.12.4 Analysis of sulfate as BaSO₄

This procedure is similar to P.12.3. However, there are significant differences in the preparation of the sample and the formation of the precipitate.

---

* Laboratory unknowns usually contain chloride salts. The instructor will tell you if this is not true of your sample.

† The reaction is

$$BaSO_4 + 4\,C \xrightarrow{\text{heat}} BaS + 4\,CO$$

If the reduction is only slight, subsequent heating in air will regenerate the sulfate

$$BaS + 2\,O_2 \xrightarrow{\text{heat}} BaSO_4$$

and no appreciable error results. The more extensive the reduction, however, the longer it will take for the crucible to reach constant weight.

‡ "Ashless" filter paper has been specially washed with HCl and HF during manufacture to remove all nonvolatile substances. The total solids remaining after ignition weight less than 0.1 mg and may be ignored.

Prepare crucibles and filtering implements as in P.12.3.

Accurately weigh out three samples of between 0.5 and 0.8 g into 400 ml beakers. Heat each sample with about 20 ml of water to dissolve it. If the sample does not dissolve at first, add concentrated HCl dropwise until it does. Acidify the solution (whether or not you have already added acid) with ten drops (~0.5 ml) of concentrated HCl and dilute to about 250 ml with water. Heat to just below boiling as in P.12.3.

To precipitate the sulfate, use a solution of approximately 0.25 F barium chloride. Assuming that your sample is 100 percent Na_2SO_4, calculate the maximum volume of barium chloride solution equivalent to each sulfate sample. While stirring, add barium chloride *very slowly* to the hot solution. Deliver the $BaCl_2$ solution from a buret clamped at an angle with the tip just touching the wall of the beaker. After about two-thirds of the maximum $BaCl_2$ volume has been added, allow the precipitate to settle and test the supernatant liquid with a drop or two of $BaCl_2$ to see if precipitation is complete. If not, add a few more ml of solution and test again until no $BaSO_4$ precipitates.

Allow the precipitate to digest for an hour or more as described in P.12.3. The formation of a granular precipitate is of more importance in this analysis than in the previous one, because the danger that barium salts of other anions will coprecipitate is greater. Test for completeness of precipitation once more before filtering.

The procedure for collecting and igniting the precipitate is identical to that given in P.12.3 except that the wash liquid is hot water delivered in a stream from a wash bottle. Do not add any sulfuric acid! Once the precipitate has been transferred to the filter, it will probably require as many as ten washings before the filtrate gives a negative AgCl test.

When all samples have been ignited to constant weight, calculate the percent of sulfate in the original unknown.

chapter 13

Acid-Base Titrations

Acid-base titrations are probably the most frequently performed of all analytical procedures. Sample preparation is normally quite simple and "neutral" solutes rarely interfere.

In this chapter we shall limit ourselves to acid-base titrations employing color indicators for endpoint detection; potentiometric methods will be described in Chapter 18. We shall consider:

1. The choice of a color indicator, and the determination of the "endpoint blank."

2. The preparation and standardization of strong-acid titrant and strong-base titrant in water.

3. The analysis of a variety of unknowns by acid-base titration in water.

4. Acid-base titration in the nonaqueous solvent methanol.

Since this is the first chapter dealing with the practice of titration, the procedures are perhaps a little more detailed than they need be. Moreover, in most cases there are valid alternatives to the procedures—indeed, your instructor may suggest some. For example, in P.13.3 the sodium hydroxide solution is standardized using potassium acid phthalate as the standard acid and phenolphthalein as the endpoint indicator. However, it is clear from Chapters 4 and 9 that there are other acidimetric standards and other color indicators that will work equally well.

CHOICE AND TEST OF INDICATOR

In Chapters 4 and 9 we have described the desired features of color endpoint indicators for acid-base titrations: The endpoint signal must occur very near the equivalence point, and it must be readily detected by the analyst. Consequently, the indicator must be chosen so that its transition range includes the pH at the equivalence point, or at least falls within the "break" on the pH titration curve near the equivalence point.

While it is often possible to select an indicator (from a list such as Table 9.2) on the basis of transition range alone, the criterion of an *easily detected* endpoint signal is a subjective matter and should be tested. Some indicators undergo more subtle color changes than others. In some cases, a third color which occurs just before the endpoint causes confusion. Many people have difficulty seeing certain color changes and, of course, those with severe color blindness will find the choice of indicators greatly reduced. In addition, there are cases where particular indicators cannot be used, perhaps because they are unstable or not sufficiently soluble in the titration medium. It is advisable, therefore, that the selection (and testing) of an indicator be a part of each new experimental situation.

The selection of an endpoint indicator should take place under conditions identical to those at the equivalence point in a titration. For example, if you are looking for an indicator for the titration of $\sim 0.1\ N$ HCl with $\sim 0.1\ N$ NaOH, start out with a solution of $\sim 0.05\ N$ NaCl. Refer to Table 9.2 or a more comprehensive reference and select some possible indicators on the basis of the proper transition range and (of course) availability. Indicators will probably be supplied to you dissolved in an appropriate solvent, ready for use. If you wish to make up your own indicator solution, a common concentration is 0.5–1.0 g of solid indicator per liter of solution, and a common solvent is ethyl alcohol (ethanol) or an alcohol-water mixture.

P.13.1 Test of endpoint indicator

Add a few drops of the chosen indicator solution to your reaction flask. If the color corresponds to the acid form of the indicator, add one drop of base (e.g., NaOH for the titration of HCl with NaOH). You may use your buret, a small pipet, or even an eye dropper. Note the color. Now add a drop or two of your acid solution (HCl in the present example). A suitable indicator will show a sharp color change each time the solution is made to contain an excess of one drop *or less* of acid or base. Try a variety of indicators, if necessary, until you find one that is acceptable to you.

For convenience, we will suggest indicators for most of the titrations that follow. However, we cannot guarantee that they will work for you, nor is it necessary that you accept our suggestion if you find a suitable indicator that you like better.

P.13.2 Measurement of indicator blank

Sometimes it is necessary to work with an endpoint signal that does not coincide precisely with the equivalence point. In that case, an empirical endpoint correction, called the *indicator blank*, is determined as follows.

Make up a solution that simulates your titration mixture at the equivalence point, either from substrate and titrant or from the pure reaction products. Add the indicator and note the color. Then determine how much titrant must be added past the equivalence point to give a recognizable endpoint signal. This amount is the indicator blank and should be subtracted from the titration results. If the endpoint signal occurs prematurely, before the equivalence point, determine the indicator blank by a similar procedure but add substrate instead of titrant.

In acid-base titration it is usually possible to find an indicator whose transition range includes the equivalence point. The indicator blank then is quite negligible, except when the substrate is so dilute ($< 0.001\ F$) that the amount of titrant required to produce the color change of the indicator becomes significant.

Endpoint color reference

Usually, the titration endpoint is taken as the point of sharpest color *change*. However, when the innate titration error becomes greater than about 1 percent, the color change near the equivalence point becomes quite gradual. To help you recognize the equivalence point, you may wish to use an *endpoint color reference*, that is, a solution whose composition is identical to that at the equivalence point and to which the requisite amount of indicator has been added. By keeping the color reference beside the titration flask, it becomes possible to compare the colors of the two solutions and to take the endpoint when the colors match. This procedure is especially helpful with indicators, such as bromothymol blue, that go through a series of color changes shortly before the equivalence point.

STANDARD SOLUTIONS OF ACID AND BASE

Because of the importance of acid-base titrations in chemical analysis, most laboratories find it convenient to maintain a steady supply of well-standardized tenth normal HCl and NaOH. If stored properly, such solutions are stable enough to maintain their normality to within 0.1 percent for several months, and they can be prepared easily from readily available reagent-grade chemicals. In the following, we shall describe some typical procedures for preparation and standardization.

P.13.3 Preparation of 0.1 *N* sodium hydroxide

Sodium hydroxide pellets normally contain several percent of water and about one percent of sodium carbonate. To prepare a solution of general utility* it is necessary to remove the carbonate from the base and keep CO_2 out during storage.

To make up 1 liter of carbonate-free 0.1 *N* NaOH solution,† dissolve ~10 g (rough weighing) of sodium hydroxide pellets in 10 ml of distilled water. (Careful! The final solution will be about 19 *F* in NaOH!) Sodium carbonate, which is insoluble in concentrated NaOH. will precipitate out.‡ Allow the precipitate to settle (this may take time), then draw off about 5.2–5.6 ml with a graduated (Mohr) pipet and deliver into a polyethylene bottle§ containing about 1 liter of cool, recently boiled (to expel CO_2) water. Stopper tightly and mix well. The resulting solution should have a normality somewhat greater than 0.100 *N*. Thorough mixing is essential

---

* Since Na_2CO_3 is itself a base, its presence is not always harmful. For example, a solution of NaOH containing carbonate can be assigned two normalities, one for titration of strong acids to the methyl orange endpoint, the other for titration of strong or weak acids to the phenolphthalein endpoint. The concentration of carbonate is equal to the difference of the two normalities. (Why?) If the carbonate accounts for no more than 1–2 per cent of the total base normality, the loss in titration accuracy owing to the presence of carbonate is usually tolerable.

† This procedure may be modified simply by scaling up or scaling down if you wish to prepare a different volume or concentration of solution than is specified.

‡ Because of the higher solubility of K_2CO_3 in KOH, this method cannot be used to prepare carbonate-free potassium hydroxide solutions. In this case, the carbonate is precipitated as $BaCO_3$ by the addition of a slight excess of $BaCl_2$ solution to a 0.1 *N* KOH solution followed by filtration.

§ A polyethylene bottle is preferred to a glass bottle because NaOH slowly attacks glass, and as a result the base normality may change. If a glass bottle must be used, cap it with a rubber stopper. A ground-glass stopper would soon "freeze."

owing to the high viscosity of the concentrated NaOH. Be sure to clean the pipet used for the transfer as soon as possible.

For convenience of handling, and to minimize contamination by atmospheric CO_2, carbonate-free base is sometimes stored in a bottle equipped with a siphon and drying-tube assembly, as shown in Fig. 13.1.

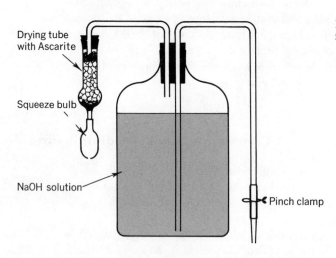

Drying tube
with Ascarite

Squeeze bulb

NaOH solution

Pinch clamp

Fig. 13.1. Storage assembly for carbonate-free base.

Because solutions of strong base pick up CO_2 so readily, titrations involving this reagent should always be done using NaOH as the *titrant*. After filling the buret, cover it with a small beaker for added protection.

P.13.4 Standardization of NaOH solution

The NaOH solution prepared in P.13.3 will be standardized using potassium acid phthalate (KHP) as primary standard. The solid acid should be dried for 1–2 hours at 110°C. Longer drying can cause decomposition if temperature regulation is not just right.

The equivalent weight of KHP is 204.22. Hence, 1.021 g of KHP is equivalent to 50.00 ml of 0.1000 N base. By the "method of successive dumping" (difference weighing) accurately weigh out 0.9–1.0 g of KHP into each of three 250 ml Erlenmeyer flasks. Dissolve in about 50 ml of CO_2-free water,* add two drops of phenolphthalein indicator, and titrate

---

* Distilled water, if stored with protection from atmospheric CO_2, need not be treated further. If necessary, boil the laboratory distilled water, or purge it of CO_2 by bubbling pure nitrogen gas through it for a brief spell. Your instructor will tell you if such treatment is necessary.

until the pink color of phenolphthalein persists for at least fifteen seconds. The pink color may fade eventually as CO_2 (a weak acid!) is absorbed from the air.

P.13.5 Preparation of 0.1 N hydrochloric acid

Concentrated reagent grade hydrochloric acid has a density of $\sim$1.19 g up per cc and is about 38 percent (by weight) HCl. Accordingly, the HCl normality is approximately 12.4 N.

To prepare 1 liter of 0.1 N HCl, deliver about 8.1 ml of the concentrated acid into 500 ml of CO_2-free water, mix well, dilute to 1 liter (the quantities need not be precise since the solution will be standardized), and mix again. Store the solution in a glass-stoppered one liter bottle.

P.13.6 Standardization of HCl by comparison with standard NaOH

The previously standardized NaOH solution (P.13.4) will be used to determine the normality of the hydrochloric acid solution prepared in P.13.5. If the solutions were prepared as directed above, the (known) NaOH normality will be somewhat greater than the (as yet unknown) HCl normality. Thus, either 50.00 ml or 25.00 ml of HCl (delivered with a pipet) can be titrated without the need for refilling the 50 ml buret with NaOH during the titration.

Pipet 25.00 or 50.00 ml of the HCl solution into each of three 250 ml Erlenmeyer flasks and add two drops of indicator to each. Use the same indicator as in the standardization of the NaOH solution. Titrate with standard NaOH and calculate the normality of the HCl solution.

P.13.7 Standardization of HCl with sodium carbonate

Sodium carbonate (equivalent weight 53.00) can be obtained in a highly pure state that is quite suitable for use as a primary standard. However, the innate titration error for the transfer of the second proton,

$$H_3O^+ + HCO_3^- \rightleftharpoons H_2CO_3 + H_2O$$

is larger than we can tolerate for standardization with four-place accuracy.*

---

* Using Eq. (8.9) with $c_S = 0.025$ M, we calculate the innate titration error as

$$\% \text{ error} = \frac{180}{(0.025 \times 2.2 \times 10^6)^{1/2}} = 0.8\%$$

Fortunately, the situation can be remedied by titrating *almost* to the endpoint, boiling to remove CO_2 (leaving essentially a NaCl solution), and completing the titration. This way we get a standardization of the desired accuracy.

Dry the Na_2CO_3 for one hour at 140°C and cool it in the desiccator. Calculate the weight of Na_2CO_3 equivalent to 30 ml of your acid and weigh out three or more samples of approximately this weight into 250 ml Erlenmeyer flasks. Dissolve in ~50 ml of water and add two drops of indicator.

The most suitable indicators for this titration have been found to be either *modified methyl orange*, or *bromocresol green-methyl red mixed indicator*. These indicators take advantage of the fact that two complementary colors cancel each other to give a resultant gray, thus sharpening the endpoint. Modified methyl orange is a mixture of methyl orange and the blue dye xylene cyanole FF and gives a sharp base-acid transition from yellow to gray. The bromocresol green-methyl red mixed indicator gives two color changes: (1) from blue-green to gray at the bromocresol green transition, and (2) from gray to red at the methyl red transition. The former gives warning that the endpoint is near; the latter *is* the endpoint.

Titrate until the indicator appears just ready to change color (the *first* perceptible change for modified methyl orange or the *slightest* pink color for the bromocresol green-methyl red mixture). Interrupt the titration. Be sure the buret is not leaking; then wash the tip, and collect the washings in your flask. Boil the solution for 1–2 minutes to expel dissolved CO_2. Be careful not to lose any solution by spattering. Then cool the titration flask in tap water.

If you have not overrun the endpoint, the indicator will return to the basic form as CO_2 escapes. Continue the titration to the appropriate endpoint color.

If an indicator blank is to be determined, note that the equivalence point in this titration corresponds to ~0.05 F NaCl, since NaCl remains when CO_2 is expelled.

SOME APPLICATIONS OF AQUEOUS ACID-BASE ANALYSIS

With standard solutions of strong acid and strong base at hand, one has the capability of solving a myriad of problems that often confront the analytical chemist. We present here only a few examples of analysis employing acid-base titration.

If you have prepared standard solutions of both HCl and NaOH, you might want to titrate with *two* burets, one containing standard acid and the other standard base. This way, you can *back-titrate* if by accident you overstep the endpoint or simply to be sure you have not gone too far. In your calculations, be sure to subtract the amount of back titrant (in milliequivalents) from the number of milliequivalents of titrant used in the experiment.

P.13.8 Analysis of an unknown sample of potassium acid phthalate

You will be given a solid containing potassium acid phthalate (KHP) mixed with a neutral salt and asked to determine the percent KHP in the sample. Assuming the sample to be 100 percent KHP (equiv. wt. 204.22), calculate the weight that would be equivalent to 50 ml of your standard NaOH solution. Weigh out *one* sample of this weight and titrate as in P.13.3 to a phenolphthalein endpoint. Now, based on the actual amount of base required, recalculate the weight of sample which would need 30 ml of base for neutralization.

Weigh out three samples of this newly calculated weight (two if you can use the first result) and titrate as before. Report your results as percent KHP (formula weight = equivalent weight) in the original unknown.

P.13.9 Equivalent weight of an acid

You are to determine the equivalent weight of a weak acid which will be given to you as a solid of high purity.

As in P.13.8 you must first determine the amount of sample to use so that 25–40 ml of standard base will be needed for each titration. As a first guess, assume an equivalent weight of 50. This will most likely be too low so that less than 40 ml of base will be needed for the first, rough titration. Phenolphthalein is usually a good indicator for titration of a weak acid with NaOH since the equivalence point will come at $pH > 7$.

Now, based on the actual amount of base required in the first titration, find the appropriate sample weight for the experiment and then perform the analysis in triplicate. Report your result as the *equivalent weight* of the weak acid. See if you can guess the identity of the acid by comparison with the substances in Table A.1. Confirm your guess by taking a melting point.

P.13.10 Determination of replaceable hydrogen

This analysis is virtually the same as P.13.9 except that the unknown is a *mixture* of one or more acids with inert material. Follow the instructions of P.13.9 to determine the number of grams of *replaceable* (acidic) hydrogen (at. wt. 1.008) per gram of sample. Report this as the percent of replaceable hydrogen in the unknown.

P.13.11 Analysis of acetic acid in vinegar

Vinegar normally contains about 5 percent acetic acid, the amount varying with the particular batch and, of course, the type of vinegar. The total acid content (there are often small amounts of other acidic substances) is easily found by titration with strong base. As a classroom exercise, it is interesting for different students to analyze a variety of vinegars—white, wine, cider, tarragon (which should be filtered), etc.—and compare results.

The following procedure should be performed twice to minimize errors in the preparation of the stock solution.

Prepare a stock solution by pipetting 25 ml of vinegar into a 250 ml volumetric flask and diluting to the mark with water. Pipet two or three 50.00 ml samples of this stock solution into 250 ml Erlenmeyer flasks. Dilute further, if necessary, to decrease the color so that the indicator (phenolphthalein or thymol blue) endpoint signal will not be obscured by the sample itself.

Titrate and report your results as formal concentration of acetic acid in the *undiluted* vinegar.

To find the percent of HAc by weight, we recall that

$$\% \text{ of HAc} = (\text{wt. of HAc per 100 g of solution})$$

$$= (\text{wt. of HAc per kg of solution})/10$$

$$= (\text{wt. of HAc per liter of solution})(\text{liters per kg})/10$$

$$= (\text{wt. of HAc per liter of solution})/(\text{density}) \cdot 10$$

$$(\text{wt. of HAc per liter}) = (\text{formal concentration}) \cdot (\text{formula weight})$$

The density of the undiluted vinegar, whose value is required for this calculation, may be measured as follows.

With a pipet or buret, carefully dispense a convenient amount of the undiluted vinegar into each of three tared weighing bottles. Reweigh each bottle (with the tops on, of course) and, from the resultant weight of vinegar, calculate its density. Using this information, along with the previously determined concentration of HAc and its formula weight, find the percent by weight of acetic acid in the vinegar.

P.13.12 Sulfuric acid in battery fluid

The "state of charge" of a lead-acid storage battery (e.g., automobile battery) is usually determined by measuring the specific gravity of the electrolyte with a device known as a *hydrometer*. Since the "active ingredient" in battery fluid is H_2SO_4,* this method is really an implicit measure of sulfuric acid concentration. In this exercise you will directly analyze for the acid by titration with standard base.

Obtain a 15 ml sample of battery acid either from your instructor or by *carefully* removing a few milliliters from each cell of a car battery. (The acid taken from the battery should subsequently be replaced with the same volume of H_2SO_4 solution of specific gravity 1.28.) Table 13.1 gives

COMPOSITION OF AUTOMOBILE BATTERY ACID[a]   **Table 13.1**

Percent charge	Density (g/ml)	Percent H_2SO_4[b]	Normality[b]
100	1.280	37.4	9.79
75	1.250	33.8	8.63
50	1.180	25.2	6.08
25	1.130	18.8	3.84
0	1.080	12.0	2.65

[a] These figures apply to U.S. automobile batteries. General purpose and foreign car batteries often have different specifications.
[b] *Acid* normality assuming H_2SO_4 is the only solute. Note that the overall battery reaction (see footnote in text) involves the sulfate moiety as well as protons.

concentration data for battery acid as a function of percent charge. Thus if your sample comes from a fully charged battery, dilution of 10 ml of sample to one liter with distilled water will produce a solution that is close to tenth normal in acid. This solution will be titrated with the NaOH solution standardized in P.13.4.

Prepare a stock solution by delivering 10 ml of battery acid from a pipet or buret into a 1000 ml volumetric flask about half full of water. Allow plenty of time for drainage of the viscous liquid. Mix well, dilute and mix again, finally bringing the solution up to the mark. This method of preparation is not the most accurate one. Your instructor will give an alternate procedure if better than 1 percent accuracy is desired. (*Precision* should still be within 2–4 parts per thousand for this method.)

Titrate 50.00 ml portions of the stock solution with standard NaOH to a phenolphthalein endpoint. Calculate the normality of H_2SO_4 in the original *undiluted* battery fluid and compare with the figures in Table 13.1. As a further check, measure the specific gravity of the fluid in the battery (this should be done *before* you replace the acid removed for analysis). A more complete table of density as a function of percent composition can be found in the Handbook of Chemistry and Physics.

* The overall reaction in the lead-acid storage battery is

$$Pb + PbO_2 + 2\,H_2SO_4 \underset{charge}{\overset{discharge}{\rightleftharpoons}} 2\,PbSO_4 + 2\,H_2O$$

P.13.13 Determination of sodium carbonate in soda ash

You will be given a solid sample of soda ash (impure sodium carbonate) to be analyzed for Na_2CO_3 by titration with standard HCl. The analysis may be carried out according to the instructions of P.13.7. A preliminary experiment will serve to determine the weight of sample required for the actual analysis. Assume ~ 50 percent purity for this rough titration.

If you have access to two burets and have standard NaOH on hand, the titration can be modified as follows. Titrate with acid until the endpoint signal begins to appear. Then add about 1 ml of acid in excess. Boil and cool as in P.13.7 and back-titrate with standard base. Since a sharp endpoint will result, ordinary methyl orange may be used as indicator. No indicator-blank correction should be made. (Why?)

P.13.14 Analysis of a solution containing NaOH and Na₂CO₃

You will be given a solution with a total base strength of approximately 0.1 N. To determine both NaOH and Na_2CO_3, you need to do two determinations. We suggest the following: (1) Analyze one 50.00 ml portion for *total* base by titration with standard 0.1 N HCl to the methyl orange endpoint, according to either P.13.7 or P.13.13. In this titration, NaOH reacts as a monoacid base, and Na_2CO_3 reacts as a diacid base (two equivalents per formula weight). (2) To a second 50.00 ml portion, add an excess of 10 percent $BaCl_2$ or $SrCl_2$ solution. Without filtering the resulting precipitate, titrate with standard HCl to a phenolphthalein endpoint, taking care to guard against absorption of carbon dioxide from the atmosphere. If possible, allow nitrogen gas to pass slowly over the solution throughout the titration. Take the endpoint when the phenolphthalein color changes to very light pink. The result of this titration measures the NaOH concentration only, since $BaCO_3$ or $SrCO_3$ remains quite insoluble at the pH of the phenolphthalein endpoint.

Repeat the analysis and report the concentrations of NaOH and Na_2CO_3, in the unknown solution.

NONAQUEOUS TITRATIONS WITH METHANOL AS SOLVENT

Acid-base titrations in methanol have been discussed in Chapter 6. The strongest acidity that can be produced by a given acid normality in methanol is that of the methyloxonium ion $CH_3OH_2^+$, which is a slightly stronger acid than H_3O^+. The strongest basicity is that of the methoxide ion CH_3O^-, which is a slightly weaker base than OH^-. If we use methanol as our solvent, we may make use of its lower dielectric constant (compared to water) to gain increased accuracy in the titration of cation acids (such as NH_4^+) with methoxide ion, of anion bases (such as Ac^-) with methyloxonium ion, and of dicarboxylic acids with methoxide ion to the first equivalence point.

Methanol also offers other advantages for titration: The commercial reagent is quite free from acidic and basic impurities and relatively inexpensive. It has a conveniently low viscosity and drains well, even in a microburet. The thermal coefficient of expansion is not excessive (0.09 percent per °C at 25°C, compared to 0.02 percent per °C for water at 25°C), although for best accuracy the temperature of the reagents should be monitored. Avoid grasping glassware tightly with your hands since this might warm the solutions. The error can be appreciable. Minor disadvantages are a relatively high volatility and some toxicity, so that pipetting should not be done by sucking with your mouth. (Methanol is the "wood alcohol" sometimes present in illegal "moonshine" whiskey and can cause blindness or even death when consumed in large quantities.) Always use a pipet bulb!

The selection of an indicator for nonaqueous titrations is usually more of a task than selection of one for aqueous titrations because the analyst, more often than not, has little experience with the particular solvent. In addition, he may not have access to pK data for substrate, titrant, or indicators. Thus, the indicator tests outlined in P.13.1 and P.13.2 become of critical importance.

Fortunately for us, the approximate pK values for several indicators in methanol are known and some are listed in Table 9.3. The "neutrality" point for methanol occurs at $[CH_3OH_2^+] = [CH_3O^-] \sim 4 \times 10^{-9}\ M$ ($K_{IP} = 1.2 \times 10^{-17}$). Since we will be titrating weak acids with methoxide ion, we need consider only indicators with basic transition ranges, that is, indicators with $pK > 10$. There are a number of indicators which satisfy this criterion listed in Table 9.3.

P.13.15 Standardization of 0.1 N sodium methoxide solution

You will be provided with a $\sim 0.1\ N$ solution of $NaOCH_3$ which has been prepared by the direct reaction of methanol with sodium metal.

Benzoic acid (HBz, equiv. wt. 122.1) makes an excellent primary acidimetric standard. Prepare an 0.08 F standard solution of benzoic acid by dissolving the appropriate weight (known exactly) in methanol in a 250 ml volumetric flask. Swirl the flask until the acid is completely dissolved. Solution of the acid will cause a decrease in temperature which later results in an *increase* in volume as the solution warms up to ambient temperature again. Therefore, fill the flask to 3–4 ml *below* the mark, mix, and let it stand for about 10–15 minutes. Then bring the solution up to volume and mix again. Alternatively, if you have a 250 ml Cassia flask, dilute initially to one of the lower graduations on the flask, wait 10–15 minutes for equilibration and read the final volume. Calculate the concentration of your standard solution. Well-stoppered solutions of benzoic acid in methanol retain their concentration at 25°C to ±0.2 percent for one or two weeks. However, on prolonged standing at room temperature or on heating, the concentration may decrease because benzoic acid tends to react with methanol.

Using a 0.05 *F* solution of sodium benzoate in methanol (as well as the HBz and $NaOCH_3$ solution), select an endpoint indicator for the titration of HBz with $NaOCH_3$ according to P.13.1.

Pipet three 50.00 ml samples of standard HBz solution into Erlenmeyer flasks, add indicator, and titrate with sodium methoxide. Calculate and report the normality of the methoxide solution.

P.13.16 Titration of a weak acid in methanol

You will be provided with either a solid or a solution of a mono- or dicarboxylic acid to analyze by titration with standard methoxide solution. Follow the guidelines on sample size selection outlined in previous procedures along with your instructor's suggestions. If the sample is a dicarboxylic acid which can be titrated selectively, you will be provided with additional help, but this analysis is best done potentiometrically (Chapter 18).

Perform the analysis in triplicate using the appropriate sample size and an indicator of your own choice. Report the normality of the unknown if it is a solution, or the equivalent weight if it is a solid.

14chapter

Precipitation Titrations with Silver Nitrate: Volumetric Chloride Analysis

Volumetric methods involving the formation of insoluble silver halide precipitates (argentometric titrations) have been surveyed in Chapter 4. The three methods described, Mohr's, Volhard's, and Fajans', are typical of all precipitation titrations that do not require some sort of instrumental endpoint detection. In this chapter we offer procedures for the determination of chloride ion by each of these methods.

Mohr's method of chloride analysis is the oldest of the three, and the least satisfactory. The color change is hard to see and the indicator blank (P.13.2) is significant. Nevertheless, with skill and practice, accurate results can be obtained.

By comparison, Volhard's and Fajans' methods deserve an "excellent" rating. The endpoints are sharp and precise. However, the indicator blanks are not quite negligible in all cases, and it is advisable to "standardize" the silver nitrate solution by titration with standard sodium chloride, using the same method and conditions as in the analysis of the unknown substrate. This way, endpoint errors will be similar in the standardization and in the titration of the unknown substrate and will tend to cancel out in the calculation of the final result.*

P.14.1 Preparation of standard 0.1 N silver nitrate solution

To prepare 500 ml of a 0.1 N $AgNO_3$ solution, first dry about 8.5 g of solid $AgNO_3$ (formula weight 169.87) in a weighing bottle at 105–110°C for about 2 hours. Pulverizing the sample with a mortar and pestle before placing it in the oven will improve the drying process. Weigh the cool bottle (with the stopper on), transfer the contents to a small beaker, and weigh again to determine the sample size. In all weighing operations, be careful not to spill any of the solid onto metallic surfaces as $AgNO_3$ is corrosive.† Dissolve the salt in water, transfer the solution quantitatively into a 500 ml volumetric flask (don't forget to rinse the beaker), dilute, mix, bring to the mark, and mix again. Calculate the normality of the solution.

If the $AgNO_3$ is to be standardized with sodium chloride, prepare 500 ml of 0.1 N NaCl solution as follows. Into a small beaker weigh out the appropriate quantity of dried (at 110°C for 2 hours) reagent-grade NaCl. Dissolve the solid in water, transfer quantitatively to a 500 ml volumetric flask (don't forget to rinse the beaker), and bring to the mark with proper mixing. Calculate the actual NaCl concentration. This solution may also be used for back-titration in case an endpoint is overrun.

For standardization of $AgNO_3$, follow the appropriate procedure given below, substituting a known volume of standard NaCl solution for the solid unknown.

MOHR'S METHOD

The pH of the titration mixture must be between 6.5 and 10.5. The endpoint is taken at the first appearance of a red precipitate of silver chromate.

---

* The cancellation of endpoint errors becomes exact as the numbers of equivalents of NaCl and of the unknown substrate become equal. However, it is sufficient, in practice, if these equivalents are of the same order of magnitude.

† Silver nitrate and skin don't go together either, so avoid getting any on your hands. If you do come in contact with $AgNO_3$, rinse the affected area with a bit of dilute KI solution followed by plenty of water. Otherwise the reduction of Ag^+ to silver metal will result in a brown stain that persists for several days.

P.14.2 Mohr's chloride analysis

Dry your solid unknown for 2 hours at 110°C. Weigh out samples of the appropriate size into 250 ml Erlenmeyer flasks, dissolve and dilute to about 50 ml. Add 1 ml of 5 percent potassium chromate solution.*

Titrate with standard $AgNO_3$ until a slight but distinct color change indicates the formation of solid Ag_2CrO_4. If standard NaCl has been used to standardize the silver nitrate titrant, the endpoint colors in standardization and titration should match exactly. Do not apply a correction for the indicator blank, as the endpoint errors will tend to cancel.

If the $AgNO_3$ is being used without comparison with standard NaCl, an indicator blank may be determined as follows. To about 90 ml of water, add approximately 2 g of solid $CaCO_3$ (to simulate the AgCl precipitate) and 1 ml of 5 percent potassium chromate solution. Titrate until the color of the solid phase matches that of the substrate solution at the endpoint. Note that for greatest accuracy, the indicator blank is measured *after* the substrate has been titrated. This is because of the tendency of the $CaCO_3$ reference mixture to change color. Allow for the indicator blank in your calculations.

Report your results as percent chloride in the solid unknown.

VOLHARD'S METHOD

The Volhard method is the most widely applicable of the argentometric methods because it is actually a method for determining silver ion by titration with thiocyanate. Thus any anion that forms an insoluble silver salt can be determined by adding an excess of $AgNO_3$, removing the precipitate if necessary, and back-titrating with standard thiocyanate. The back-titration must be done below pH 3 to avoid precipitation of the Fe(III) endpoint indicator as ferric hydroxide, and to minimize the hydrolysis of Fe(III). In practice, the solution is acidified with chloride-free reagent-grade 6 N nitric acid to a pH less than 1. The nitric acid must be absolutely colorless. If there is even a trace of color, the acid must be boiled to remove oxides of nitrogen which interfere in the titration. The endpoint is taken when the color of the solution changes sharply to rust-orange owing to the formation of $FeSCN^{2+}$.

In the procedure described below, the back-titration with thiocyanate

---

* If the sample contains an acidic chloride (ask your instructor), add $\sim$0.5 g of $NaHCO_3$ to adjust the pH.

is done so that less than 1 ml of thiocyanate is required. The standard thiocyanate solution is prepared directly from the appropriate weight of KSCN after drying at 110°C for two hours. 100 ml of solution, prepared carefully in the same manner as previous standard solutions, should be sufficient. If desired, the concentration of thiocyanate can be checked by titration with standard silver nitrate according to the procedure given below for back-titration.

Two burets are needed for this analysis. We recommend that a 50 ml buret be used for the $AgNO_3$ solution, and a 5 or 10 ml buret for back-titration with KSCN.

P.14.3 Chloride analysis by Volhard's method

Dry the solid unknown for 2 hours at 110°C. Accurately weigh out samples of the appropriate size into 250 ml Erlenmeyer flasks, dissolve, and dilute to ~50 ml. Add 5 ml of 6 N (1:1) HNO_3 (which has been boiled to remove NO_2 if necessary).

Deliver $AgNO_3$ solution from the buret until the precipitate begins to coagulate. This is a sign that the equivalence point is near. Swirl the flask and then let it stand. Test for completeness of precipitation by adding $AgNO_3$ to the solution a little at a time until you are certain that Ag^+ is in excess. No more than one ml of excess $AgNO_3$ should be added.

Add about 4 ml of nitrobenzene and swirl vigorously to coat the precipitate. Now add 1 ml of saturated ferric ammonium sulfate solution which is 1 F in HNO_3.

Back-titrate with standard thiocyanate solution until one drop of titrant produces a sharp color-change to rust-orange. If in doubt, record the buret reading and add another drop until you are sure you have passed the endpoint.

From the volumes of $AgNO_3$ and KSCN, calculate the *net* amount of Ag^+ required to precipitate all the chloride ion. Report the percentage of chloride in the solid unknown.

FAJANS' METHOD

The unknown chloride sample is titrated directly with a standard solution of silver nitrate at pH 4–10 in the presence of the organic dye, dichlorofluorescein. The endpoint is taken when the color of the AgCl precipitate changes sharply to pink, owing to the adsorption of dichlorofluorescein ions. In the procedure described below, dextrin is added to promote the formation of many fine crystals and to prevent their coagulation at the equivalence point.* In this way, the color change to pink will appear to take place throughout the entire solution.

Dichlorofluorescein sensitizes (speeds up) the photoreduction of AgCl, but interference from this process is usually not serious. Nevertheless, avoid exposing the titration flask to bright light. If the precipitate turns gray (owing to photoreduction) so that the endpoint is obscured, do a rough titration first, to locate the equivalence point approximately, and thereafter add the indicator only just before the equivalence point. For best results, it is advisable to "standardize" the silver nitrate solution by titration with standard NaCl.

P.14.4 Chloride analysis by Fajans' method

Dry the solid unknown at 110°C for two hours. Weigh out three samples of appropriate size into Erlenmeyer flasks and dissolve in ~50 ml of water. Add one ml of a buffer solution that is $0.5 F$ each in acetic acid and sodium acetate. Dissolve about 0.2 g of dextrin in the solution. Add five drops of dichlorofluorescein indicator solution† and titrate with standard silver nitrate, swirling the flask vigorously (this is important!) until the color of the precipitate changes to light pink.

Report the percent chloride in the solid unknown.

---

* If dextrin is not available, add a few ml of methanol instead.

† The indicator is prepared by dissolving 0.1 g of dichlorofluorescein in 100 ml of 70 percent ethanol.

Complexometric Titrations

Analytical methods involving the formation of complexes between metals and the Y^{4-}* anion of ethylenediamine tetraacetic acid (EDTA) were first proposed as recently as 1945. Since that time a wealth of procedures has been developed which place EDTA among the most versatile of all analytical reagents.

In Chapter 4 we described the acid-base and complexing properties of EDTA and introduced the most commonly used conditions for analysis with EDTA. In this chapter we provide details for the complexometric determination of a few metals and some further information to illustrate the feasibility of a multitude of additional EDTA titrations.

---

* We shall continue to use the nomenclature and abbreviations introduced in Chapter 4.

STANDARD REAGENTS

P.15.1 Preparation of a standard 0.01 *F* EDTA solution

Analytical-reagent grade disodium EDTA dihydrate, $Na_2H_2Y \cdot 2 H_2O$ (formula weight 372.2) can be used as a primary standard for EDTA solutions. For all work involving EDTA, it is important that your "pure" water be free from polyvalent metal ions. Distilled water of questionable quality should be passed through a cation exchange resin to remove metals that form strong EDTA complexes.

For one liter of solution, dry about 3.8 g of $Na_2H_2Y \cdot 2 H_2O$ overnight or longer at 80°C.* Weigh (by difference) the appropriate amount of solid directly into a 1 liter volumetric flask with the aid of a powder funnel. Carefully rinse all particles into the flask, fill partially with water, and mix. The salt is somewhat slow to dissolve, so be patient! When all traces of solid are gone, bring the solution up to the mark.

Use the actual weight of salt in the solution to calculate (to four significant figures) its concentration.

Glass bottles should not be used for storing EDTA solutions as metal ions in the glass are leached out and complexed, often resulting in a considerable decrease in titer.† Store EDTA solutions and, wherever practical, other reagents for these analyses in plastic containers.

P.15.2 Eriochrome black T and *p*H 10 buffer solutions

As noted in Chapter 4, Eriochrome Black T (Erio T) is a good endpoint indicator when titrating at *p*H 10 in the presence of magnesium either directly or in a back-titration. Other metals such as zinc and cadmium can be determined by *direct titration* with EDTA to an Erio T endpoint.

To prepare the Erio T indicator solution, dissolve 0.1 g of the dye in about 100 ml of water. Add 0.2 g of ascorbic acid (to prevent air oxidation of Erio T) and 1 ml of *p*H 10 buffer (next paragraph). The indicator should

---

* The optimum period has been reported to be four days.

† We introduced the concept of *titer* in Problem 25 of Chapter 4. A more general definition is that the titer of a solution is equal to the weight of a particular substrate that is equivalent to exactly 1 ml of solution. Thus a 0.01000 *F* EDTA solution has a calcium carbonate titer of 1.001 milligram of $CaCO_3$ (formula weight 100.1) per ml.

be stable for at least two weeks, but if poor endpoints result sooner, a fresh batch should be prepared.

The pH 10 ammonia-ammonium chloride buffer should contain 70 g NH_4Cl and 570 ml of concentrated ammonia per liter of solution.

In an analysis, one ml of this concentrated buffer solution provides sufficient buffer strength if the metal concentration does not exceed 0.01 M and the sample solution was originally neutral. Too much buffer is to be avoided as the formation of metal-ammonia complexes will reduce the sharpness of the endpoint signal.

P.15.3 Standard 0.01 F Mg(II) solution

Although high purity $MgSO_4 \cdot 7\ H_2O$ (formula weight 246.49) is available, it is advisable to standardize the solution against standard EDTA to give practice in titration with EDTA and Erio T and to guard against endpoint errors in subsequent analyses.

Without drying, weigh out the amount of $MgSO_4 \cdot 7\ H_2O$ necessary to prepare 1 liter of 0.01 F solution. Transfer the solid to a 1 liter volumetric flask. Dissolve, dilute, and mix to make up 1000 ml of solution.

Standardize the $MgSO_4$ solution according to P.15.4. If the appropriate weight of solid was used, 25.00 ml of this solution should require approximately 25 ml of EDTA.

DIRECT TITRATION WITH EDTA

The procedures described in this section are for the direct titration of Mg^{2+}, Zn^{2+}, and Cd^{2+} with 0.01 F EDTA to the Erio T endpoint. For best results, the formal concentrations of substrate (c_S) at the equivalence point should be below the following values:

$$Mg^{2+}, \quad c_S < 0.01\ F\ (0.24\ g\ Mg\ per\ liter)$$
$$Zn^{2+}, \quad c_S < 0.004\ F\ (0.25\ g\ Zn\ per\ liter)$$
$$Cd^{2+}, \quad c_S < 0.0045\ F\ (0.50\ g\ Cd\ per\ liter)$$

P.15.4 Liquid unknowns

If your unknown substrate is a liquid solution, do a preliminary titration to find the approximate concentration of metal ion. Then choose a pipet size so that one pipetful will require 20–50 ml of 0.01 F EDTA for titration, and calculate how much water must be added to the titration flask so that the formal concentration (c_S) at the equivalence point will be below the recommended maximum.

Titration procedure. Following the above guidelines, deliver one pipetful of the substrate solution into each of three 250 ml Erlenmeyer flasks, and add the requisite amounts of water. Just before each titration, add pH 10 buffer (P.15.2; use 1 ml of buffer for each 50 ml of solution), 1–2 drops of Erio T indicator, and $\sim$0.2 g (half a small spatula full) of ascorbic acid to help prevent air oxidation of the indicator. Titrate with your standard 0.01 F EDTA solution, adding titrant slowly as the endpoint begins to appear. The solution will turn from red (the metal-Erio T complex) through magenta to blue (Erio T). The endpoint should be taken at the disappearance of the *last trace* of magenta.

Report the concentration of metal ion in the liquid unknown.

P.15.5 Formula weight of an unknown salt

You will be given a relatively pure salt of zinc, cadmium, or magnesium for analysis. Your instructor will tell you which metal is involved.

Before attempting any quantitative analysis, find out how to get the salt into solution. Test whether 0.1 g will dissolve in 50 ml of water. If not, add 0.5 ml of 6 N HCl and see if it dissolves now. If it still will not go into solution, add 1.5 ml of 6 N ammonia and see if it dissolves in the resulting ammonia-ammonia chloride buffer. If all fails, consult your instructor.

After you know how to get the sample into solution, prepare 250.0 ml of an approximately 0.008 F stock solution in a volumetric flask.* (Assume that the formula weight of the salt is 150.) Deliver 50.00 ml of this stock solution into a 250 ml Erlenmeyer flask, add 1 ml† of pH 10 buffer (P.15.2), and swirl. Then add 1–2 drops of Erio T indicator, $\sim$0.2 g (half a small spatula full) of ascorbic acid, and titrate with your standard 0.01 F EDTA solution as described in P.15.4. Repeat the titration, modifying the volume of stock solution (if necessary) so that 20–50 ml of EDTA solution is used in titration.

Calculate the formula weight of the unknown salt. On the basis of formula weight and solubility, suggest its probable (or possible) identity.

---

* If your sample dissolves in HCl but not in water, weigh the sample into a beaker, add 100 ml of water, then add HCl dropwise until solution is accomplished. Transfer the solution to the volumetric flask and make up to volume. If your sample dissolves in NH_3-NH_4Cl buffer, proceed similarly but add buffer (dropwise) instead of HCl.

† Use 2 ml of buffer if your sample has been dissolved in HCl.

SUBSTITUTION TITRATION OF CALCIUM(II) WITH EDTA

The procedure described in this section is for the titration of Ca^{2+} with 0.01 F EDTA in an ammonium chloride-ammonia buffer at pH 10 to the Erio T endpoint by a method called *substitution titration*. Ca^{2+} by itself cannot be titrated to the Erio T endpoint because the association constant for the formation of the red calcium-Erio T complex is too small.* However, a very satisfactory endpoint can be obtained simply by adding about 0.05 milliequivalents of magnesium-EDTA complex (MgY^{2-}) to the titration flask. Since calcium forms a more stable complex with EDTA than does magnesium (Table 4.6), a substitution reaction ensues and Mg^{2+} is liberated.

$$Ca^{2+} + MgY^{2-} \rightleftharpoons Mg^{2+} + CaY^{2-}$$

When this equilibrium mixture is subsequently titrated with EDTA, reaction takes place approximately in two steps, with Ca^{2+} reacting first and Mg^{2+} second. The endpoint is taken when *both* ions have reacted. The endpoint signal is therefore essentially the same as if 0.05 milliequivalents of Mg^{2+} were titrated with EDTA (without any Ca^{2+} present) and is readily detected with Erio T indicator. At the same time, the equivalents of EDTA added are precisely equal to the unknown equivalents of Ca^{2+}. To appreciate this, compare (in the following table) the state of the

Ion	State in which ion is added	State of ion at endpoint	Net reaction
Magnesium	MgY^{2-}	MgY^{2-}	None
Calcium	Ca^{2+}	CaY^{2-}	Forms one-to-one complex with EDTA

calcium and magnesium species as they are added and as they exist at the titration endpoint. The table shows that the net reaction is simply the conversion of Ca^{2+} to CaY^{2-}.

P.15.6 Determination of Ca^{2+} by substitution titration

The procedure consists of two parts: (1) preparation of a magnesium-EDTA solution, and (2) titration of the unknown substrate.

* $K_{assoc} = $ [Ca-Erio T]/[Ca^{2+}][Erio T] $= 10^{5.4}$ at pH 10.

Mg-EDTA solution. Pipet 50.00 ml of your standard $0.01\,F$ $MgSO_4$ solution (P.15.3) into a small plastic storage bottle or 125 ml Erlenmeyer flask. Now add from a buret the *exact* amount of standard EDTA solution required to quantitatively complex all the Mg^{2+}. Mix well.

Test a 5 ml portion of the solution by adding three drops of pH 10 buffer and one drop of Erio T indicator which has been diluted by a factor of 10. The solution should be dull violet. A fractional drop of EDTA should turn the solution blue while the same amount of $MgSO_4$ should cause a red color. If the solution is not at the equivalence point as shown by these tests, add either Mg^{2+} or EDTA solution and test again.

Titration of unknown substrate. The formal concentration (c_S) of Ca^{2+} at the titration endpoint should not exceed $0.01\,F$ (0.40 g per liter of calcium). Ask your instructor for the approximate calcium concentration in your liquid unknown, or do a preliminary experiment to find it. Then, using a buret or pipet, deliver 0.25–0.40 milliequivalents of Ca^{2+} into each of three 250 ml Erlenmeyer flasks. To each flask add the requisite amount of water to make a total volume of 40–60 ml, 1 ml of ammonium chloride-ammonia buffer (P.15.2), and 10 ml of the Mg-EDTA solution prepared above. Just before titration add approximately 0.2 g of ascorbic acid and 1–2 drops of Erio T indicator, then titrate with your standard $0.01\,F$ EDTA solution. The color of the solution will change from red through magenta to blue. The endpoint should be taken at the disappearance of the last trace of magenta.

Report the concentration of Ca^{2+} in your liquid unknown.

DETERMINATION OF COBALT(II) BY BACK-TITRATION

Cobalt(II) cannot be titrated directly with EDTA to the Erio T endpoint because cobaltous hydroxide precipitates from solution above pH 8. In the following procedure, an excess of EDTA is added to the unknown Co^{2+} solution at low pH, the pH is increased to pH 10, and the solution is back-titrated with standard $0.01\,F$ Mg^{2+} (P.15.3) to the Erio T endpoint. Since you will be dealing with an unknown solution, a preliminary run should be done to determine the approximate Co^{2+} concentration. In planning the preliminary run, assume that $[Co^{2+}] \sim 0.01\,F$. Your unknown solution will be at a low pH.

P.15.7 Co^{2+} by back-titration

Deliver an appropriate pipetful (to contain 0.2–0.3 milliequivalents of Co^{2+}) of your liquid unknown to each of three 250 ml Erlenmeyer flasks, then add 50.00 ml (0.5 milliequivalents) of your standard $0.01\,F$ EDTA solution and swirl. Add 3 ml of ammonium chloride-ammonia buffer of pH 10, about 0.2 g of ascorbic acid, and a few drops of Erio T indicator. Titrate with your standard $0.01\,F$ magnesium solution until the color changes from blue to magenta. From the difference between the added equivalents of EDTA and the equivalents of Mg^{2+} needed to reach the Erio T endpoint, determine the concentration of Co^{2+} in the unknown solution.

DETERMINATION OF WATER HARDNESS

In most areas of the world, drinking water (tap water, well water, river, or stream water) contains significant quantities of metal ions, most notably Ca^{2+} and Mg^{2+}. The *total hardness* of water is normally taken as the sum of Ca(II) and Mg(II) concentrations and is usually reported as ppm (parts per million by weight) of calcium carbonate, as if calcium (the predominant metal) were responsible for all the hardness. (See Chapter 4, Problem 25.)

P.15.8 describes the analysis of water for total hardness using EDTA titration to an Erio T endpoint. Metal ions such as aluminum, nickel, iron, and copper do not interfere with the stoichiometry of the titration because they are present only in trace amounts and/or form insoluble hydroxides. However, such ions can obscure the sharpness of the endpoint by combining with the indicator.*

In P.15.9 we describe the older method (remember, EDTA has only recently been introduced to analytical chemistry) of determining water hardness by titration with a standard soap solution. Although the endpoint is not as good as with color indicators, the soap titration is an interesting alternative. It is accurate to *ca.* ± 5 percent.

Water samples for hardness determination may come from your city or town water supply or from a more remote origin. Your instructor will advise you of the best method to use if you wish to test a sample of water from a source of your own choosing.

P.15.8 Total hardness of water by EDTA titration

Since water hardness varies considerably depending on its source, the volume of water for each titration should be determined by an initial experiment with a 50.00 ml sample. For rather soft water, the standard

---

* Elimination of all interfering metal ions necessitates the addition of KCN, a modification we don't recommend for student experiments.

0.01 *N* EDTA solution may be (quantitatively) diluted by up to a factor of ten.

In most cases, the water being tested will contain enough Mg^{2+} ion that *direct titration* (P.15.4) is possible. For the sharpest endpoint, the sample should be acidified slightly with HCl, boiled (to remove CO_2), and neutralized with NaOH (to a methyl red endpoint) before titration.

If you do not get a sharp endpoint in the direct titration, the water may be relatively free from magnesium. In that case, perform the analysis for calcium by substitution titration (P.15.6), adding a few ml of a solution of MgY^{2-} before titration with EDTA.*

From the (average) "hardness normality" determined by titration, calculate the total water hardness as ppm calcium carbonate.

P.15.9 Water hardness by titration with standard soap solution

The semi-quantitative determination of water hardness by titration with soap is probably more of a precipitation titration than a complexation titration since the interaction between soap and calcium (or magnesium) ion produces a solid curdy substance. We offer it here, however, as a possible substitute for P.15.8. This procedure consists of (1) preparation of a standard $CaCl_2$ solution, (2) preparation and standardization of the soap solution, and (3) titration of the unknown.

Standard calcium chloride solution. If you have already analyzed a solution containing calcium (in P.15.6), you may use it after diluting (quantitatively) so that the calcium concentration is $\sim$0.002 *F*. Otherwise dissolve $\sim$0.2 g (accurately weighed) of dry $CaCO_3$ in a small amount of dilute HCl. Be careful not to let the solution bubble too vigorously. Evaporate to dryness in a (clean) evaporating dish a few times, using about 5 ml of water for each dissolution. This drives off excess HCl. (The soap titration will not work if the *p*H is less than 5.) Dissolve the solid and transfer quantitatively to a 1 liter volumetric flask. Dilute and mix well.

Soap solution. Use the best grade of soap available—99.44 percent purity is fine. Dissolve about 10 g (rough weighing) in 80 percent ethanol to make about 100 ml of solution. This solution should be allowed to stand for a few days. Then dilute 20 ml of solution to 250 ml with 80 percent ethanol.

Titration procedure. For standardization use 25.00 ml of $CaCl_2$ solution and $\sim$25 ml of distilled water. For an unknown determination use

---

* If acceptable results are still not obtained, the cause is probably interference from other metal ions as noted above. A few crystals of Na_2S added before the indicator often helps "mask" the effects of interfering ions.

50.00 ml of the unknown water sample.* Put the sample into a 250 ml glass container with a well-fitting ground glass stopper. A volumetric flask is acceptable, but be careful not to lose any titrant. Deliver the soap solution in small portions from your buret. Add about 1 ml at a time for the first, rough titration.

After each addition of soap, stopper the container and shake. The endpoint occurs at the first sign of a persistent lather. After a rough titration, subsequent determinations should give an endpoint that is reproducible to about 0.1–0.2 ml of titrant.

Sometimes a premature endpoint is noted in this titration. This is called the magnesium endpoint and its origin is somewhat obscure. In any case, when you think you have reached the endpoint, read the buret. Then add about 0.5 ml more of soap solution and shake the flask. If the lather disappears, continue the titration until the true endpoint is reached.

From the results of the standardization, determine the titer of the soap solution in terms of milligrams of $CaCO_3$ equivalent to 1 ml of titrant. Then calculate the hardness of your water sample as ppm calcium carbonate.

OTHER EDTA TITRATIONS

In this chapter we have given procedures for only a few applications of titration with standard EDTA, all involving titration at pH 10 to an Erio T endpoint. As is apparent from Table 4.6, the magnitude of association constants for metal-EDTA complexes is such that direct titration should be feasible for most of the metals of the periodic chart. Indeed, direct titration at low pH is possible for many metals that precipitate (as oxides or hydroxides) in the high pH region required for the Erio T endpoint. There is a large variety of indicators and instrumental methods of detection available to make these titrations practical. In many instances, direct titration can be used at high pH by adding an auxiliary complexing agent to prevent hydroxide formation. Thus, for example, in P.15.4 and P.15.5, zinc is easily determined at pH 10 (NH_3-NH_4Cl buffer) because the formation of $Zn(NH_3)_4^{2+}$ keeps the metal in solution, while Mn^{2+}, which does not form strong ammonia complexes, requires the addition of tartrate ion.

Figure 15.1 indicates actual pH ranges for EDTA titration of a large number of metals. Note the wide selection of endpoint indicators that are

---

* The optimum conditions occur when less than 7 ml of soap are required for 50 ml of solution. Thus, if larger volumes of titrant are required, the sample should be diluted accordingly. If only 7 ml or so are required for a titration, use of a 10 ml buret might improve accuracy but the error inherent in getting a good endpoint is probably larger than the error in reading a 50 ml buret.

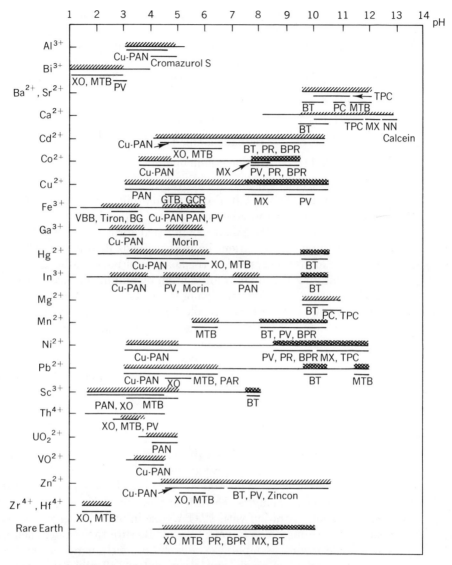

- ———— pH region where EDTA reacts with metal ion quantitatively
- ////////// pH region where metal indicators can be used to determine the end point of EDTA titration
- \\\\\\\\\\ pH region where auziliary complexing agents should be used to prevent hydroxide precipitation

BG Bindscheidler's green, leuco base	MX Murexide	PR Pyrogallol red
BPR Bromopyrogallol red	NN Patton & Reeder's Dye; 2-Hydroxy-1-(2-hydroxy-4-sulfo-1-napthylazo)-3- naphthoic acid	PV Pyrocatechol violet
BT Eriochrome black T		TPC Thymolphthalein complexone
GCR Glycinecresol red	PAN Pyridylazonaphthol	VBB Variamine blue B, base
GTB Glycinethymol blue	Cu-PAN PAN plus Cu-EDTA	
MTB Methylthymol blue	PC o-Cresolphthalein complexone	XO Xylenol orange

Fig. 15.1. Conditions for metal-EDTA titrations. [Reprinted, with permission, from K. Ueno, *J. Chem. Educ.*, **42**, 432 (1965).]

suitable for these titrations. Although many of the metals can be titrated at high *p*H in the presence of auxiliary complexing agents, under proper conditions each can be determined without the need for additional ligands.

Hydrochloric or nitric acid is used in reactions at very low *p*H. Otherwise appropriate buffer systems are chosen as follows:

*p*H 2–4	Glycine-glycine hydrochloride
*p*H 4–6.5	Acetic acid-sodium acetate
*p*H 6.5–8	Triethanolamine-triethanolamine hydrochloride
*p*H 8–11	Ammonia-ammonium chloride

As in the case of zinc mentioned above, the buffers serve in many cases not only to control the *p*H but also as complexing agents for the metal ions, thus preventing the formation of insoluble hydroxides.

Metals (such as the alkali metals) that form weak EDTA complexes can be determined by various indirect methods involving EDTA. EDTA titration can also be applied to the analysis of nonmetallic anions, usually by precipitating the anion from solution by adding an excess of a suitable metal, followed by back-titration of the metal with EDTA.

With your standard EDTA solution and the proper indicator in hand, you are prepared to perform a large number of simple analyses that were unknown before 1945. For further details of particular procedures refer to G. Schwarzenbach and H. Flaschka, *Complexometric Titrations*, 2nd Eng. Ed., Trans. by H. M. N. H. Irving (London: Methuen & Co., Ltd., 1969).

P.15.10 Devise your own procedure

For this analysis you will be given a solid or liquid unknown containing a constituent that can be titrated with EDTA. The identity of this constituent (and of possible interferences) will be disclosed by your instructor in advance of the analysis. Devise your own procedure for analysis, which should include titration with EDTA. Use information given in this book and reference materials in your library. Be sure to consider such matters as sample drying, sample size, dissolution, *p*H, temperature, endpoint indicator, indicator blank, and other matters affecting the analysis. Go over your plan of attack with the instructor before finalizing the details.

Perform the analysis and report the results.

16 *chapter*

Oxidation-Reduction Titrations

In Chapter 4 we have surveyed the application of oxidation-reduction (redox) reactions to chemical analysis. In this chapter we shall describe the use of two outstandingly important titrating agents: potassium permanganate in acid solution and the iodine-iodide couple. Potassium permanganate is a strong oxidizing agent that reacts rapidly and quantitatively with many substrates. The iodine-iodide couple is used as a mild oxidizing agent and as a selective reducing agent.

Since the stoichiometry of redox reactions can get quite complicated, be sure to write down all pertinent equations in preparing for an analysis. Otherwise you might be surprised to find that 10 ml or 250 ml of a titrant are required rather than the 50 ml you had planned on. For help in this matter, refer to Tables 4.4, 4.5, and A.5.

STANDARD POTASSIUM PERMANGANATE SOLUTION

A 0.02000 F KMnO$_4$ solution contains 3.1605 grams of solute per liter and is 0.1000 N when used as an oxidant in acidic media where MnO$_4^-$ is reduced to Mn^{2+}.* (Verify this.)

The permanganate ion is such a good oxidizing agent that it reacts with dust, rubber, cork and any organic impurities that might be present in the distilled water supply. Thus, only glass- or inert plastic-stoppered vessels should be used for storage. Since a certain amount of contamination is unavoidable, a freshly prepared permanganate solution should be allowed to stand for at least 24 hours before it is standardized. Water is also oxidized by MnO$_4^-$ ions, but in neutral solution this reaction is slow if care is taken to protect the solution from light and remove any MnO$_2$ precipitate. Because of the possibility of decomposition, solutions which are to be used over long periods of time should be restandardized periodically.

P.16.1 Preparation of 0.1 N KMnO$_4$

Dissolve about 3.2 g (rough weighing) of potassium permanganate crystals in hot water, dilute to one liter in a large beaker, and heat to boiling. Keep the solution hot for several minutes. Then cover the beaker with a large watch glass and set aside in a dark place until the next laboratory period.

Using a wad of glass wool packed firmly in a glass funnel, filter the solution directly into your storage vessel. Be sure that the container is absolutely clean and free from grease around the ground glass joint.

Standardize the solution with arsenious oxide (P.16.2) or sodium oxalate (P.16.3). Store it in a dark place when not in use.

P.16.2 Standardization of permanganate with As$_2$O$_3$

Dry some primary-standard grade As$_2$O$_3$ in a small weighing bottle at 110°C for one to two hours. Weigh out (by successive dumping) three samples of 0.20–0.24 g each into 250 ml Erlenmeyer flasks. Dissolve each

---

* Since we shall be concerned only with these conditions, we will refer to the permanganate solution prepared in P.16.1 as "0.1 N" without repeatedly specifying normality with respect to what.

sample with 10 ml of water and ∼1.5 g of sodium hydroxide. Acidify the solution with 15 ml of 6 N HCl, dilute to ∼100 ml, and add one drop of KIO_3 catalyst.*

Titrate with your permanganate solution until a faint pink color persists for at least 10 seconds, indicating an excess of MnO_4^-. The titrant should be added dropwise during the last milliliter or so of the titration. Allow time for complete decoloration before introducing the next drop.

Determine an endpoint blank correction (refer to P.13.2) by titrating 150 ml of water plus one drop of catalyst to the same faint pink color.

P.16.3 Standardization of permanganate with sodium oxalate

The direct reaction between permanganate and oxalate ions is quite slow at ordinary temperatures. Fortunately, the process is catalyzed by the Mn^{2+} produced by the reduction of MnO_4^- and becomes sufficiently rapid near the equivalence point to provide an easily detectable endpoint signal.

Dry the primary-standard grade $Na_2C_2O_4$ at 110°C for one hour. Accurately weigh out three samples of 0.25–0.30 g each into 250 ml Erlenmeyer flasks. (Convince yourself that this is the correct quantity!)

Dissolve each sample in about 75 ml of 0.8 F H_2SO_4 solution. Just before each titration heat the solution to 85–90°C. Don't forget to rinse your thermometer into the titration flask before removing it.

Titrate slowly with your permanganate solution. Stop the addition of titrant occasionally and allow complete decoloration before proceeding. It is especially important that the flask be constantly in motion to prevent a local deficiency of hydrogen ions. Reheat the solution if the temperature falls below 60°C. The endpoint should be taken as the first pink color that persists for more than 20 seconds.

Run an endpoint blank with 100 ml of hot (80°C) 1 N H_2SO_4. Subtract the volume of $KMnO_4$ solution required to produce a faint pink color from the volume needed for standardization.

From the results of three titrations, calculate the normality of your permanganate solution.†

---

* A solution of ∼0.05 g KIO_3 per 100 ml.

† This method of standardization (often called the McBride procedure) contains a small systematic error of ∼0.1– 0.3 percent due to slight air oxidation of oxalic acid. However, this error is small enough to be negligible for most purposes.

PERMANGANATE TITRATIONS IN ACID MEDIA

P.16.4 Devise your own procedure

For this analysis, you will be given a solid or solution of unknown composition to analyze for a particular reducing agent. Your instructor will announce the identity of the unknown (selected from the list on p. 84) in advance of the analysis. Devise your own procedure for the determination. Use information from Chapter 4 and from reference material in your library. Be sure to include such points as sample drying, size of sample, dissolution, acidity, and special conditions for the titration (temperature, reduction of the substrate, catalyst, etc.). Go over your plan of attack with the instructor before finalizing the details.

Perform the analysis and report the results.

P.16.5 Determination of iron in an oxide or ore

In this analysis the sample is dissolved in HCl, reduced to the ferrous state by stannous chloride, and titrated with a standard solution of permanganate in the presence of Zimmerman-Reinhardt "preventive solution." The chemistry of these steps is described in Chapter 4. Be sure to study it before you proceed.

The following solutions should be prepared in advance.

1. $SnCl_2$: Dissolve $\sim$3 g of $SnCl_2 \cdot 2\ H_2O$ in 10 ml of concentrated HCl and dilute to $\sim$20 ml with water. (Prepare just before use.)

2. Zimmerman-Reinhardt solution: Prepare with care! Dissolve 20 g of $MnSO_4 \cdot 4\ H_2O$ in 100 ml of water. In a separate vessel mix (caution!) 120 ml of water, 40 ml of concentrated H_2SO_4, and 40 ml of 85 percent H_3PO_4. Cool the acid mixture and combine with the $MnSO_4$ solution.

3. $HgCl_2$: 5 g in 100 ml of water.

Dissolution of sample. If the unknown is not already in powder form, pulverize it with mortar and pestle (preferably agate). Dry the solid for $\sim$2 hours at 110°C and weigh out samples of appropriate size* into 150 ml beakers.

To each sample add 10 ml of water and 20 ml of concentrated HCl. Cover loosely with a watch glass, and heat to just below boiling on a hot plate or steam bath *in the hood* until all dark colored material has dissolved, usually within a half hour. Add additional acid if necessary to maintain

---

* Your instructor can provide you with a rough estimate of the percent iron in the sample to use as a guide. It is wise to start with four samples to insure against one being ruined.

the original volume of solution. A colorless residue of silica may remain, but the presence of this solid does not interfere with the analysis.*

Reduction. By the time the iron is all in solution, there should be enough Fe(III) to give a definite yellow color. If the solution is not yellow, add a small crystal of $KMnO_4$ to oxidize some of the ferrous iron. This will allow you to see a definite color change with the addition of tin(II) solution.

Reduce all the Fe(III) to Fe(II) by adding $SnCl_2$ solution dropwise from a pipet to the *hot* solution in the beaker. When the last trace of yellow has gone (the resulting solution will be colorless or pale green) add one or two drops of $SnCl_2$ in excess.†

Allow the solution to cool to room temperature and quickly (important!) dump in 20 ml of $HgCl_2$ solution. A small amount of pure white precipitate (Hg_2Cl_2) should form. If the precipitate is gray or if no solid forms, this sample should be discarded.

Let the solution stand for five minutes. Then quantitatively transfer the contents of the beaker into a 500 ml Erlenmeyer flask and dilute to $\sim$300 ml with water.

Titration. Add 25 ml of the Zimmerman-Reinhardt preventive solution and titrate immediately with standard permanganate solution. The titrant should be added slowly with constant swirling of the flask. The endpoint is signaled by the characteristic faint pink color of excess permanganate.

To determine the endpoint correction, take a 500 ml Erlenmeyer flask and add 5 ml of concentrated HCl, then 2 drops of $SnCl_2$, then the quantities of $HgCl_2$, water, and preventive solution specified above. Titrate with permanganate to a faint pink. Subtract this endpoint correction from each titration volume.

Calculate the percent iron in the iron oxide or ore.

THE IODINE-IODIDE COUPLE: STANDARD SOLUTIONS

When iodine is used as a quantitative oxidizing agent (iodimetry), the unknown substrate is titrated directly with a standard solution of iodine.

---

* Samples given to students for analysis can almost always be treated in this manner. If the HCl treatment is not effective in dissolving all the dark colored iron compounds, try a 1:1 HCl-HNO₃ mixture with heating. When the solid has dissolved, carefully evaporate the solution to dryness to remove nitrates and redissolve the solid in HCl as in the original treatment. If solution is still unsatisfactory, consult your instructor for help.

† To minimize air oxidation of iron back to ferric ion, this reduction step should *not* be done simultaneously for all samples, but rather reduction and titration should be performed for each sample without delay.

When iodide is used as a quantitative reducing agent (iodometry), the unknown substrate is treated with a large excess of iodide ion and the liberated iodine is titrated with a standard solution of sodium thiosulfate. Unfortunately, neither of these standard solutions is reliably stable and both titrants need frequent restandardization.

An approximately 0.1 N solution of iodine, to be used as titrant, will be made up by dissolving pure solid iodine in 0.2 F aqueous KI. The familiar reaction, $I_2 + I^- \rightleftharpoons I_3^-$, serves not only to raise the solubility of iodine, but also to "buffer" the iodine normality against loss of I_2 by volatilization during handling or storage.

Solutions of thiosulfate in water are vulnerable to contamination by bacteria (called thiobacteria) which feed on $S_2O_3^{2-}$, converting it to a mixture of SO_4^{2-}, SO_3^{2-} and sulfur. An approximately 0.1 N solution will be made up, for use as titrant, by dissolving solid $Na_2S_2O_3 \cdot 5\ H_2O$ in freshly distilled or boiled distilled water, thus reducing the bacterial count in the freshly prepared solution. Moreover, two or three drops of chloroform will be added to destroy bacteria that are left behind or creep into the solution. Chloroform is immiscible with water and settles to the bottom of the container.

To standardize the preceding solutions we shall use a 0.0167 F (0.1 N) aqueous solution of potassium iodate as our primary standard. Iodate ion (IO_3^-) reacts quantitatively with iodide and hydrogen ions to produce the stoichiometric amount of iodine which, in turn, is used to standardize the thiosulfate solution. The reactions in this procedure are:

$$IO_3^- + 8\ I^- + 6\ H^+ \longrightarrow 3\ I_3^- + 3\ H_2O$$

$$\frac{3\ I_3^- + 6\ S_2O_3^{2-} \longrightarrow 9\ I^- + 3\ S_4O_6^{2-}}{IO_3^- + 6\ H^+ + 6\ S_2O_3^{2-} \longrightarrow I^- + 3\ S_4O_6^{2-} + 3\ H_2O}$$

Note that in the overall reaction iodate is reduced all the way down to I^-. Hence the equivalent weight of KIO_3 is one-sixth its formula weight. Note also that large quantities of hydrogen ions are needed for the reaction. Iodide ions are needed in the first reaction and are regenerated in the second.

If a standard solution of iodine is required, we shall prepare an approximately 0.1 N solution of iodine in aqueous KI and determine the precise normality by direct titration with the standard thiosulfate solution.

P.16.6 Preparation of 0.0167 F (0.1 N) KIO₃

Prepare 250.0 ml of a standard solution by quantitatively transferring the appropriate weight of reagent-grade KIO_3 (dried at 110°C) to a volumetric flask, dissolving in water and diluting to the mark.

P.16.7 Preparation of 0.1 F (0.1 N) sodium thiosulfate

This solution should be prepared about one week in advance, as there is an unavoidable initial decomposition that should be allowed to occur before standardization. To make an approximately 0.1 F solution of thiosulfate, start with about one liter of *freshly* distilled or boiled distilled water. Dissolve about 25 g of $Na_2S_2O_3 \cdot 5\ H_2O$ in some of the water, after first crushing any large crystals. Transfer this solution to a volumetric flask or storage bottle, dilute to ∼1 liter, and add a few drops of chloroform.*

P.16.8 Standardization of 0.1 F (0.1 N) sodium thiosulfate

Pipet 25.00 ml of the standard KIO_3 solution (P.16.6) into a 250 ml Erlenmeyer flask and add ∼2 g of KI. After the potassium iodide has dissolved, acidify with 10 ml of dilute (∼1 N) sulfuric acid and titrate without delay with thiosulfate. When the iodine color has faded appreciably, add 3 ml of starch solution.† It is important not to add starch until nearly all of the iodine has reacted; otherwise, slowly reacting starch-iodine complexes are formed. Continue to titrate until the blue color just disappears.

Repeat the procedure twice, rapidly adding thiosulfate to within 1 ml of the endpoint before introducing the starch indicator.

Since KI is slightly susceptible to air oxidation and sometimes contains small amounts of iodate, a control experiment should be run. Dissolve ∼2 g of KI in 50 ml of water, add 10 ml of dilute HCl and 3 ml of starch solution. If a blue color results, titrate with $S_2O_3{}^{2-}$ and subtract this amount from the volume used in each standardization.

Calculate the concentration of your thiosulfate solution based on three determinations. The solution should be restandardized at weekly intervals.

---

* Chloroform is volatile. Check periodically for the presence of a chloroform layer at the bottom of the bottle. Add a few drops if necessary.

† Starch solution is prepared by mixing thoroughly about 2 g of soluble starch with 100 ml of cold water and pouring the suspension into 200 ml of rapidly boiling water. The boiling is stopped immediately after the addition. If visible particles remain, the solution must be filtered through a very porous paper. After the solution has cooled, dissolve in it 6 g of potassium iodide, and add a crystal of mercuric iodide the size of a pinhead.

P.16.9 Preparation of 0.05 *F* (0.1 *N*) iodine solution

Prepare 500 ml of approximately 0.05 *F* iodine by dissolving about 6.5 g of iodine and 15 g of potassium iodide (KI) in ~40 ml of warm water in a small beaker. After making certain that the iodine is completely dissolved, transfer the solution to a 500 ml volumetric flask and dilute to the mark. It is essential that the solution be preserved in a glass-stoppered flask, as both rubber and cork are attacked by iodine.

P.16.10 Standardization of 0.05 *F* (0.1 *N*) iodine solution

The iodine solution may be standardized by direct titration against thiosulfate. The easiest method is to use two 50 ml burets. Run about 40 ml of thiosulfate solution (record initial volume) into the titration flask and add 3 ml of starch solution. Titrate with your 0.05 *F* iodine solution until you see the intense blue color of the starch iodine complex. Back-titration with thiosulfate to the disappearance of color will give a reliable endpoint without the need for a blank correction.

TITRATIONS INVOLVING THE IODINE-IODIDE COUPLE

P.16.11 Determination of arsenic(III) or antimony(III)

Trivalent arsenic and antimony can be determined by titration with iodine if the *p*H is kept within the range 5 to 9. Below *p*H 5 the pseudo-equilibrium constant for titration becomes unsatisfactorily small, and above *p*H 9 the titrant becomes unstable. In addition, antimony salts are prone to precipitate from solution at *p*H values above neutrality. It is therefore recommended to add potassium sodium tartrate, which forms soluble tartrate complexes with both Sb(III) and Sb(V).

You will be given a solid unknown. Following guidelines provided by your instructor, determine the appropriate sample size. Dry the solid (at 110°C) and weigh out three samples for analysis into 250 ml Erlenmeyer flasks. Dissolve the solid by warming it with small amounts of 2 *F* sodium hydroxide. If your sample contains antimony(III), add ~2 g of potassium sodium tartrate. If necessary, add some water to dissolve inert material. Cool, and add a drop of methyl orange or methyl red indicator solution.

Cautiously acidify with dilute hydrochloric acid. Immediately add ~5 g of sodium bicarbonate and dissolve. Dilute to ~100 ml. Titrate with standard iodine solution to the starch-iodine endpoint. Do not be surprised if bubbles of CO_2 gas escape from the solution during the titration.

Report the percentage of As_2O_3 or Sb_2O_3 in the solid unknown.

P.16.12 Devise your own procedure

For this analysis you will be given a solid or liquid unknown to analyze for a particular component. Your instructor will announce the identity of that component in advance of the analysis. Devise your own procedure so as to make use of the iodine-iodide couple. Use information from Chapter 4 and from reference materials in your library. Be sure to consider such matters as sample drying, sample size, dissolution, pH, temperature, and other conditions affecting the analysis. Go over your plan of attack with the instructor before finalizing the details.

Perform the analysis and report the results.

P.16.13 Determination of copper in brass

Brass is an alloy consisting mainly of copper and zinc, with possible traces of tin, lead, and iron. In the following procedure, the unknown brass sample is brought into solution by oxidation with hot nitric acid and the copper in the sample is thus oxidized to Cu(II). Upon subsequent addition of an excess of potassium iodide to the solution, the following reaction takes place:

$$2 \, Cu^{2+} + 5 \, I^- \; \rightleftharpoons \; Cu_2I_2(s) + I_3^-$$

The iodine liberated in the reaction is then titrated with thiosulfate to a starch-iodine endpoint.

Of the other metals present in brass, iron is the only one that would interfere with this analysis because Fe(III) can be reduced by iodide ion. The interference is eliminated by the addition of phosphate, which converts any Fe(III) to a stable phosphate complex that is not reduced under the conditions of the analysis.

Another potential source of difficulty is *adsorption* of iodine on the surface of the Cu_2I_2 precipitate. It is believed that iodine molecules interact with iodide ions on the surface of the crystals, forming an ion analogous to I_3^- whose desorption is quite slow. As a result, the starch-iodine endpoint is not sharp. The problem is solved by addition of potassium thiocyanate to the solution. Since cuprous thiocyanate is less soluble than cuprous iodide, the thiocyanate ions quickly replace I^- and I_3^- ions in the surface layers of the crystals and thus return the adsorbed iodine into solution.*

Dissolution of the brass sample. Using guidelines provided by the instructor, determine the appropriate sample size (usually 0.2–0.5 g) and weigh three or four brass samples into 250 ml Erlenmeyer flasks.

Dissolve each sample in 4–5 ml of concentrated nitric acid. When the sample has dissolved (tin will remain undissolved as a white solid), add 10 ml of concentrated sulfuric acid. Rinse the sides of the flask with the acid and evaporate *under a hood.* Use low heat and swirl the flask to avoid bumping and spattering. Continue the evaporation until dense white fumes of sulfur trioxide are seen in the flask. (Do not confuse the fumes with water vapor.) Cool, add 20 ml of water and, after all of the copper salt has dissolved, cool again.

All traces of nitric acid are removed by the fuming. Tin precipitates initially as an insoluble tin(IV) oxide; lead precipitates as lead sulfate. Only copper(II), zinc(II), and iron(III) will go into solution. The Zn^{2+} ions will not interfere, and the iron will be converted to an unreactive phosphate complex.

Titration. Add concentrated ammonium hydroxide until the solution just turns dark blue. Discharge the deep blue color with dilute acetic acid, and add 3–4 ml of glacial acetic acid in excess. Add 2 ml of concentrated phosphoric acid; cool. Now add 2–3 g of potassium iodide and stir well.

Titrate the liberated iodine immediately with standard thiosulfate solution. When most of the iodine has disappeared (leaving brownish cuprous iodide), add starch indicator. Continue the titration until the starch-iodine color has nearly disappeared. Then add about 2 g of potassium thiocyanate and dissolve by swirling. The blue color will return. Titrate very carefully until the disappearance of the blue color signals the endpoint.

If a significant endpoint correction was determined in the control experiment in P.16.8, subtract this amount from the final volume of titrant.†

Report the percentage of copper in the brass sample.

---

* We do not know whether thermodynamic equilibrium between solid cuprous iodide, solid cuprous thiocyanate, and ions in solution is reached during the short time required to reach the titration endpoint.

† The thiosulfate used in this analysis may be standardized against pure copper metal following the procedure given here for brass. If this is done, no endpoint correction need be made. While standardization by the same procedure as in the actual analysis is usually preferred because systematic errors tend to cancel out, we recommend the standardization with iodate (P.16.8) because it is easier to do and because it will teach you an important and different method.

17 chapter

Chemical Kinetics

One of the uses of quantitative analysis is to monitor the progress of chemical reactions and to establish their rate laws. In the "batch method" of measuring reaction rates, one starts the reaction at a known time by suddenly mixing solutions of the reactants at the desired temperature. Then, at a known later time, one suddenly "stops" the reaction, by dropping the temperature or by inactivating a reactant or catalyst, and determines its progress by titration.

Besides the familiar variables of titration, two additional variables now become important: the time (t), and the temperature (T). Because precision of timing is important, you must discipline yourself to work quickly and purposefully. Do not dally!

Precise control of the temperature is important only during the time that the reaction is actually in progress, from the time of mixing the reactant to the time the reaction is "stopped." However, during that time the temperature should be known to 0.1°C or better, because reaction rates studied by the batch method typically have temperature coefficients of 5–10 percent per degree C. In practice, reaction mixtures are placed in a constant-temperature bath whose temperature fluctuates by less than 0.1°C and (barring power failures) remains steady indefinitely. Unfortunately, constant-temperature baths of that quality are rarely available in the introductory analytical laboratory. For this reason we shall study a reaction that is fast enough so that adequate kinetic results can be obtained during a continuous two-hour period at room temperature, hoping that changes in ambient temperature will thus be tolerably small. However, except for the inadequate temperature control, the experiment will be conducted in a thoroughly professional manner. Because of the need for fast action, and because of the substantial amount of glassware required, we suggest that you do this experiment in collaboration with a partner.

REACTION TO BE STUDIED

In the following experiment, you will study the kinetics of the reaction between ethyl lactate (EtL) and hydroxide ion (OH⁻) in aqueous solution. As shown in Eq. (17.1), the products are ethyl alcohol (EtOH) and lactate ion (L⁻).

$$
\underset{\substack{\text{Ethyl} \\ \text{lactate}}}{CH_3CH_2O\overset{\displaystyle O}{\overset{\|}{C}}\underset{\substack{| \\ OH}}{C}HCH_3} + \underset{\substack{\text{Hydroxide} \\ \text{ion}}}{OH^-} \longrightarrow \underset{\substack{\text{Ethyl} \\ \text{alcohol}}}{CH_3CH_2OH} + \underset{\substack{\text{Lactate} \\ \text{ion}}}{{}^-O\overset{\displaystyle O}{\overset{\|}{C}}\underset{\substack{| \\ OH}}{C}HCH_3} \quad (17.1)
$$

This reaction can be followed conveniently by acid-base titration for hydroxide ion, which is consumed as the reaction progresses. Ethyl lactate and ethyl alcohol are neutral solutes in water, and lactate ion is a very weak base ($pK_B = 10.14$ at 25°C). Bromothymol blue is a convenient endpoint indicator for this titration.

In previous kinetic studies, the reaction of ethyl lactate with hydroxide ion was found to be second-order: first-order in each reactant. The reaction half-life was found to be on the order of 10 minutes at 25°C when the initial concentrations of the reactants were 0.001 M. In the following kinetic measurements we shall likewise use initial concentrations of 0.001 M and "stop" the reaction (by adding acid) after 5, 10, 15, and 20 minutes. Our data will enable us to test the second-order rate law and to calculate the rate constant.

ANALYTICAL METHOD

The reaction is started by mixing a dilute solution of ethyl lactate with a dilute solution of sodium hydroxide. As the reaction proceeds, hydroxide ion is consumed and the hydroxide concentration therefore decreases with time. The hydroxide concentration is determined by back-titration: At a time of t minutes after mixing, the reaction is "stopped" by adding excess hydrochloric acid to neutralize the hydroxide ion. The amount of acid in excess is then determined by back-titration with sodium hydroxide. Titration must be done at once, because ethyl lactate hydrolyzes slowly in acid. The endpoint is that point at which the color due to the bromothymol blue indicator changes sharply from greenish blue to blue. The blue endpoint color will, in time, revert to green, because ethyl lactate reacts further at the alkaline pH corresponding to blue. But there should be no trouble in identifying the endpoint. Efficient stirring during the titration is advisable to avoid local regions of high alkalinity in the titration flask.

In addition to following the hydroxide concentration as a function of time, we also need to know the initial concentration of ethyl lactate. However, no other concentrations need be measured. To *calculate* the concentrations of the other solutes as a function of time, we apply the principle of mass balance. In the following, the subscripts zero and t indicate that $t = 0$ and $t = t$, respectively. The experiment is done so that $[\text{EtOH}]_0 = [\text{L}^-]_0 = 0$. The equations of mass balance are therefore (17.2) through (17.4).

Conservation of ethyl (CH_3CH_2) groups

$$[\text{EtL}]_0 = [\text{EtL}]_t + [\text{EtOH}]_t \tag{17.2}$$

Conservation of lactate (L) groups:

$$[\text{EtL}]_0 = [\text{EtL}]_t + [\text{L}^-]_t \tag{17.3}$$

Conservation of hydroxyl (OH) groups:

$$[\text{OH}^-]_0 = [\text{OH}^-]_t + [\text{EtOH}]_t \tag{17.4}$$

These equations are readily rearranged to express the concentrations of the other solutes as a function of the measured hydroxide concentrations and of $[\text{EtL}]_0$.

$$[\text{EtOH}]_t = [\text{L}^-]_t = [\text{OH}^-]_0 - [\text{OH}^-]_t \tag{17.5}$$

$$[\text{EtL}]_t = [\text{EtL}]_0 + [\text{OH}^-]_t - [\text{OH}^-]_0 \tag{17.6}$$

REAGENTS AND CONTROL EXPERIMENTS

P.17.1 0.01 *N* HCl and 0.01 *N* NaOH

If stock solutions of standard 0.01 *N* HCl and 0.01 *N* NaOH are available in the laboratory, obtain 500 ml of each for use in all kinetic experiments. Otherwise prepare such solutions by quantitative dilution of 0.1 *N* solutions prepared previously. In either case, standardize the 0.01 *N* NaOH solution by titrating 25.00 ml of 0.01 *N* HCl with the NaOH solution, using 3–4 drops of bromothymol blue indicator. The endpoint is taken as the sharp color-change to blue. Do the standardization in triplicate and, from the average result, calculate the normality of the NaOH solution.

P.17.2 Test of distilled water

Because the solutions used in the kinetic experiments are rather dilute, it is important that the distilled water be free from acidic and basic impurities. Deliver 50.00 ml of the laboratory distilled water into a 250 ml Erlenmeyer flask, add 3–4 drops of bromothymol blue indicator, and note the color. If the indicator turns blue, one drop of 0.01 *N* HCl (delivered from a medicine dropper) should cause a distinct color change to blue-green or green; if not, consult your instructor. More likely, the indicator will turn green or greenish yellow. In that case, titrate with 0.01 *N* NaOH until the color changes to blue. If more than 0.10 ml of NaOH is required, take a fresh 50.00 ml sample of distilled water and expel dissolved CO_2 by boiling or by bubbling a slow stream of nitrogen gas through it, with gentle swirling, for about two minutes. Then titrate again to the bromothymol blue endpoint. If the volume of NaOH required is still more than 0.10 ml, consult your instructor.

If this test indicates that the laboratory distilled water contains too much carbon dioxide, prepare one liter of CO_2-free distilled water for use in the following procedures (by boiling or by sweeping out the CO_2 with nitrogen) and store in a glass-stoppered bottle.

P.17.3 0.01 *F* solution of ethyl lactate

Pure ethyl lactate is a relatively nonvolatile liquid at room temperature, with a density of 1.030 g per ml at 25°C; boiling point 154.5°C (1 atm); formula weight 118.13.

To prepare 500 ml of 0.01 *F* solution, deliver (with a graduated pipet)

0.59 g (0.57 ml) of pure ethyl lactate into a tared glass-stoppered weighing bottle, being careful not to wet the ground glass neck. Replace the stopper and weigh again. The net weight *must* be between 0.54 and 0.64 g. If it is outside that range, add or remove ethyl lactate as required. Place a funnel into the neck of a 500 ml volumetric flask and, with the aid of water from your wash bottle, transfer the ethyl lactate quantitatively from the weighing bottle to the volumetric flask, then add water (with proper mixing) to the 500 ml mark. All water used in this procedure should be able to pass the test of P.17.2.

The solution of ethyl lactate should be sensibly neutral. To test for neutrality, titrate 50.00 ml of your solution with 0.01 N NaOH to the bromothymol blue endpoint. If more than 1 ml of NaOH is required, consult your instructor.

P.17.4 Equivalent weight of ethyl lactate

Deliver 30.00 ml of 0.01 F ethyl lactate (P.17.3; use a 25 ml pipet plus a 5 ml pipet) into a 250 ml Erlenmeyer flask. Add 50.00 ml of 0.01 N NaOH solution, cover the flask, and allow to stand for at least 30 minutes. Under these conditions, reaction (17.1) will go practically to completion.*

Add 3–4 drops of bromothymol blue indicator and 25.00 ml of 0.01 N HCl. Swirl and note the color. If the solution is still basic (that is, if the color is still blue), add another 5.00 ml of HCl. Back-titrate the acidified solution with 0.01 N NaOH to the bromothymol blue endpoint.

Calculate the normality of ethyl lactate. The volume (V) is 30.00 ml or 0.030 liter. The number of equivalents is equal to the difference between the equivalents of NaOH and those of HCl. (Why?)

Calculate the equivalent weight according to Eq. (1.16). [*Hint:* Rewrite the equation in the form, equivalent weight = (weight per liter)/ (normality).]

KINETIC EXPERIMENTS

P.17.5 Hydroxide concentration as a function of time

The following experiments require a stopwatch, or a watch or clock with a second hand.

Into each of three 250 ml Erlenmeyer flasks, deliver 75.0 ml of tested distilled water, 10.00 ml of 0.01 F ethyl lactate, and 4 drops of bromothy-

---

* At these concentrations ([EtL]$_0$ $\sim$ 0.004 F; [OH$^-$]$_0$ $\sim$ 0.006 F), 30 minutes is more than ten half-lives, even though we noted earlier that at initial concentrations of 0.001 F the half-life is approximately 10 minutes.

mol blue indicator. Stopper the flasks, or cover them with an inverted beaker or plastic wrap, and allow to come to room temperature.

To start the reaction, deliver 10.00 ml of 0.01 N NaOH into one of the Erlenmeyer flasks. Record "zero time" as the time when about half of the pipetful has been delivered. Swirl continuously while the NaOH is being introduced to avoid the accumulation of base in regions of high local concentration. After 10.00 ml of base has been delivered, swirl for another second, then stopper or cover the flask and allow it to stand. Record the room temperature.

After approximately 10 minutes, deliver (with continuous swirling) 10.00 ml of 0.01 N HCl to "stop" the reaction. Record "stop time" as the time when about half of the pipetful has been delivered. Back-titrate at once with 0.01 N NaOH, swirling vigorously, until the indicator changes to green. Then continue the titration, adding one drop at a time and swirling (do not split drops; it is too slow for best results here) until the blue endpoint color persists for at least one or two seconds.

Do a second experiment in which the reaction time is about 15 minutes. After that has been done, do a third experiment in which the reaction time is about 20 minutes, and a fourth experiment in which the reaction time is about 5 minutes.

Calculate the hydroxide concentration in the reaction mixture (1) at zero time, by recalling that 10.00 ml of 0.01 N NaOH is diluted to a reaction volume of 95.0 ml, and (2) at "elapsed time" t (where t = "stop time" − "zero time"), from the titration results. Then summarize your results for the four experiments in the form of a table:

Run #	"Zero time"	"Stop time"	t	$[OH^-]_t$	$[OH^-]_0 - [OH^-]_t$	Room temperature
	—	—	zero	$[OH]_0$	zero	—
1			t_1	etc.	etc.	T_1
	(initial	(final				
2			t_2			T_2
	clock	clock				
3			t_3			T_3
	reading)	reading)				
4			t_4			T_4

Test of rate law and calculation of rate constant

You are now ready to test the second-order rate law (17.7) and to calculate the rate constant k.*

* A more detailed treatment of this material may be found in any elementary physical chemistry text as well as in many (often without use of the calculus) general chemistry texts.

$$-\frac{d[OH^-]}{dt} = k[OH^-][EtL] \tag{17.7}$$

The experiments have been designed so that $[OH^-]_0 \approx [EtL]_0 \approx 1.0 \times 10^{-3}\ M$. It is therefore convenient to define an average "zero-time" concentration a according to (17.8) and to express the change in concentration by x, which is defined in (17.9).

$$a = \frac{([OH^-]_0 + [EtL]_0)}{2} \tag{17.8}$$

$$x = ([OH^-]_0 - [OH^-]_t) \tag{17.9}$$

On introducing these quantities in (17.6) and (17.7) and assuming that $[OH^-]_0 = [EtL]_0$, we transform Eq. (17.7) to (17.10).

$$\frac{dx}{dt} = k(a - x)^2 \tag{17.10}$$

Integration of Eq. (17.10) between the limits $t = 0$ and $t = t$ leads to (17.11).

$$\frac{1}{a - x} - \frac{1}{a} = kt \tag{17.11}$$

Mathematical analysis shows that Eq. (17.11) is an excellent approximation even if $[EtL]_0$ and $[OH^-]_0$ are not exactly equal, provided that their difference is less than 20 percent and that a is defined according to (17.8).

Test Eq. (17.11) by means of your data as follows: Construct a graph of your data in which you plot $y = 1/(a - x)$ versus t. Include the point $y = 1/a$ when $t = 0$. If the second-order rate law is correct, the experimental points will accurately define a straight line.

Calculate the rate constant k from the slope of the plot of your data. What are the units? What is the average reaction temperature?

chapter **18**

Potentiometric Titrations

Chemical analysis by potentiometry has become a powerful tool of scientists and engineers in many fields of endeavor. For example, by using specific sensing electrodes such as those discussed in Chapter 10 (Tables 10.3, 10.4, 10.5), water quality control personnel can determine selectively the concentrations of a host of natural impurities or additives (e.g., fluoride in drinking water). Environmental engineers can detect pollutants. Oceanographers can measure the concentrations of specific ions in sea water. Soil scientists can detect ions in moist soil or mud. Ionic composition of blood, bone, and tissue samples can be determined by medical researchers. Potentiometric measurements are made routinely in many industrial plants to maintain (often automatically) proper composition of solutions used in, for example, food processing, metal-plating, and pharmaceutical manufacture.

The most common of all potentiometric measurements is, of course, that of pH. Thus direct-reading electronic potentiometers are often called "pH meters" by their manufacturers.

The procedures given in this chapter are designed to illustrate the operation of the pH meter and its application to pH, redox, and precipitation titrations employing one of each of the three types of sensing electrodes described in Chapter 10.

The type of reference electrode used for a titration is simply a matter of availability. You will probably use either a saturated calomel (S.C.E.) or a Ag-AgCl electrode.

OPERATION OF THE pH METER

Since this is probably your first experience with a pH meter, let the instructor explain its operation. Then study the manufacturer's directions until you are sure you understand the function of every switch and dial.

pH meter

A pH meter is essentially a precision potentiometer designed for high-resistance circuits.* The particular pH meter that you will be using is powered by the laboratory's regular 115 V ac supply. The typical pH meter will have the following physical features.

1. A large meter which reads directly in pH units. Some meters also have a millivolt scale (and a knob which selects scales). Either scale can be used as "arbitrary units" for any potentiometric titration.

2. A "standby-read" knob. The meter should always be left on "standby" except when measurements are being taken. *Never* remove the electrodes from solution with the switch in the "read" position. Serious damage to the instrument or electrodes might result.

3. A "standardize" or "calibrate" knob. When accurate pH measurements are desired, the instrument is calibrated by immersing the electrodes in a buffer solution of known pH and setting the proper value on the pH scale. For a potentiometric titration where calibration is not needed, use this knob to set the meter to any convenient reading such that the needle will not run off scale during the course of the titration.

---

* For pH measurements with a glass electrode, electric current flowing in the cell must pass through the glass membrane. The internal resistance of the cell is, therefore, very large, on the order of 10^8 ohm. (By comparison, the internal resistance of a small $1\frac{1}{2}$ V dry-cell is about 100 ohm.)

4. Input jacks for sensing and reference electrodes. Many instruments also have an electrode holder.

5. A "temperature compensation" knob (on more expensive models). This dial should be set to the temperature of the sample (room temperature except in special cases).

Electrodes

The electrode compartments are more fragile than most glass apparatus, and, unfortunately, quite expensive. Handle them gently. In particular, avoid subjecting the glass envelopes to large torques. (Remember the formula: torque = force × lever arm. If you *must* push or pull hard, place your hand close to the trouble spot where the bind is, so that, if you don't push straight, the torque will be small because the lever arm is small.) Read the manufacturer's directions and consult your instructor about how to mount and prepare the electrodes for use.

Since the electrodes will be in continuous use over several weeks, they should be stored between laboratory periods with their tips immersed in buffered aqueous KCl. This is especially important for the glass electrode, for if the glass membrane is allowed to dry out, the electrode may lose its easy reversibility and may not regain it until after many hours of soaking in aqueous buffer.

P.18.1 Potentiometric titration procedure

The step-by-step outline below gives a general method for potentiometric titration using a pH meter. Specific instructions and modifications will be supplied by your instructor, the pH meter manual, and your own experience.

1. Set up the meter, electrodes, buret, and mechanical stirrer in an area where it will be safe from human traffic, spills, and fumes from corrosive chemicals. Use a beaker large enough to accommodate sensing electrode, reference electrode or salt bridge, stirrer, and buret tip as well as substrate and titrant solution. Be sure the meter is on "standby," then plug it in and allow it to warm up.

2. Rinse off the electrodes thoroughly and carefully with distilled water and blot them dry with a lint-free tissue.

3. If the meter is to be standardized, place the electrodes in a buffer of known pH, switch the meter to "read," and set the scale with the "standardize" dial. Return the meter to "standby," raise the electrodes, and clean them as in step (2).

4. Pipet the appropriate quantity of substrate into the titration beaker,

dilute if necessary,* insert the mechanical stirrer and lower the electrodes.†
If the *p*H meter scale is to be used without calibration, switch the instrument to "read" and set the scale to some convenient reading with the "standardize" knob. If the instrument has been standardized, simply take an initial reading. Record this "zero ml of titrant" reading as the first entry in a list such as that in Table 10.6. The mechanical stirrer may be left on throughout the titration unless it seems to cause the meter to fluctuate.‡

5. Add some titrant from the buret, wait for the meter to stabilize, and record buret and meter reading. Repeat this process, adding titrant in successively smaller increments as the endpoint approaches and $\Delta E/\Delta V$ becomes larger. Continue the titration until well past the endpoint, that is, until $\Delta E/\Delta V$ has clearly passed through a maximum, making appropriate adjustments in the amount of titrant added between readings.

6. Repeat the titration. After the last run of the day, clean the electrodes as in step (2). Dismantle and store all equipment as directed by the instructor.

7. Determine the endpoint volume (V_e) for each titration according to the maximum-slope method described in Chapter 10. In some cases you may also wish to plot your data.

AQUEOUS ACID-BASE TITRATIONS

As explained in Chapter 10, the *p*H of aqueous solutions is measured conveniently with an electrochemical cell employing a glass sensing electrode and an appropriate reference electrode. The *p*H meter is usually standardized before each series of titrations. Your instructor will provide the appropriate buffer solution.

P.18.2 Procedures from Chapter 13

Any of the aqueous titrations described in Chapter 13 can be adapted to potentiometric study. Your instructor will assign you one or more of these experiments. Follow the appropriate directions for preparing

* The liquid level must be high enough to allow immersion of the electrode tips without interference from the magnetic stirring bar (if one is used). This often requires the addition of some solvent.

† For aqueous *p*H titrations (where no salt bridge is needed) a single "combination electrode" containing glass and reference electrodes in the same unit may be used.

‡ Magnetic stirrers often induce meter drift or sluggishness, especially near the end point of a titration. If this happens, simply shut off the stirrer each time you take a reading.

standard solutions,* dissolving samples, etc., and titrate (no indicator, of course) according to P.18.1.

For experience in potentiometric titration, it is a good idea to construct the entire titration curve for a few cases, as well as determine the endpoint by the maximum-slope method.

If you titrate a weak acid with NaOH, determine the pK_A of the acid (equal to pH when $V = \frac{1}{2} V_e$) along with your other results.

NONAQUEOUS ACID-BASE TITRATIONS

In Chapters 5 and 6 we have shown how variations in acidity, basicity, and dielectric constant of the solvent can lead to sizable changes in the equilibrium constant for titration. These principles are dramatically illustrated in Figs. 7.3 and 7.4 which compare the titration of sulfuric acid with strong base in water and in 95 percent acetone. The latter titration results in the discrete, stepwise removal of the two H_2SO_4 protons; the titration in water shows only one break. In the procedure that follows, you will perform similar titrations to determine the composition of a solution containing both H_2SO_4 and p-toluenesulfonic acid, $C_7H_7SO_3H$.

There is a marked similarity in the molecular structures of un-ionized sulfuric acid and p-toluenesulfonic acid.

$$
\begin{array}{cc}
\overset{\displaystyle :\ddot{O}:}{H:\ddot{O}:\overset{}{\underset{\displaystyle :\ddot{O}:}{\ddot{S}}}:\ddot{O}:H} & \overset{\displaystyle :\ddot{O}:}{H:\ddot{O}:\overset{}{\underset{\displaystyle :\ddot{O}:}{\ddot{S}}}:C_7H_7}
\end{array}
$$

Sulfuric acid   *p*-**Toluenesulfonic acid (HTs)**

p-Toluenesulfonic acid (HTs) is a strong monobasic acid—not as strong as H_2SO_4, but decidely stronger than HSO_4^-. As a result, the potentiometric titration curve of a mixture of H_2SO_4 and HTs with strong base in 95 percent acetone will show two discrete steps: The first equivalence point measures *two* reactions, $H_2SO_4 \rightarrow HSO_4^-$ *and* HTs $\rightarrow$ Ts$^-$, while the second equivalence point measures only the reaction, $HSO_4^- \rightarrow SO_4^{2-}$.

P.18.3 Potentiometric titration of a sulfuric acid-toluenesulfonic acid mixture

Your unknown sample will be an aqueous solution of sulfuric and p-toluenesulfonic acids with a total acid strength of about $1\ N$. In aqueous

---

* To save time standard titrant solutions may be supplied to you or you may use previously prepared solutions.

solution, the three protons (two from H_2SO_4, one from $C_7H_7SO_3H$) are titrated nonselectively by strong base. When the sample is titrated with strong base in an acetone medium, there are two breaks in the titration curve: the first resulting from the neutralization of one proton from each acid, the other from the titration of the second sulfuric acid proton.

Titration in water.* Accurately dilute 5.00 ml of the unknown sample to 100.0 ml with water in a volumetric flask. Pipet 25.00 ml of this solution into a beaker and titrate with standard 0.05 N NaOH† by the method described in P.18.1.

No salt bridge is necessary to separate the glass electrode from the reference half-cell. You should see only one break in the titration curve, corresponding to the total acid normality.

Nonaqueous titration. Pipet 5.00 ml of the original, undiluted, unknown solution into a 100 ml volumetric flask and dilute to the mark (mix well!) with acetone. The solvent is now 95 percent acetone and 5 percent water. Your instructor will provide you with a *rationed* volume of standard 0.06 N tetrabutylammonium hydroxide ($Bu_4N^+OH^-$) in methanol to be used as titrant.‡

Because acetone interferes with the reference electrode (both Ag-AgCl and S.C.E.), the titration must be carried out with a salt bridge. The appropriate glassware will be supplied to you.

Pipet 25.00 ml of the acetone-acid solution into the titration vessel and insert electrodes and stirrer. Although the glass electrode can be used for this titration, the pH scale of the pH meter will read in arbitrary units. Use the "standardize" knob to bring the meter needle to a point on the scale from which the entire titration can be followed.§ Alternatively, if your meter reads in millivolts, set the switch to "+MV" and take your first reading. Change to "−MV" as required.

Perform the titration as described in P.18.1. Use the positions of the first and second endpoints to determine the total acid normality‖ and the concentration of each acid in the unknown sample.

---

* If time is short, this part may be omitted since it yields quantitative information that is also obtained in the nonaqueous titration. It does, however, serve to provide experience if you have not previously done a pH titration, illustrate the difference between the aqueous and nonaqueous titrations, and check the result of the acetone titration.

† Use either the solution prepared in P.13.3 or a standard solution provided by the instructor.

‡ The titrant is prepared by dilution of 25 percent tetrabutylammonium hydroxide in methanol (Eastman EK7774). Be frugal in your use of this titrant—it is expensive. Rather than using large amounts of titrant to rinse out your buret, you may use methanol except for the last rinsing.

§ Refer to Fig. 7.4, but remember that higher numbers on the pH scale correspond to decreasing emf.

‖ If this value agrees with that from the aqueous titration, you may assume that the solvents do not contain acidic or basic impurities. If there is a *significant* discrepancy,

POTENTIOMETRIC REDOX TITRATIONS

Half-cells that are sensitive to both members of a redox couple are discussed in Chapter 10 and some examples are given in Table 10.4. Platinum is by far the most widely used electrode material because of the ease with which reactions occur on a properly prepared platinum surface. The platinum sensing electrode is often called an *inert* electrode, because platinum does not (except as an active surface) participate in the redox reaction. The electrode itself may be a piece of platinum foil or simply a short length of platinum wire sealed into a glass tube and connected either by solder or by liquid mercury to the electrical lead.

There are a large number of redox titrations that can easily be performed with the platinum and reference electrodes in the same beaker. We suggest the titration of ferrous iron with cerium(IV) as a student experiment.

P.18.4 Titration of Fe(II) with Ce(IV)

You will determine the percent of iron(II) in a water-soluble sample by potentiometric titration with a standard cerium(IV) solution.

Titrant. If a standard cerium(IV) solution is not provided, prepare a standard $0.1\ N$ solution by direct dissolution of dry primary-standard grade ceric ammonium nitrate, $(NH_4)_2Ce(NO_3)_6$. The solution should be made $\sim 1\ F$ in H_2SO_4.*

Substrate. Following your instructor's guidelines, weigh out three samples into dry beakers. Just before each titration, dissolve one of the samples in about 100 ml of $1\ F\ H_2SO_4$. Ferrous ion is easily oxidized to ferric ion by oxygen. You should, therefore, be careful to avoid unnecessary exposure to air. A small amount (~ 0.1 g) of solid $NaHCO_3$ added to the substrate solution will remove dissolved O_2.

Titration. Titrate each sample *immediately* after preparation. Intro-

---

check the acetone reagent used in diluting your unknown sample to make sure it is neutral. A convenient procedure is to take 5.00 ml of the acetone-acid solution (instead of 25.00 ml), add 1.0 ml of distilled water and 19 ml of acetone, and titrate potentiometrically with $0.06\ N\ Bu_4N^+OH^-$ as above. The two endpoint volumes now should be one-fifth of the endpoint volumes found before. If not, apply the appropriate endpoint corrections.

* If pure ceric ammonium nitrate is not available, an approximately $0.1\ N$ solution can be prepared and standardized with As_2O_3 (see P.16.2), using OsO_4 or ICl as catalyst and ferroin (ferrous 1,10-phenanthroline sulfate) as indicator.

duce platinum* and reference (S.C.E. or Ag-AgCl) electrodes and mechanical stirrer. Switch the pH meter to "read" and set the scale. (The voltage will increase as titrant is added.) Titrate with standard ceric solution according to P.18.1 and report your results.

POTENTIOMETRIC DETERMINATION OF MIXED HALIDES

In the procedure that follows, a silver-silver halide electrode is used for the potentiometric titration of a mixture of halide salts with silver nitrate. The electrode is simply a silver wire that has been "sensitized" by treatment with nitric acid to provide an active surface layer of silver(I). When such a wire is dipped into a solution containing a halide salt, the silver wire becomes coated with a microscopic deposit of solid silver halide. If several halide salts are present, the composition of the silver halide coating is such that the electrode exists in equilibrium with the solution. Thus, as silver nitrate is added and first I^-, then Br^-, and then Cl^- are removed from solution, the electrode changes successively from Ag-AgI to Ag-AgBr to Ag-AgCl. Accordingly, the potentiometric titration curve will show three discrete breaks corresponding to the selective titration of first I^-, then Br^-, and finally Cl^-.

Unfortunately, coprecipitation and the formation of solid solutions prevent optimum accuracy from being achieved (unless special precautions are taken or empirical corrections applied). The titration normally results in errors no greater than ± 0.5 percent for iodide, ± 3 percent for bromide, and ± 4 percent for chloride.† However, the bromide and chloride endpoint errors tend to be of similar magnitude and opposite sign so that the total titrant volume is close to the theoretical amount.

P.18.5 Potentiometric titration of a halide mixture

You will be asked to determine the composition of a mixture of KI, KBr, and KCl.‡

Prepare 250 ml of a solution that would be $\sim 0.1\ N$ if the sample were 100 percent potassium iodide.§

---

* A few seconds in aqua regia (3:1 concentrated HCl to concentrated HNO_3) will regenerate a contaminated Pt surface.

† D. Jaques, *J. Chem. Educ.*, **42**, 429 (1965).

‡ If only two salts are present, you will be told which ones.

§ Because of the rather large inaccuracy of the method, it is not necessary to dry the sample.

Prepare a standard $0.1\,N$ $AgNO_3$ solution as described in P.14.1. Sensitize the silver wire electrode by dipping it in $6\,N$ HNO_3 containing a little $NaNO_3$ until bubbles begin to form. Connect the reference electrode to the titration vessel via a salt bridge. (Why is this necessary?)

Pipet 50.00 ml of the unknown solution into the titration beaker. Then immerse the electrodes and stirrer, set the pH meter scale, and titrate to all three endpoints.

Use the titration results to calculate the percentage of each potassium halide in the solid unknown. If the total does not add up to 100 percent, is the discrepancy experimentally significant?

CONCLUDING REMARKS

We began this chapter by mentioning a few applications of potentiometric measurements currently employed in industry and research. Many of these determinations are done quickly and easily using ion-selective membrane electrodes. However, because such electrodes are expensive to buy and maintain, the procedures described in this chapter make use only of the more common glass and metallic electrodes. If your laboratory is set up with some of the more exotic electrodes so that you can undertake the titration (or estimation from a calibration curve) of fluoride, cadmium, etc., you will be provided with special instructions. In any case, we hope that your experiments will demonstrate the simplicity and scope of modern potentiometric methods of analysis.

19 *chapter*

Some Additional
Instrumental Methods

In this chapter we add three new techniques of instrumental analysis to the potentiometric titration methods described in Chapter 18. The first two methods—spectrophotometry and conductometry—have been discussed in Chapters 1, 7, and 10. The third—amperometric titrations—will probably be new to you. We shall therefore present the principles of this technique as well as describe laboratory procedures.

SPECTROPHOTOMETRIC ANALYSIS

In Chapter 10 we demonstrated that spectrophotometric measurements can be used to determine stoichiometry and equilibrium constants as well as to follow the course of titrations. In actual practice we often use the high specificity and reliability of absorption data in the ultraviolet and visible region to determine the concentration of a solute *directly* from the absorbance at a particular wavelength and a previously measured extinction coefficient. (See Eq. (1.2).) However, for each case this procedure must be established as valid before routine measurements may be begun. The preliminary experimentation must answer such questions as: Does Beer's law hold in the concentration range of interest? Is the unknown sample likely to contain other solutes that absorb appreciably at the given wavelength? Was the extinction coefficient measured with an instrument having the same optical characteristics as the one used to analyze the unknown sample?* The consideration of these questions will point out some of the pitfalls that must be avoided if accurate results are to be obtained.

The procedures described in this section are designed to provide you with experience in the operation of the spectrophotometer and with some applications to analytical problems.

Since many types of spectrophotometers (the simpler instruments are often called *photometers* or *colorimeters*) are available for student use, we shall leave the manipulative details to your instructor, who will demonstrate the instrument used in your laboratory. Become thoroughly familiar with the instrument before you need to use it for your first important measurement.

P.19.1 Determination of the pK_A of an acid-base indicator

In planning your work for this laboratory, be sure you understand the principles of determining equilibrium constants from spectral absorption data and can derive Eq. (10.24). You will be given a solid sample or liquid solution of one of the indicators listed in Table 9.2 with a pK_A between 3 and 9.

Solutions. You will need the following solutions:†

(a) $1\,F$ HNO$_3$. This solution must be colorless.
(b) $1\,F$ NaOH. This solution must be colorless.
(c) Three buffer solutions (the particular buffer system will depend on the transition range of your indicator) with $pH \approx pK_A + 0.4$, $pH \approx pK_A$, and $pH \approx pK_A - 0.4$, where pK_A is the acid dissocia-

---

* A spectrophotometer of high resolution will give an absorbance reading corresponding to a very narrow wavelength region. An instrument of lower quality will give a reading that is an average of absorbances over a wider wavelength region.

† To save time, solutions may be prepared for you. Otherwise, you will be given the identity of your indicator in advance so you can determine the composition of your buffer solutions before coming to class. The total ionic strength of each buffer solution should be about $1\,M$.

tion constant of the indicator. These solutions must be colorless.

(d) A solution of your indicator (usually $\sim$0.05 g in 100 ml of ethanol or water) such that the spectrophotometer does not go off scale at any wavelength used for recording spectra (next paragraph). The formal concentration of indicator need not be known unless you are asked to report accurate values of extinction coefficients for HIn and In.

Pipet 1.00 ml of your indicator solution into each of five 50 ml volumetric flasks. Add 10 ml of 1 F HNO$_3$ to one, 10 ml of 1 F NaOH to the second, and 10 ml of the three buffer solutions, respectively, to the last three flasks. Dilute to the mark with water and mix. Measure the pH of the three buffered solutions with a pH meter.

Spectra. After you have learned to use the spectrophotometer, measure the absorption spectra of the acidic (in HNO$_3$) and basic (in NaOH) indicator solutions between 4000 and 6500 Å. Measurements should be at 200 Å intervals except near absorption maxima, where they should be at 50 Å intervals. The "blanks" (whose absorbance is zero, by definition) may be pure water. If the absorbance goes off scale at or near a maximum, prepare a 1:10 dilution of the acidic or basic indicator and remeasure the region in which the previous absorbance reading had been greater than 1.00.

Plot absorbance versus wavelength. The plots will represent the spectra of HIn and unprotonated In, respectively. Compare the two absorption spectra. Pick one wavelength at which the absorbance of HIn is much greater than that of In, and one wavelength at which the absorbance of HIn is much smaller than that of In. These wavelengths are to be used in the subsequent measurements. If such a pair of wavelengths cannot be found, choose two wavelengths in different parts of the spectrum such that (for each) one absorbance is much smaller than the other.

***p*K$_A$ determination.** Measure the absorbance of each of the three buffered indicator solutions at both wavelengths selected for analysis, using pure water as "blank."

Derive an equation analogous to Eq. (10.23) from which you can calculate the degree of dissociation, α, corresponding to each of the six absorbances (three solutions at two wavelengths).* The two values of α obtained for each solution should agree within the range permitted for experimental error.

Now use the average value of α and the pH for each solution to calculate pK$_A$ for your indicator. Compare the final result with the pK$_A$ of the indicator listed in Table 9.2.

---

* Use extinction coefficients if you know c_{HIn}; otherwise you may write the equation in terms of absorbances since all your solutions contain the same total indicator concentration.

P.19.2 Determination of manganese in steel

In this experiment, you will analyze a steel sample for its manganese content by dissolving the steel in nitric acid, oxidizing Mn^{2+} with periodate ion, and measuring the absorbance of the resulting permanganate ion at its absorption maximum.

The absorption spectrum of permanganate ion is shown in Fig. 1.2. The absorbance is greatest at 5250 Å. Since the condition and quality of your spectrophotometer may be different from the instrument used to construct Fig. 1.2, you will need to select a wavelength for measurement and plot a calibration curve of absorbance versus concentration using a standard MnO_4^- solution. A single absorbance measurement, made on the treated unknown sample, will then be used to calculate the percent manganese in the solid steel.

Spectrum and calibration curve. For these measurements, use either a standard $KMnO_4$ solution provided by your instructor or a recently standardized solution prepared as in P.16.1.

If necessary, carefully dilute this solution so that you have a *standard* solution about $4 \times 10^{-4} \, F$ in $KMnO_4$. Using a water "blank" to set zero absorbance for each wavelength, determine the spectrum of this solution between 4000 and 6000 Å. Take as many measurements as you think are necessary to construct a good plot and locate the absorption maximum to within 50 Å.*

Once you have determined the wavelength of maximum absorption (it should be near 5250 Å), measure the absorbance of four more solutions—containing 10, 20, 30 and 40 ml of the $\sim 4 \times 10^{-4}$ solution diluted to 50 ml in a volumetric flask—at this wavelength. Use the five data points to plot absorbance versus concentration for this wavelength. Is Beer's law obeyed?

Treatment of the sample. Steel drillings and shavings are sometimes contaminated by machine oil. If contamination is suspected, wash your sample with ether before weighing. *Do not use ether in a room where flames are present!*

Insert at least three 0.8 g samples into 250 ml beakers. The weights should vary by no more than 4% as one sample will be used (without oxidation by periodate) as an absorbance blank for the others.

Add 50 ml of 6 $NHNO_3$ to each sample and heat until boiling to dissolve the metal. If the steel does not dissolve, add water to make a total liquid volume of about 75 ml and swirl. If this doesn't help, consult your instruc-

---

* Unless you are using a rather high quality instrument, your spectrum may show one broad peak instead of the two peaks and two shoulders displayed in Fig. 1.2.

tor.* After the sample has dissolved, boil the solution for another minute to expel nitrogen oxides.

Add 1 g of ammonium peroxydisulfate ($(NH_4)_2S_2O_8$, usually called *ammonium persulfate* for short) and boil for 10 minutes. The persulfate ion will oxidize carbon (always present in steel) to CO_2. Boiling destroys the excess reagent.

If a brown precipitate (MnO_2) or purple color (MnO_4^-) appears, add a few crystals of sodium sulfite or sodium bisulfite and boil for another one or two minutes. You should now have a clear, yellowish solution.

Dilute each of the sample solutions to about 100 ml and add 10 ml of concentrated (15 F) H_3PO_4. This reagent forms complexes with the iron and should remove all but a slight hint of color.

Transfer *one* of the solutions to a 250 ml volumetric flask. Dilute with water, cool to room temperature, and fill to the mark. This is your absorbance blank.

To each of the remaining solutions, add 0.4 g of KIO_4. Boil for about 3 minutes to oxidize Mn^{2+} to MnO_4^-.† Quantitatively transfer (with rinsing) the solutions to 250 ml volumetric flasks. Dilute with water, cool to room temperature, fill to the mark, and mix well; then allow the solutions to stand until all undissolved silica has settled out.

Absorbance measurement. You are now ready to determine the absorbance of the permanganate ion in each oxidized solution. Use the steel solution that was not treated with periodate as your "blank" to set zero absorbance. This will correct your absorbance readings for possible contributions due to ions such as Co^{2+}, Ni^{2+}, and $Cr_2O_7^{2-}$ that might be present.‡

Use your MnO_4^- calibration curve to determine the concentration of permanganate in each oxidized sample solution. Then translate this concentration into the *weight of manganese* in the sample and finally the *percent manganese* in the steel.

Report the results of individual determinations and the average.

CONDUCTOMETRIC TITRATIONS

The application of conductance measurements to quantitative analysis has been illustrated by the linear titration curves in Figs. 7.2 and 10.4. Of

---

* Do not be concerned if a fine precipitate of silica remains.
† The reaction is

$$2\ Mn^{2+} + 5\ IO_4^- + 3\ H_2O \longrightarrow 2\ MnO_4^- + 5\ IO_3^- + 6\ H^+$$

‡ The absorbance of these ions is small enough, compared to that due to permanganate ion, so that a 4% difference in sample weights causes no appreciable error.

special note in Fig. 10.4 is the accuracy that can be attained despite an innate titration error of 25 percent!

Procedures 19.3 and 19.4 demonstrate the conductometric method as applied to acid-base and precipitation titrations. Because of the wide variety of conductance instrumentation that can be used, we shall not give step-by-step instructions for these experiments. We shall, however, make some general remarks about the measurement of conductance.

Measurement of conductance

By now you are probably familiar with the qualitative aspects of electrical conduction by ions. Beside the fact that strong electrolytes conduct electricity more readily than do weak electrolytes at equivalent concentrations, it can be observed that each electrolyte has its own characteristic charge-carrying ability, known as the *equivalent conductance*. Thus, in the conductometric titration of *para*-nitrophenol with NaOH (Fig. 7.2), the conductivity increases with a steeper slope after the equivalence point than before, because the equivalent conductance of NaOH is greater than that of sodium phenoxide. For a detailed discussion of the equivalent conductance of electrolytes and their relationship to the migration of ions, consult a textbook on general or physical chemistry.

Since electrical conductance is defined simply as the reciprocal of electrical resistance, measuring the conductance of a solution is similar to measuring the resistance of a metallic resistor. However, an alternating current (ac) is used in preference to direct current (dc) because the passage of direct current through an ionic solution is complicated by electrolysis at the electrodes. The frequency of the ac is usually between 1000 and 10,000 cycles per second, and the voltage is 10 V or less. Under such conditions, complications from electrode phenomena are effectively avoided.

Figure 19.1 is a schematic diagram of a simple ac circuit suitable for measuring the conductance of solutions. The measuring apparatus is called a *Wheatstone bridge*. R_1 is a known, fixed resistor. R_2 and R_3 are variable resistors ($R_2 + R_3$ equals a constant). The solution is placed between two platinized platinum electrodes in the cell. An ac signal is fed into the bridge and the ratio of R_2 to R_3 is varied until no current flows through the detector, D, indicating that the bridge is *balanced*. On simple circuits the detector is usually a telephone earpiece. At the balance point, the resistance of the cell is related to the other three resistances by

$$R_c = \frac{R_1 \times R_3}{R_2}$$

The conductance is then simply

$$\text{Conductance} = \frac{1}{R_c}$$

More sophisticated circuits have been designed that incorporate (1) a switch that permits the value of the known resistor R_1 to be varied; (2) a variable capacitor in parallel with R_1, which is used to balance the capacitance of the cell and leads; (3) a more sensitive detector, such as an oscilloscope or a vacuum-tube voltmeter. Convenient "black boxes" called *conductance bridges*, which employ a tuning-eye detector and read directly in conductance, are also based on the ac Wheatstone bridge.

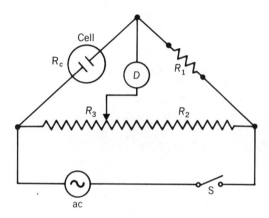

Fig. 19.1. AC resistance bridge. See text for identification of symbols and details of operation.

Before you begin to make conductance measurements, be sure you understand how to operate the instrument assigned to you. Your instructor will demonstrate the proper technique for handling and filling the conductance cell, as well as the best method for using the apparatus.

Since dilution of the sample during titration would tend to cause curvature in the linear portion of the titration curve, dilution must be kept to a minimum. Thus, the titrant should always be at least ten times as concentrated as the substrate. The use of 10 ml burets is advisable to retain the desired precision.

P.19.3 Conductometric acid-base titrations

You will be given separate ~0.01 N solutions of a strong acid and weak acid for titration with standard 0.1 N NaOH. The titrant will be a solution standardized for a previous experiment or one provided by the instructor.

Acquaint yourself thoroughly with the operation of the conductance bridge. Then perform duplicate titrations with *each* of the acid solutions as follows.

Pipet 50.00 ml of acid into the conductance cell and dilute to ~100 ml with water. The mechanical stirring motor may be left on during the titration, but be careful not to disturb the electrodes. Measure and record the conductance (or resistance). Add titrant in approximately 0.5 ml increments, taking buret and conductance (or resistance) readings after each addition. Use at least 9 ml of NaOH for each titration.

Plot conductance versus ml titrant for each run and determine the endpoints from the intersection of straight line segments. Compare and explain the shapes of the curves for the strong and weak acid titrations. Report your results.

P.19.4 Conductometric precipitation titration

Standard 0.05 N silver nitrate and silver acetate solutions provided by the instructor will be used to titrate a $\sim$0.01 N NaCl solution.

After learning how to measure conductance, pipet a 25.00 ml sample of the NaCl solution into the conductivity cell and dilute to about 50 ml with water. Turn on the stirring motor and measure the conductance. Titrate first with $AgNO_3$, adding at least 9 ml of titrant in approximately 0.5 ml increments and taking buret and conductance (or resistance) readings after each addition. Repeat the titration using silver acetate as the titrant.

Be sure to clean the cell well between titrations and after the last run. A bit of 1 F NH_3 may be used to dissolve the precipitate if necessary.

Plot the two titration curves and compare their differences. Determine the endpoint for each titration and report your results.

AMPEROMETRIC TITRATIONS

In an amperometric measurement, one *applies* an emf across an electrochemical cell such that a particular solute in the vicinity of the cathode will be reduced.* The reduction produces a current proportional to the concentration of the solute. Thus, measurement of current as a function of titrant volume results in a linear titration curve.†

---

* Anodic oxidation of the species of interest is less common.

† The method of amperometric titration is an extension of *polarography*—a technique in which the current through a cell is measured as a function of applied voltage. For further reading on this subject, see the instrumental analysis texts cited at the end of Chapter 10, or J. J. Lingane, *Electroanalytical Chemistry*, 2nd ed. (New York, Interscience Publishers, 1958) or any other advanced book on electrochemistry.

Amperometric measurement

The simplest arrangement for amperometry involves a dc potentiometer circuit such as the one shown in Fig. 19.2. However, instead of using the potentiometer to balance the cell voltage, as in a potentiometric emf measurement (Fig. 10.10b), a known voltage is applied in *opposition* to the spontaneous cell reaction. The cell thus becomes very much like an *electrolysis cell* since a nonspontaneous reaction (reduction of a solute that does not reduce spontaneously) is occurring. Unlike an industrial scale electrolysis, however, very little solute is reduced and the macroscopic composition of the cell remains unchanged.

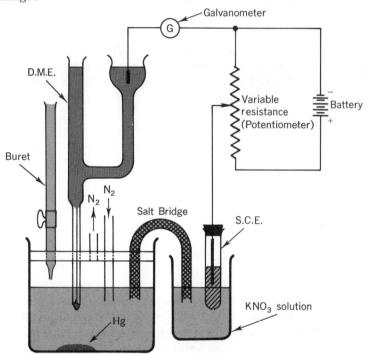

Fig. 19.2. Amperometric titration apparatus. The galvanometer, G, measures the current produced by the reduction of solute at the dropping mercury electrode (D.M.E.).

The mathematics of current production in an amperometric measurement will not be treated here. Let us just point out that if the electrochemical reduction process is very rapid with respect to diffusion of solute to the cathode, then the amount of current will be proportional to the rate of diffusion. The rate of diffusion is in turn proportional to the difference in solute concentration between the bulk solution (c) and the immediate vicinity of the electrode (c_o). If the applied potential is great enough (negative) so that reduction occurs for every solute ion (or molecule) that comes close to the electrode, then c_o will be very small compared to c and we may write

$$\text{Current } (i_d) \propto \text{rate of diffusion} \propto c - c_o \approx c$$

where i_d is the "diffusion current."

By carefully selecting the applied potential for an experiment, we may selectively reduce one solute that is more easily reduced than other components in the same solution. The ease of reduction, as might be expected, follows the same order as standard reduction potentials.

The cathode is commonly a dropping mercury electrode (D. M. E.) (Fig. 19.2) which consists of a capillary tube with a mercury reservoir above it. In operation, mercury drops are constantly formed at the capillary tip, which is immersed in the solution, and drop off at regular intervals. Since the electrode surface area is small, only very small amounts of solute are reduced—not enough to alter the concentration of solute. As each drop of mercury falls, it carries the reduced solute, often in the form of a metal amalgam. This leaves a clean, new electrode surface at the capillary.

Although any suitable reference half-cell may be used, the saturated calomel electrode has become the standard reference for amperometry (and polarography). Indeed, the established convention is to talk about applied voltages "versus S. C. E." Since the S. C. E. has a reduction potential of $+0.242$ V at 25°C, we may write

$$E_{\text{applied}} = E_{\text{DME}} - 0.242 \text{ V}$$

The lead-dichromate reaction

P.19.5 demonstrates the application of the method of amperometric titration to the precipitation reaction between lead and dichromate ions. The "black box" consisting of the electronic circuits will be set up and its operation demonstrated by the instructor.* The reaction cell (Fig. 19.2) will be a covered beaker with D. M. E., salt bridge to reference electrode, buret tip, nitrogen inlet tube† and nitrogen outlet tube all inserted in the cover.

The reaction to be studied is

$$\text{Pb}^{2+} + \tfrac{1}{2}\,\text{Cr}_2\text{O}_7{}^{2-} + \tfrac{1}{2}\,\text{H}_2\text{O} \longrightarrow \text{PbCrO}_4(\text{s}) + \text{H}^+$$

---

* The apparatus will most likely be a manual polarograph similar to those described by O. H. Muller, *J. Chem. Educ.*, **18**, 111 (1941) or I. M. Kolthoff and J. J. Ligane, *Polarography*, 2nd ed., Vol. I. (Interscience Publishers, New York, 1952), p. 297 ff, especially Fig. XVI-2.

† When dissolved oxygen would be reduced (to water) by the applied potential, it is necessary to purge the solution with N_2 before titration.

You will perform the titration at two applied voltages. At -1.0 V versus S. C. E., both substrate and titrant are reduced and produce a current; at 0 V, only the titrant is reduced. Thus linear titration curves with different appearances will result.

As in any linear titration, the titrant will be much more concentrated than the substrate and 10 ml burets will be used.

P.19.5 Amperometric titration of lead with dichromate

You will study this precipitation reaction by titrating a ~0.01 F $Pb(NO_3)_2$ solution with standard 0.05 F potassium dichromate (provided by the instructor) in an acetate buffer.

In addition to the buffer, the titration vessel should contain KNO_3 (to allow free conduction of current through the cell) and a bit of gelatin solution, which helps prevent some spurious electrochemical problems.

Prepare a "supporting electrolyte" solution that is ~0.2 F in KNO_3, 0.17 F in acetic acid and 0.06 F in sodium acetate. The pH should be ~4.2. Pipet 50.00 ml of this solution and 50.00 ml of ~0.01 F $Pb(NO_3)_2$ solution into the titration vessel and add 5 ml of 0.2 percent gelatin solution.

Cover the cell, insert all components and bubble nitrogen through the solution for 10 minutes to remove dissolved oxygen.

Adjust the flow rate on the D. M. E. so that a drop falls every 3 to 5 seconds. Be extremely careful with the electrode assembly and whenever handling mercury. Mercury is poisonous and hard to clean up if it is spilled! For protection, keep a large tray under the apparatus in case some of the mercury gets loose.

Set the applied potential to -1.0 V and reduce the nitrogen flow so that there is no apparent motion of solution in the vicinity of the D. M. E.* Read and record the current.

Add titrant in ~0.5 ml increments. Wait for a stable galvanometer reading and record current and ml titrant. Continue the titration until you are confident that you have enough data to construct a good titration curve.

Repeat the titration at 0 volts versus S. C. E. This time it is not necessary to bubble nitrogen to remove O_2, but don't forget to mix the reactants after each addition of titrant.

Construct titration curves for both titrations and compare their features. Use the endpoints to calculate the concentration of the lead nitrate solution. Report your results.

---

* Since diffusion of ions to the cathode determines the current, the solution must not be disturbed by unnecessary motion.

PART THREE

Appendix

Acid		K_A		pK_A
Acetic acid	H_3CCOOH		1.75×10^{-5}	4.76
Aluminum(III)	$Al(H_2O)_6^{3+}$		1.4×10^{-5}	4.85
Ammonium ion	NH_4^+		5.6×10^{-10}	9.26
Arsenic acid	H_3AsO_4	K_{A1}	6.0×10^{-3}	2.22
		K_{A2}	1.0×10^{-7}	7.00
		K_{A3}	3×10^{-12}	11.5
Arsenious acid	H_3AsO_3	K_{A1}	6.0×10^{-10}	9.22
		K_{A2}	3×10^{-14}	13.5
		K_{A3}	—	—
Benzoic acid	C_6H_5COOH		6.6×10^{-5}	4.18
Boric acid	H_3BO_3		5.8×10^{-10}	9.24
Carbonic acid	H_2CO_3 $(CO_2 \cdot H_2O)$	K_{A1}	4.6×10^{-7}	6.34
		K_{A2}	4.4×10^{-11}	10.36
Chloroacetic acid	$H_2ClCCOOH$		1.5×10^{-3}	2.82
Dichloroacetic acid	HCl_2CCOOH		5.0×10^{-2}	1.30
Trichloroacetic acid	Cl_3CCOOH		2.0×10^{-1}	0.70
Chromic acid	H_2CrO_4	K_{A1}	1.8×10^{-1}	0.74
		K_{A2}	3.2×10^{-7}	6.49
Citric acid	$H_3C_6H_5O_7$	K_{A1}	8.4×10^{-4}	3.08
		K_{A2}	1.8×10^{-5}	4.74
		K_{A3}	4.0×10^{-6}	5.40
Copper(II)	$Cu(OH_2)_n$		5×10^{-7}	6.30
Ethylenediaminetetraacetic acid	H_4Y (see Table 4.7)	K_{A1}	6.6×10^{-3}	2.18
		K_{A2}	1.9×10^{-3}	2.73
		K_{A3}	6.3×10^{-7}	6.20
		K_{A4}	1.0×10^{-10}	10.0
Formic acid	$HCOOH$		1.8×10^{-4}	3.74
Glycine	$^+H_3NCH_2COO^-$		1.65×10^{-10}	9.78
Hydrazoic acid	HN_3		1.9×10^{-5}	4.72
Hydrofluoric acid	HF		6.0×10^{-4}	3.22
Hydrogen cyanide	HCN		7.2×10^{-10}	9.14
Hydrogen sulfide	H_2S	K_{A1}	5.7×10^{-8}	7.24
		K_{A2}	1.2×10^{-15}	14.92
Hypochlorous acid	$HOCl$		3.0×10^{-8}	7.52
Iron(II)	$Fe(OH_2)_6^{2+}$		5.0×10^{-9}	8.30
Iron(III)	$Fe(OH_2)_6^{3+}$		6×10^{-3}	2.22
Lactic acid	$CH_3CHOHCOOH$		1.65×10^{-4}	3.81
Magnesium ion	$Mg(OH_2)_6^{2+}$		4×10^{-12}	11.4
Nitrous acid	HNO_2		4.5×10^{-4}	3.35
Oxalic acid	$H_2C_2O_4$	K_{A1}	6.3×10^{-2}	1.20
		K_{A2}	6.1×10^{-4}	3.21
Phenol	C_6H_5OH		1.3×10^{-10}	9.89
Phosphoric acid	H_3PO_4	K_{A1}	7.6×10^{-3}	2.12
		K_{A2}	6.2×10^{-8}	7.21
		K_{A3}	4.8×10^{-13}	12.32
Phosphorous acid	H_3PO_3	K_{A1}	1.0×10^{-2}	2.00
		K_{A2}	2.6×10^{-7}	6.59
Phthalic acid	$C_6H_4(COOH)_2$	K_{A1}	1.3×10^{-3}	2.89
		K_{A2}	3.9×10^{-6}	5.41
Sulfuric acid	H_2SO_4	K_{A1}	large	—
		K_{A2}	1.2×10^{-2}	1.92
Sulfurous acid	H_2SO_3	K_{A1}	1.7×10^{-2}	1.77
		K_{A2}	6.2×10^{-8}	7.21
Tartaric acid	$H_2C_4H_4O_6$	K_{A1}	9.4×10^{-4}	3.03
		K_{A2}	2.9×10^{-5}	4.54
Tin(II)	$Sn(OH_2)_3^{2+}$		2×10^{-2}	1.70
Zinc ion	$Zn(OH_2)_6^{2+}$		2.5×10^{-10}	9.60

354 *Appendix*

**Table
A.1
(Cont.)**

Base			K_B	pK_B
Ammonia	NH_3		1.85×10^{-5}	4.76
Aniline	$C_6H_5NH_2$		4.2×10^{-10}	9.38
Benzidine	$(p\text{-}NH_2C_6H_4\text{-})_2$	K_{B1}	9.3×10^{-10}	9.03
		K_{B2}	5.6×10^{-11}	10.25
Ethylamine	$CH_3CH_2NH_2$		5.6×10^{-4}	3.25
Diethylamine	$(CH_3CH_2)_2NH$		1.26×10^{-3}	2.90
Triethylamine	$(CH_3CH_2)_3N$		5.6×10^{-4}	3.25
Glycine	$^-OOCCH_2NH_3^+$		2.3×10^{-12}	11.64
Methylamine	CH_3NH_2		5×10^{-4}	3.3
Pyridine	C_5H_5N		1.4×10^{-9}	8.85

[a] Aquated metal ions are listed as the oxidation state rather than the correctly named ion. The prefixes "di-," "tri-," "p-" have been ignored in placing entries alphabetically.
[b] The dissociation constants of a conjugate acid-base pair are related by $K_A \cdot K_B = K_W$ or $pK_A + pK_B = 14.00$.
[c] pK_A values for a variety of indicators are given in Table 9.2.

pK_A VALUES OF ACIDS IN NONAQUEOUS
SOLVENTS (25°C)

(a) Methanol-Water Mixtures

Acid	Methanol, % by volume						
	0	20	40	60	80	95	100
Acetic	4.76	5.01	5.33	5.81	6.50	7.86	9.72
Benzoic	4.18	4.51	4.97	5.54	6.29	7.47	9.38
Anilinium ion	4.62	4.46	4.32	4.17	4.07	4.61	5.80
Ammonium ion	9.24	....	....	....	....	....	11.13

(b) Ethanol-Water Mixtures

Acid	Ethanol, % by weight						
	0	20	35	50	65	80	100
Acetic	4.76	5.13	5.43	5.84	6.29	6.87	10.32
Cyanoacetic	2.47	2.78	3.07	3.39	3.81	4.39	7.49
Benzoic	4.18	4.77	5.24	5.76	6.19	6.79	10.25
Anilinium ion	4.62	4.42	4.16	3.92	3.80	3.75	5.70
Ammonium ion	9.24	9.00	8.78	8.57	8.45	....	10.45

(c) Dimethylsulfoxide

Acid	*p*K_A
Acetic	11.6
HCN	12.9
Benzoic	11.0

(d) Acetonitrile

Acid	*p*K_A
Benzoic	20.7
Phenol	26.6
H_2SO_4	7.3
HCl	8.9

(See also Table 6.7.)

(e) Dimethylformamide

Acid	*p*K_A
Acetic	12.5
Benzoic	11.6
Phenol	*ca.* 18
Hydrazoic	9.9
4-Methylpyridinium ion	4.98[a]

[a] *p*K_A = 6.02 in water.

(f) Acetic Acid

Acid	*p*K_A
HCl	8.6
HNO_3	9.4
Cl_3CCO_2H	11.5
2,5-Dichloroanilinium ion	5.0[b]

[b] *p*K_A = 1.57 in water.

Table   SOLUBILITY PRODUCT CONSTANTS
A.3   SOME SALTS IN WATER AT 25°C[a]

Compound	Formula	K_{SP}
Aluminum hydroxide	$Al(OH)_3$	5×10^{-33}
Barium carbonate	$BaCO_3$	6×10^{-9}
Barium iodate	$Ba(IO_3)_2$	6×10^{-10}
Barium oxalate	BaC_2O_4	2×10^{-7}
Barium sulfate	$BaSO_4$	1×10^{-10}
Cadmium carbonate	$CdCO_3$	3×10^{-14}
Cadmium oxalate	CdC_2O_4	1×10^{-8}
Cadmium sulfide	CdS	1×10^{-28}
Calcium carbonate	$CaCO_3$	5×10^{-9}
Calcium fluoride	CaF_2	1×10^{-10}
Calcium oxalate	CaC_2O_4	2×10^{-9}
Calcium sulfate	$CaSO_4$	6×10^{-5}
Cupric hydroxide	$Cu(OH)_2$	2×10^{-19}
Cupric iodate	$Cu(IO_3)_2$	1×10^{-7}
Cupric sulfide	CuS	$\sim 10^{-40}$
Cuprous bromide	Cu_2Br_2[b]	2×10^{-13}
Cuprous chloride	Cu_2Cl_2[b]	8×10^{-11}
Cuprous iodide	Cu_2I_2[b]	1×10^{-18}
Cuprous thiocyanate	$Cu_2(SCN)_2$[b]	4×10^{-21}
Ferric hydroxide	$Fe(OH)_3$	1×10^{-36}
Ferrous hydroxide	$Fe(OH)_2$	2×10^{-14}
Ferrous sulfide	FeS	2×10^{-19}
Lanthanum iodate	$La(IO_3)_3$	6×10^{-10}
Lanthanum oxalate	$La_2(C_2O_4)_3$.	2×10^{-28}
Lead carbonate	$PbCO_3$	2×10^{-13}
Lead chloride	$PbCl_2$	1×10^{-4}
Lead chromate	$PbCrO_4$	2×10^{-14}
Lead hydroxide	$Pb(OH)_2$	3×10^{-16}
Lead iodate	$Pb(IO_3)_2$	2×10^{-13}
Lead oxalate	PbC_2O_4	3×10^{-11}
Lead sulfate	$PbSO_4$	2×10^{-8}
Lead sulfide	PbS	5×10^{-28}
Magnesium carbonate	$MgCO_3$	2×10^{-5}
Magnesium hydroxide	$Mg(OH)_2$	1×10^{-11}
Magnesium oxalate	MgC_2O_4	9×10^{-5}
Manganous carbonate	$MnCO_3$	9×10^{-11}
Manganous hydroxide	$Mn(OH)_2$	4×10^{-14}
Manganous sulfide	MnS	5×10^{-16}
Mercuric oxide	HgO	1×10^{-26}
Mercuric sulfide	HgS	$\sim 10^{-52}$
Mercurous bromide	Hg_2Br[b]	1×10^{-22}
Mercurous chloride	Hg_2Cl_2[b]	1×10^{-18}
Mercurous iodide	Hg_2I_2[b]	1×10^{-28}
Mercurous sulfate	Hg_2SO_4[b]	6×10^{-7}
Mercurous sulfide	Hg_2S[b]	$\sim 10^{-54}$
Silver arsenate	Ag_3AsO_4	1×10^{-22}
Silver bromide	$AgBr$	6×10^{-13}
Silver carbonate	Ag_2CO_3	7×10^{-12}
Silver chloride	$AgCl$	2×10^{-10}
Silver chromate	Ag_2CrO_4	2×10^{-12}
Silver cyanide	$AgCN$	3×10^{-12}
Silver hydroxide	$AgOH$	2×10^{-8}
Silver iodate	$AgIO_3$	2×10^{-8}
Silver iodide	AgI	1×10^{-16}

Compound	Formula	K_{SP}
Silver oxalate	$Ag_2C_2O_4$	7×10^{-12}
Silver sulfide	Ag_2S	1×10^{-50}
Silver thiocyanate	$AgSCN$	1×10^{-12}
Strontium carbonate	$SrCO_3$	2×10^{-9}
Strontium oxalate	SrC_2O_4	5×10^{-8}
Strontium sulfate	$SrSO_4$	3×10^{-7}
Thallous bromide	$TlBr$	2×10^{-6}
Thallous chloride	$TlCl$	2×10^{-4}
Thallous iodate	$TlIO_3$	2×10^{-6}
Thallous iodide	TlI	3×10^{-8}
Thallous sulfide	Tl_2S	1×10^{-22}
Stannous hydroxide	$Sn(OH)_2$	5×10^{-26}
Stannous sulfide	SnS	8×10^{-29}
Zinc hydroxide	$Zn(OH)_2$	2×10^{-14}
Zinc oxalate	ZnC_2O_4	5×10^{-9}
Zinc sulfide	ZnS	4×10^{-24}

[a] In calculating solubilities from K_{SP} values and vice versa it is important to realize that hydrolysis, incomplete or stepwise dissociation, complex-ion formation, and ion pairing can cause large deviations from the approximation that salts are completely dissociated in solution. However, for purposes of calculating concentrations to determine if a precipitation titration is feasible or to get an idea of the composition of a solution this approximation, while not really justified, can usually be tolerated. (*Warning:* Small changes in pH can drastically affect the solubility of some compounds.) The K_{SP} values given here have been selected from among a variety of sources which in some cases differ greatly from each other.

[b] Cuprous ion exists as Cu^{2+} and mercurous ion exists as Hg_2^{2+} in solution. K_{SP} values are based on these formulas

Table A.4   ASSOCIATION CONSTANTS OF METAL COMPLEXES IN WATER AT 25°C[a,b,c]

Complex	K_{assoc}
$Ag(CN)_2^-$	1×10^{21}
$Ag(en)_2^+$ [d]	5×10^7
$Ag(NH_3)_2^+$	1.4×10^7
$Ag(S_2O_3)_2^{3-}$	1×10^{13}
$Cu(NH_3)_4^{2+}$	4×10^{12}
$Cu(en)^{2+}$ [d]	4×10^{19}
$Cu(trien)^{2+}$ [e]	2.5×10^{20}
$Fe(CN)_6^{4-}$	1×10^{24}
$Fe(CN)_6^{3-}$	1×10^{31}
$FeSCN^{2+}$	2×10^2
$Fe(SCN)_2^+$	2×10^4
$HgCl_2$	1.5×10^{13}
$Ni(NH_3)_4^{2+}$	2×10^7
$Ni(CN)_4^{2-}$	2×10^{31}
$Zn(NH_3)_4^{2+}$	1×10^9
$Zn(CN)_4^{2-}$	5×10^{16}
$Zn(en)_2^{2+}$	2.5×10^{10}

[a] Values are *overall* equilibrium constants for formation of the species shown. For example, for $Ag(CN)_2^-$,

$$K_{assoc} = \frac{[Ag(CN)_2^-]}{[Ag^+][CN^-]^2}$$

[b] The *caveats* of footnote a, Table A.3 apply here too.
[c] See Table 4.6 for metal-EDTA constants.
[d] "en" is ethylenediamine, $NH_2CH_2CH_2NH_2$, a bidentate chelating ligand.
[e] "trien" is triethylenetetramine $NH_2(CH_2)_2NH_2(CH_2)_2NH_2(CH_2)_2NH_2$, a tetradentate ligand.

STANDARD REDUCTION POTENTIALS[a]

Couple (half-reaction)	E°_{red} (volts)
$F_2 + 2\ H^+ + 2\ e^- = 2\ HF(aq)$	+3.06
$S_2O_8{}^{2-} + 2\ e^- = 2\ SO_4{}^{2-}$	+2.01
$Ag^{2+} + e^- = Ag^+$	+1.98
$HN_3 + 3\ H^+ + 2\ e^- = NH_4{}^+ + N_2$	+1.96
$Co^{3+} + e^- = Co^{2+}$	+1.82
$H_2O_2 + 2\ H^+ + 2\ e^- = 2\ H_2O$	+1.77
$Ce^{4+} + e^- = Ce^{3+}$ (perchlorate media)	+1.70
$MnO_4{}^- + 4\ H^+ + 3\ e^- = MnO_2 + 2\ H_2O$	+1.695
$PbO_2 + SO_4{}^{2-} + 4\ H^+ + 2\ e^- = PbSO_4 + 2\ H_2O$	+1.685
$Ce^{4+} + e^- = Ce^{3+}$ (nitrate media)	+1.61
$H_5IO_6 + H^+ + 2\ e^- = IO_3{}^- + 3\ H_2O$	+1.6
$HBrO + H^+ + e^- = \frac{1}{2}\ Br_2 + H_2O$	+1.59
$BrO_3{}^- + 6\ H^+ + 5\ e^- = \frac{1}{2}\ Br_2 + 3H_2O$	+1.52
$MnO_4{}^- + 8\ H^+ + 5\ e^- = Mn^{2+} + 4\ H_2O$	+1.51
$Mn^{3+} + e^- = Mn^{2+}$	+1.51
$HO_2 + H^+ + e^- = H_2O_2$	+1.5
$Au^{3+} + 3\ e^- = Au$	+1.50
$PbO_2 + 4\ H^+ + 2\ e^- = Pb^{2+} + 2\ H_2O$	+1.455
$HIO + H^+ + e^- = \frac{1}{2}\ I_2 + H_2O$	+1.45
$Ce^{4+} + e^- = Ce^{3+}$ (sulfate media)	+1.44
$Cl_2 + 2\ e^- = 2\ Cl^-$	+1.36
$Cr_2O_7{}^{2-} + 14\ H^+ + 6\ e^- = 2\ Cr^{3+} + 7\ H_2O$	+1.33
$PdCl_6{}^{2-} + 2\ e^- = PdCl_4{}^{2-} + 2\ Cl^-$	+1.29
$ClO_2 + H^+ + e^- = HClO_2$	+1.28
$N_2H_5{}^+ + 3\ H^+ + 2\ e^- = 2\ NH_4{}^+$	+1.28
$Tl^{3+} + 2\ e^- = Tl^+$	+1.25
$MnO_2 + 4\ H^+ + 2\ e^- = Mn^{2+} + 2\ H_2O$	+1.23
$O_2 + 4\ H^+ + 4\ e^- = 2\ H_2O$	+1.23
$IO_3{}^- + 6\ H^+ + 5\ e^- = \frac{1}{2}\ I_2 + 3\ H_2O$	+1.195
$Br_2 + 2\ e^- = 2\ Br^-$	+1.065
$ICl_2{}^- + e^- = 2\ Cl^- + \frac{1}{2}\ I_2$	+1.06
$AuCl_4{}^- + 3\ e^- = Au + 4\ Cl^-$	+1.00
$HNO_2 + H^+ + e^- = NO + H_2O$	+1.00
$NO_3{}^- + 4\ H^+ + 4e^- = NO + 2\ H_2O$	+0.96
$NO_3{}^- + 3\ H^+ + 2\ e^- = HNO_2 + H_2O$	+0.94
$FeO_4{}^{2-} + 2\ H_2O + 3\ e^- = FeO_2{}^- + 4\ OH^-$	+0.9
$ClO^- + H_2O + 2\ e^- = Cl^- + 2\ OH^-$	+0.89
$HO_2{}^- + H_2O + 2\ e^- = 3\ OH^-$	+0.88
$Cu^{2+} + I^- + e^- = CuI$	+0.86
$Ag^+ + e^- = Ag$	+0.80
$Hg_2{}^{2+} + 2\ e^- = 2\ Hg$	+0.789
$Fe^{3+} + e^- = Fe^{2+}$	+0.771
$BrO^- + H_2O + 2\ e^- = Br^- + 2\ OH^-$	+0.76
$H_3IO_6{}^{2-} + 2\ e^- = IO_3{}^- + 3\ OH^-$	+0.7
$HN_3 + 11\ H^+ + 8\ e^- = 3\ NH_4{}^+$	+0.69
$O_2 + 2\ H^+ + 2\ e^- = H_2O_2$	+0.682
$UO_2{}^+ + 4\ H^+ + e^- = U^{4+} + 2\ H_2O$	+0.62
$BrO_3{}^- + 3\ H_2O + 6\ e^- = Br^- + 6\ OH^-$	+0.61
$MnO_4{}^{2-} + 2\ H_2O + 2\ e^- = MnO_2 + 4\ H^+$	+0.60
$MnO_4{}^- + 2\ H_2O + 3\ e^- = MnO_2 + 4\ OH^-$	+0.588
$MnO_4{}^- + e^- = MnO_4{}^{2-}$	+0.564
$2\ AgO + H_2O + 2\ e^- = Ag_2O + 2\ OH^-$	| 0.57
$H_3AsO_4 + 2\ H^+ + 2\ e^- = H_3AsO_3 + H_2O$	+0.56
$Cu^{+2} + Cl^- + e^- = CuCl$	+0.538

**Table
A.5
(Cont.)**

Couple (half-reaction)	E°_{red} (volts)
$I_3^- + 2\ e^- = 3\ I^-$	+0.536
$I_2 + 2\ e^- = 2\ I^-$	+0.5355
$4\ H_2SO_3 + 4\ H^+ + 6\ e^- = S_4O_6{}^{2-} + 6\ H_2O$	+0.51
$OI^- + H_2O + 2\ e^- = I^- + 2\ OH^-$	+0.49
$O_2 + 2\ H_2O + 4\ e^- = 4\ OH^-$	+0.401
$Ag(NH_3)_2{}^+ + e^- = Ag + 2\ NH_3$	+0.373
$Fe(CN)_6{}^{3-} + e^- = Fe(CN)_6{}^{4-}$	+0.36
$Cu^{2+} + 2\ e^- = Cu$	+0.337
$UO_2{}^{2+} + 4\ H^+ + 2\ e^- = U^{4+} + 2\ H_2O$	+0.334
$IO_3^- + 3\ H_2O + 6\ e^- = I^- + 6\ OH^-$	+0.26
$AgCl + e^- = Ag + Cl^-$	+0.222
$HSO_4^- + 3\ H^+ + 2\ e^- = H_2SO_3 + H_2O$	+0.17
$Cu^{+2} + e^- = Cu^+$	+0.153
$Sn^{4+} + 2\ e^- = Sn^{2+}$	+0.15
$S + 2\ H^+ + 2\ e^- = H_2S$	+0.141
$TiO^{2+} + 2\ H^+ + e^- = Ti^{3+} + H_2O$	+0.1
$AgBr + e^- = Ag + Br^-$	+0.095
$S_4O_6{}^{2-} + 2\ e^- = 2\ S_2O_3{}^{2-}$	+0.08
$HCOOH + 2\ H^+ + 2\ e^- = HCHO + H_2O$	+0.06
$CuBr + e^- = Cu + Br^-$	+0.033
$NO_3^- + H_2O + 2\ e^- = NO_2^- + 2\ OH^-$	+0.01
$2\ H^+ + 2\ e^- = H_2$	0.000
$MnO_2 + H_2O + 2\ e^- = Mn(OH)_2 + 2\ OH^-$	−0.05
$O_2 + H_2O + 2\ e^- = HO_2^- + OH^-$	−0.076
$Cu(NH_3)_2{}^{2+} + 2\ e^- = Cu + 2\ NH_3$	−0.12
$Pb^{2+} + 2\ e^- = Pb$	−0.126
$O_2 + H^+ + e^- = HO_2$	−0.13
$CrO_4{}^{2-} + 4\ H_2O + 3\ e^- = Cr(OH)_3 + 5\ OH^-$	−0.13
$Sn^{2+} + 2\ e^- = Sn$	−0.136
$AgI + e^- = Ag + I^-$	−0.151
$CuI + e^- = Cu + I^-$	−0.185
$CO_2 + 2\ H^+ + 2\ e^- = HCOOH$	−0.196
$HO_2^- + H_2O + e^- = OH + 2\ OH^-$	−0.24
$V^{3+} + e^- = V^{2+}$	−0.26
$Co^{2+} + 2\ e^- = Co$	−0.28
$Tl^+ + e^- = Tl$	−0.34
$PbSO_4 + 2\ e^- = Pb + SO_4{}^{2-}$	−0.36
$Cd^{2+} + 2\ e^- = Cd$	−0.40
$Cr^{3+} + e^- = Cr^{2+}$	−0.41
$Fe^{2+} + 2\ e^- = Fe$	−0.440
$Ni(NH_3)_6{}^{2+} + 2\ e^- = Ni + 6\ NH_3$	−0.47
$2\ CO_2 + 2\ H^+ + 2\ e^- = H_2C_2O_4$	−0.49
$Fe(OH)_3 + e^- = Fe(OH)_2 + OH^-$	−0.56
$Cr^{3+} + 3\ e^- = Cr$	−0.74
$Zn^{2+} + 2\ e^- = Zn$	−0.76
$Fe(OH)_2 + 2\ e^- = Fe + 2\ OH^-$	−0.877
$PbS + 2\ e^- = Pb + S^{2-}$	−0.95
$Mn^{2+} + 2\ e^- = Mn$	−1.18
$ZnS + 2\ e^- = Zn + S^{2-}$	−1.44
$Al^{3+} + 3\ e^- = Al$	−1.66
$\frac{1}{2}H_2 + e^- = H^-$	−2.25
$Mg^{2+} + 2\ e^- = Mg$	−2.37
$Ce^{3+} + 3\ e^- = Ce$	−2.48

Couple (half-reaction)	E°_{red} (volts)
$Na^+ + e^- = Na$	-2.71
$Ca^{2+} + 2\,e^- = Ca$	-2.87
$Ba^{2+} + 2\,e^- = Ba$	-2.90
$K^+ + e^- = K$	-2.92
$Li^+ + e^- = Li$	-3.04
$\frac{3}{2} N_2 + H^+ + e^- = HN_3$	-3.09

[a] Taken from *Oxidation Potentials*, by W. M. Latimer, 2nd Ed., Prentice-Hall, Inc., copyright 1952. Couples involving OH^- or NH_3 are for reactions which occur in basic solution. Other reactions occur in acid solution. (For simplicity H^+ is written rather than the hydronium ion H_3O^+.) E°_{red} values for couples not involving H^+ or OH^- should be independent of pH, but because of additional reactions, such as metal hydrolysis, care must be exercised when considering these reactions under varying conditions.

Table A.6   LOGARITHMS OF NUMBERS

Natural numbers	0	1	2	3	4	5	6	7	8	9	1	2	3	4	5	6	7	8	9
10	0000	0043	0086	0128	0170	0212	0253	0294	0334	0374	4	8	12	17	21	25	29	33	37
11	0414	0453	0492	0531	0569	0607	0645	0682	0719	0755	4	8	11	15	19	23	26	30	34
12	0792	0828	0864	0899	0934	0969	1004	1038	1072	1106	3	7	10	14	17	21	24	28	31
13	1139	1173	1206	1239	1271	1303	1335	1367	1399	1430	3	6	10	13	16	19	23	26	29
14	1461	1492	1523	1553	1584	1614	1644	1673	1703	1732	3	6	9	12	15	18	21	24	27
15	1761	1790	1818	1847	1875	1903	1931	1959	1987	2014	3	6	8	11	14	17	20	22	25
16	2041	1068	2095	2122	2148	2175	2201	2227	2253	2279	3	5	8	11	13	16	18	21	24
17	2304	2330	2355	2380	2405	2430	2455	2480	2504	2529	2	5	7	10	12	15	17	20	22
18	2553	2577	2601	2625	2648	2672	2695	2718	2742	2765	2	5	7	9	12	14	16	19	21
19	2788	2810	2833	2856	2878	2900	2923	2945	2967	2989	2	4	7	9	11	13	16	18	20
20	3010	3032	3054	3075	3096	3118	3139	3160	3181	3201	2	4	6	8	11	13	15	17	19
21	3222	3243	3263	3284	3304	3324	3345	3365	3385	3404	2	4	6	8	10	12	14	16	18
22	3424	3444	3464	3483	3502	3522	3541	3560	3579	3598	2	4	6	8	10	12	14	15	17
23	3617	3636	3655	3674	3692	3711	3729	3747	3766	3784	2	4	6	7	9	11	13	15	17
24	3802	3820	3838	3856	3874	3892	3909	3927	3945	3962	2	4	5	7	9	11	12	14	16
25	3979	3997	4014	4031	4048	4065	4082	4099	4116	4133	2	3	5	7	9	10	12	14	15
26	4150	4166	4183	4200	4216	4232	4249	4265	4281	4298	2	3	5	7	8	10	11	13	15
27	4314	4330	4346	4362	4378	4393	4409	4425	4440	4456	2	3	5	6	8	9	11	13	14
28	4472	4487	4502	4518	4533	4548	4564	4579	4594	4609	2	3	5	6	8	9	11	12	14
29	4624	4639	4654	4669	4683	4698	4713	4728	4742	4757	1	3	4	6	7	9	10	12	13
30	4771	4786	4800	4814	4829	4843	4857	4871	4886	4900	1	3	4	6	7	9	10	11	13
31	4914	4928	4942	4955	4969	4983	4997	5011	5024	5038	1	3	4	6	7	8	10	11	12
32	5051	5065	5079	5092	5105	5119	5132	5145	5159	5172	1	3	4	5	7	8	9	11	12
33	5185	5198	5211	5224	5237	5250	5263	5276	5289	5302	1	3	4	5	6	8	9	10	12
34	5315	5328	5340	5353	5366	5378	5391	5403	5416	5428	1	3	4	5	6	8	9	10	11
35	5441	5453	5465	5478	5490	5502	5514	5527	5539	5551	1	2	4	5	6	7	9	10	11
36	5563	5575	5587	5599	5611	5623	5635	5647	5658	5670	1	2	4	5	6	7	8	10	11
37	5682	5694	5705	5717	5729	5740	5752	5763	5775	5786	1	2	3	5	6	7	8	9	10
38	5798	5809	5821	5832	5843	5855	5866	5877	5888	5899	1	2	3	5	6	7	8	9	10
39	5911	5922	5933	5944	5955	5966	5977	5988	5999	6010	1	2	3	4	5	7	8	9	10
40	6021	6031	6042	6053	6064	6075	6085	6096	6107	6117	1	2	3	4	5	6	8	9	10
41	6128	6138	6149	6160	6170	6180	6191	6201	6212	6222	1	2	3	4	5	6	7	8	9
42	6232	6243	6253	6263	6274	6284	6294	6304	6314	6325	1	2	3	4	5	6	7	8	9
43	6335	6345	6355	6365	6375	6385	6395	6405	6415	6425	1	2	3	4	5	6	7	8	9
44	6435	6444	6454	6464	6474	6484	6493	6503	6513	6522	1	2	3	4	5	6	7	8	9
45	6532	6542	6551	6561	6571	6580	6590	6599	6609	6618	1	2	3	4	5	6	7	8	9
46	6628	6637	6646	6656	6665	6675	6684	6693	6702	6712	1	2	3	4	5	6	7	7	8
47	6721	6730	6739	6749	6758	6767	6776	6785	6794	6803	1	2	3	4	5	5	6	7	8
48	6812	6821	6830	6839	6848	6857	6866	6875	6884	6893	1	2	3	4	4	5	6	7	8
49	6902	6911	6920	6928	6937	6946	6955	6964	6972	6981	1	2	3	4	4	5	6	7	8
50	6990	6998	7007	7016	7024	7033	7042	7050	7059	7067	1	2	3	3	4	5	6	7	8
51	7076	7084	7093	7101	7110	7118	7126	7135	7143	7152	1	2	3	3	4	5	6	7	8
52	7160	7168	7177	7185	7193	7202	7210	7218	7226	7235	1	2	2	3	4	5	6	7	7
53	7243	7251	7259	7267	7275	7284	7292	7300	7308	7316	1	2	2	3	4	5	6	6	7
54	7324	7332	7340	7348	7356	7364	7372	7380	7388	7396	1	2	2	3	4	5	6	6	7

Natural numbers	0	1	2	3	4	5	6	7	8	9	Proportional parts								
											1	2	3	4	5	6	7	8	9
55	7404	7412	7419	7427	7435	7443	7451	7459	7466	7474	1	2	2	3	4	5	5	6	7
56	7482	7490	7497	7505	7513	7520	7528	7536	7543	7551	1	2	2	3	4	5	5	6	7
57	7559	7566	7574	7582	7589	7597	7604	7612	7619	7627	1	2	2	3	4	5	5	6	7
58	7634	7642	7649	7657	7664	7672	7679	7686	7694	7701	1	1	2	3	4	4	5	6	7
59	7709	7716	7723	7731	7738	7745	7752	7760	7767	7774	1	1	2	3	4	4	5	6	7
60	7782	7789	7796	7803	7810	7818	7825	7832	7839	7846	1	1	2	3	4	4	5	6	6
61	7853	7860	7868	7875	7882	7889	7896	7903	7910	7917	1	1	2	3	4	4	5	6	6
62	7924	7931	7938	7945	7952	7959	7966	7973	7980	7987	1	1	2	3	3	4	5	6	6
63	7993	8000	8007	8014	8021	8028	8035	8041	8048	8055	1	1	2	3	3	4	5	5	6
64	8062	8069	8075	8082	8089	8096	8102	8109	8116	8122	1	1	2	3	3	4	5	5	6
65	8129	8136	8142	8149	8156	8162	8169	8176	8182	8189	1	1	2	3	3	4	5	5	6
66	8195	8202	8209	8215	8222	8228	8235	8241	8248	8254	1	1	2	3	3	4	5	5	6
67	8261	8267	8274	8280	8287	8293	8299	8306	8312	8319	1	1	2	3	3	4	5	5	6
68	8325	8331	8338	8344	8351	8357	8363	8370	8376	8382	1	1	2	3	3	4	4	5	6
69	8388	8395	8401	8407	8414	8420	8426	8432	8439	8445	1	1	2	2	3	4	4	5	6
70	8451	8457	8463	8470	8476	8482	8488	8494	8500	8506	1	1	2	2	3	4	4	5	6
71	8513	8519	8525	8531	8537	8543	8549	8555	8561	8567	1	1	2	2	3	4	4	5	5
72	8573	8579	8585	8591	8597	8603	8609	8615	8621	8627	1	1	2	2	3	4	4	5	5
73	8633	8639	8645	8651	8657	8663	8669	8675	8681	8686	1	1	2	2	3	4	4	5	5
74	8692	8698	8704	8710	8716	8722	8727	8733	8739	8745	1	1	2	2	3	4	4	5	5
75	8751	8756	8762	8768	8774	8779	8785	8791	8797	8802	1	1	2	2	3	3	4	5	5
76	8808	8814	8820	8825	8831	8837	8842	8848	8854	8859	1	1	2	2	3	3	4	5	5
77	8865	8871	8876	8882	8887	8893	8899	8904	8910	8915	1	1	2	2	3	3	4	4	5
78	8921	8927	8932	8938	8943	8949	8954	8960	8965	8971	1	1	2	2	3	3	4	4	5
79	8976	8982	8987	8993	8998	9004	9009	9015	9020	9025	1	1	2	2	3	3	4	4	5
80	9031	9036	9042	9047	9053	9058	9063	9069	9074	9079	1	1	2	2	3	3	4	4	5
81	9085	9090	9096	9101	9106	9112	9117	9122	9128	9133	1	1	2	2	3	3	4	4	5
82	9138	8143	9149	9154	9159	9165	9170	9175	9180	9186	1	1	2	2	3	3	4	4	5
83	9191	9196	9201	9206	9212	9217	9222	9227	9232	9238	1	1	2	2	3	3	4	4	5
84	9243	9248	9253	9258	9263	9269	9274	9279	9284	9289	1	1	2	2	3	3	4	4	5
85	9294	9299	9304	9309	9315	9320	9325	9330	9335	9340	1	1	2	2	3	3	4	4	5
86	9345	9350	9355	9360	9365	9370	9375	9380	9385	9390	1	1	2	2	3	3	4	4	5
87	9395	9400	9405	9410	9415	9420	9425	9430	9435	9440	0	1	1	2	2	3	3	4	4
88	9445	9450	9455	9460	9465	9469	9474	9479	9484	9489	0	1	1	2	2	3	3	4	4
89	9494	9499	9504	9509	9513	9518	9523	9528	9533	9538	0	1	1	2	2	3	3	4	4
90	9542	9547	9552	9557	9562	9566	9571	9576	9581	9586	0	1	1	2	2	3	3	4	4
91	9590	9595	9600	9605	9609	9614	9619	9624	9628	9633	0	1	1	2	2	3	3	4	4
92	9638	9643	9647	9652	9657	9661	9666	9671	9675	9680	0	1	1	2	2	3	3	4	4
93	9685	9689	9694	9699	9703	9708	9713	9717	9722	9727	0	1	1	2	2	3	3	4	4
94	9731	9736	9741	9745	9750	9754	9759	9763	9768	9773	0	1	1	2	2	3	3	4	4
95	9777	9782	9786	9791	9795	9800	9805	9809	9814	9818	0	1	1	2	2	3	3	4	4
96	9823	9827	9832	9836	9841	9845	9850	9854	9859	9863	0	1	1	2	2	3	3	4	4
97	9868	9872	9877	9881	9886	9890	9894	9899	9903	9908	0	1	1	2	2	3	3	4	4
98	9912	9917	9921	9926	9930	9934	9939	9943	9948	9952	0	1	1	2	2	3	3	4	4
99	9956	9961	9965	9969	9974	9978	9983	9987	9991	9996	0	1	1	2	2	3	3	3	4

Index

Absorbance, 5
Absorbancy index (*see* Extinction coefficient)
Absorption, of gases, 14
Absorption spectrum, 6, 222
  equilibrium constants from, 229, 232, 235
  for analysis of mixtures, 232, 233
  isosbestic point, 231, 233
  measurement, 341, 342
  stoichiometry from, 231
Acetic acid:
  acid-base properties, 128, 129, 135, 137, 139, 144, 145
  as pH buffer, 39
  as solvent, 104, 139
    ionic equilibria in, 109, 110, 111, 159
    titrations in, 139
  in carbon tetrachloride, 119
  in water, 19, 156, 163

Acetic acid (*contd.*)
  in water-dioxane mixtures, 113
  titration of, 25, 61, 182, 205
    in presence of phenol, 53, 55, 59, 185
Acetolysis, 159
Acetone, 104, 127, 129, 137, 153, 227, 334
Acetonitrile, 104, 117, 129, 137, 144, 145
Acid, Brønsted, 18, 26, 125
  strength of, 126
Acid, Lewis, 18, 120
Acid-base indicators, 200 (*see also* Color indicators)
  concentration of, 285
  for acid-base titration, 70, 72, 203, 205, 206, 285
  for pH, 72, 201, 202, 203
  for selective titration, 207
  indicator blank, 286
  in nonaqueous solvents, 212

Acid-base indicators (*contd.*)
  measurement of pK_A, 340
  molecular structure, 207
  practical test, 206, 285, 286
  transition range, 72, 203, 213
Acid-base reactions, 18, 26, 49
  in acetonitrile, 117
  in carbon tetrachloride, 119
  in water-dioxane mixtures, 113
Acid-base titration in nonaqueous solvents, 137, 153
  conductometric, 225
  in acetic acid, 139
  in benzene, 226
  in dioxane-water mixtures, 183
  in methanol, 141, 226, 295
  indicators for (*see* Acid-base indicators)
  and ion product of the solvent, 138
  of dibasic acids, 142
  pH at equivalence point, 206

Acid-base titration in nonaqueous solvents (*contd.*)
  potentiometric, 334
  prediction of pK_A for, 143
  of salts of carboxylic acids, 142
Acid-base titration in water, 67, 284
  conductometric, 345
  indicators for (*see* Acid-base indicators)
  innate titration error, 68, 69
  pH at equivalence point, 70, 185, 205, 206
  of polybasic acids, 57
  potentiometric, 333
  standard solutions for, 287
  standards for, 74
  strong acid and strong base, 67, 70
  titration curves, 62, 70, 204
  weak acid and strong base, 68, 71, 182, 183
  weak base and strong acid, 68, 71, 140
Acid dissociation constant, 26, 28, 49, 53, 130
  and equivalence point, 69, 71
  and selective titration, 56, 59
  and titration error, 68, 183
  effect of solvent on, 143, 183
  Tables, 144, 145, 213, 355
  for color indicators, 201, 204, 213
  for common solvents, *Table*, 135
  for dibasic acids, 28
  for EDTA, 93
  spectrophotometric determination of, 340
  Tables, 135, 353, 354, 355
Acidity of solutions, 132, 135
  leveling of, 133
Activity, 25
Adsorption indicator, 79
Air oxidation, 84, 86
  of Erio T, 95, 303
  of iodine, 87
Algebraic equations:
  by successive approximation, 161
  for titration curves, 156
  solution of, 159
Algebraic order, 178
Alizarin yellow, 204
Aluminum(III), 92, 120
  as Brønsted acid, 121
  association, 121
  hexahydrate, 120
American Society for Testing Materials, 67
Ammonia, 104, 127, 129, 137, 139, 140
  base ionization, 131
Ammonium chloride, 183
Amperometric titration, 339, 346
  apparatus, 347
  of lead with dichromate, 349
Analysis (*see specific methods*):
  gravimetric (*see* Gravimetric analysis)

Analysis (*contd.*)
  volumetric (*see* Titration)
Analytical balance, 264
  operation of, 265
Analytical samples, 274
Angstrom unit (*see Glossary*)
Anode, 237
Antimony, 84
  determination of, 320
Aprotic solvents, 117, 127
Aqua regia, 274
Argentometric titration, 75, 297
Arsenic(III), 50, 51, 81, 84, 88, 186
  determination, 320
Arsenic(V), 51
Arsenic trioxide (*see* Arsenious oxide)
Arsenious oxide, 40, 90, 314
Asbestos, 278
Ascarite, 14
Ascorbic acid, in EDTA titration, 95, 303
Ashless filter paper, 15, 282
Association constant, 27, 29
  for complex ions, *Table*, 92, 358
  ionic (*see* Ionic association constant)
Association, ionic (*see* Ionic association)
ASTM, 67
Atomic weights, *Table* (*see inside front cover*)
Autoprotolysis constant, 27, 128, 139 (*see also* Ion product, of solvent)
Average, 34
Azo dyes, 210

Back-titration, 77, 290, 307
Baking soda, 22
Balance (*see* Analytical balance)
Barium, gravimetric analysis, 280
Barium sulfate:
  gravimetric analysis, 280
  precipitation, 281, 283
Base, Bronsted, 125
  strength, 126
Base, Lewis, 18
Base dissociation constant, 26, 28, 130
  for very weak bases, 135
  Table, 354
Basicity:
  leveling of, 134
  of hydroxide in Me2SO, 119
  of solutions, 132
Battery, 237
  lead-acid, 292
Battery acid, 292
Beer's law, 6, 194, 214, 218, 340
  on molar basis, 231
Benzene, 105, 226, 229
Benzoic acid, 74, 141
  in methanol, 296
  in water, 55

Bicarbonate ion, 74
  as pH buffer, 40
Black boxes, 217
Blank, in spectrophotometry, 341
Brass analysis, 321
Bromide ion, 25, 75, 242, 245
  potentiometric determination of, 337
  titration of, 76, 77, 79
Bromine, 89, 157, 162
Bromocresol green, 204, 213, 290
Bromocresol purple, 213
  spectrum, 7
Bromophenol blue, 140, 141, 142, 204, 213
Bromophenol red, 213
Bromothymol blue, 73, 141, 204, 212, 213
Brønsted acid-base reactions (*see* Acid-base reactions)
Buffers, 39
  choice of, 41
  for chloride, 41
  for EDTA titrations, 312
  for pH, 40, 41
Bulk solution properties, 166, 217
  and linear titration curves, 167, 222
  calibration curves, 219
  equilibrium constants from, 229, 235, 236
  specificity, 222
  stoichiometry from, 226, 231
  temperature control, 219
Buret, 271
Butter yellow (*see* Methyl yellow)
Butyl alcohol (*tertiary*), 104, 131, 137
Butyl lithium (*secondary*), 227
Butyl mercaptan, 127

Cadmium(II), 242, 245
  titration, 304
Calcite, 74, 96
Calcium(II), 95
  titration, 306
Calcium carbonate, 13, 74, 96
Calcium chloride, standard solution, 309
Calcium sulfate, solubility, 111
Calomel electrode (*see* Saturated calomel electrode)
Carbonate ion, in strong base, 73
Carbon dioxide, in aqueous solution, 40
  solubility, 40
Carbonic acid, 40
Carbon tetrachloride, 105, 119, 157
Cassia flask, 269
Cathode, 237
Cell constant, 152, 241, 242
  inaccuracy, 249
Cells, electrochemical, 151, 236 (*see also* Half-cell, sensing)
  emf, equations for, 239
  for amperometric titration, 347

Cells, electrochemical (*contd.*)
  for measuring concentration, 241, 243, 248
  for measuring pH, 151
  for potentiometric titration, 249
  liquid junctions, 248
  reversibility, 239
Ceric sulfate (*see also* Cerium(IV))
  as titrant, 83
  catalysts for, 83
Cerium(IV), 81, 199, 223, 243
  E^0 in 1 N acids, 48
  potentiometric titration with, 336
  standard solution, 336
Charged-sphere model, of ions, 102, 103
  of equilibrium constants, 112, 114
  of ionic association, 106, 107
Chemical composition (*see* Composition)
Chemical equation, 45
  balancing of, 9
    by algebraic method, 10
    by half-reaction method, 11
  formal, 45, 47
  for selective titration, 59, 61, 63
  molecular, 45, 47
  stoichiometric, 45
Chemical equilibrium, 155
  algebraic equations for, 156
Chemical equivalent (*see* Equivalent)
Chemical kinetics, 323
  batch method, 323
  timing, 323, 324
  temperature, 324
Chemical methods, 4, 12 (*see also specific method*)
Chemical standards, 16
  acidimetric, 74, 140
  argentometric, 80
  EDTA, 96
  oxidimetric, 40, 89
Chloride, 75, 140, 242, 243, 245
  oxidation of, 83, 86
  titration of, 76, 77, 79, 297
    potentiometric, 337
    gravimetric, 278
  standard solution, 298
Chlorine, 89, 243
Chloroform, 105
Chlorophenol red, 204
Chromate, 50, 81, 82, 89
Cleaning solution (dichromate), 273
Cobalt(II), 81, 92, 120, 243
  in methanol, 120
  titration, 307
Cobalt(III), 81, 92, 243
Color, 192, 285
  and absorption of light, 5, 192
  and concentration, 193, 194
  and indicator ratio, 201
  hue, 192, 213

Color (*contd.*)
  of solutions, 192
  shade, 193
Colorimeter, 340
Color indicators, 16, 285
  adsorption on crystals, 79
  as stepwise reactants, 196
  at high dilution, 198
  effect of concentration, 197
  extinction coefficient, 196, 198
  for EDTA titrations, 310
  for pH (*see* Acid-base indicators)
  for acid-base titration (*see* Acid-base indicators)
  indicator blank, 286
  one-color, 201, 202, 213
  self-indicators, 195
  transition range, 72, 203, 204
  two-color, 201, 203, 213
Color reference, for endpoint, 286
Common-ion effect:
  and ionic association, 111
  on solubility, 110, 111
Complex ions:
  and pH, 49
  association constants, *Table*, 92, 358
  in redox, 48
Complexity:
  molecular, 44
  stoichiometric, 44, 51
Complexometric titrations, 302
Component, 4, 44, 154, 155, 164
  conventional formula, 46
Composition, 16
  units, 16
Concentration, 16
  by titration, 19
  dilution of, 22, 23
  formal, 16, 17, 155, 164
  from bulk solution properties, 219, 221
  from emf, 241, 243, 248
  molar, 16, 17, 155, 164, 167
  normal, 19
Conductivity, electrical, 150, 218, 222
  measurement of, 339, 344, 345
Conductometric titration, 150, 225, 343, 345
Conjugate acid, base, 126
Conservation:
  of atoms, 10, 155
  of electrical charge, 10, 155
Conventional formula, 46, 47
Conversion diagram, 52, 53
  and titration, 54
Conversion interval, 54
  and pH, 55, 57
  of color indicators (*see* Transition range)
Cooperative reaction, 56
Coordination, 120
Copper, 90
  in brass, analysis, 321
Copper(II), 45, 86, 242, 245, 321
m-Cresol purple, 204
Cresol red, 213

Crucible, 13, 15, 267, 268, 278
  Gooch, 268, 278
Crystal violet, 140, 204, 210, 211
Cyanide, 245
Cyclohexene, 157, 162
  dibromide, 157, 162
Cysteine, 88

Decomposition, 13
  of limestone rock, 13
Decomposition pressure, 13
Degree of conversion, 53
Density:
  and composition, 4, 219, 220
  optical, 5
Desiccator, 266
Determinate error, 33
Dextrin, 300
Diagram:
  conversion, 52
  principal species, 49, 50, 55
Dichlorofluorescein, 79, 300
Dichromate, 82, 85, 89, 90, 348 (*see also* Chromate)
  cleaning solution, 273
Dielectric constant, 102, 126
  and ionic association, 106, 107, 109
  effect on equilibrium constant, 113, 114, 115, 183, 209
  of liquid mixtures, 112
  of solutions, 222, 226
  Table of, 104, 105
Diffusion current, 348
Dilution, 22, 23
Dimethylformamide, 104, 129, 137, 144, 145
Dimethylsulfoxide, 104, 119, 137, 144, 145
3,5-dinitrobenzoic acid, 118
Dioxane, 105, 129, 137
Dioxane-water, 112
  ionic equilibria in, 113, 183
Diphenylamine sulfonate, 86
Diphosphopyridine nucleotide, 82
Dissociation constant, 27, 29, 131, 132 (*see also* Acid dissociation constant *or* Base dissociation constant)
D.M.E. (*see* Dropping mercury electrode)
Dropping mercury electrode, 348, 349
Drying, 266, 274
  of precipitates, 15
Drying agents, 266
Drying oven, 15, 266, 274
Dye (*see name of dye*)

EDTA, 90
  acid-base properties, 92, 94
  as complexing agent, 90, 91, 302
  association constants, 92
  as titrant, 91, 92, 94, 186, 199, 302, 310, 312

EDTA (*contd.*)
  disodium salt, dihydrate, 96, 303
  *p*H in titration, 95, 304, 311, 312
  half-cell for, 243
  standard solutions, 303
  substitution titration, 95, 306
Effective equation for titration, 61, 184
Effective substrate, 60, 184
Effective titrant, 61, 63, 184
  color indicators as, 197
  solids, 187
Effective titrant rule, 63
Electrical conductivity (*see* Conductivity, electrical)
Electrical neutrality, 155
Electrical polarization, 102
Electric charge, 102, 106
  in chemical equilibria, 114, 115
Electrochemical cells (*see* Cells, electrochemical *and* Half-cell, reference)
Electrode, 236, 241, 332 (*see also name of electrode*)
  for redox couples, 243
  for specific ions, 242
  reference, 241
Electrode potential, standard, 27, 29
  convention for, 30
  data, 31, 81
  effect of complex ion formation, 48
  Table, 359, 360, 361
Electrolytes (*see also* Ion *or* Ionic association):
  in gas phase, 106
  in liquid solution, 107, 108
  principal species, 109
  solubility of, 110, 111, 114
emf, equations for, 239
  at equivalence point, 251
Endpoint color reference, 286
Endpoint indicator (*see* Color indicators *or name of indicator*)
Eosin, 79
Equation, chemical (*see* Chemical equation)
Equilibrium, chemical (*see* Chemical equilibrium)
Equilibrium constant, 25, 28, 29 (*see also specific name, e.g.* Dissociation constant)
  algebraic order, 178
  and dielectric constant, 112
  for ionic reactions, 115, 183
  and principal species, 46
  conditional, 42
  data for, 26, 31
  effect on titration curve, 165, 169, 173
  for selective titration, 62, 184
  from bulk solution properties, 229, 236
  from E⁰, 29, 30, 31
  pseudo-, 41

Equilibrium constant (*contd.*)
  relationships among, 32, 33
  specific solvation effects, 142
  traditional symbols for, 26, 27
Equilibrium constant for titration, 25, 32
  and titration error (*see* Innate titration error)
Equivalent, chemical, 17, 164
  equations for, 18, 19
  in ionic reactions, 18
Equivalent weight, 18
Equivalence point, 18, 19
  emf at, 251
  from titration curve, 70, 165, 168, 173, 174, 176, 177, 224, 225
  general treatment, 178–182
  *p*H at, 185, 205, 206
Eriochrome Black T, 95, 199
  solution of, 303
Erio T (*see* Eriochrome Black T)
Error, experimental, 33, 35 (*see also specific name, e.g.* Determinate error)
Ethanol, 104, 137, 139, 144, 145
Ethylenediamine, 127
Ethylenediamine tetraacetate ion (*see* EDTA)
Ethyl lactate, 324
  alkaline hydrolysis, 324, 325
  rate, 327, 328, 329
  equivalent weight, 327
Expression, mass action (*see* Mass action expression)
Extinction coefficient, 6, 194
  formal, 230
  of color indicators, 195, 198

Factor, gravimetric, 21
Fajan's method, for insoluble silver salts, 79, 300
Faraday, of charge, 17, 30 (*see also* the Glossary)
Fatuous error, 33
  test for, 35
Ferric ammonium sulfate, 77, 78, 300
Ferric ion (*see* Iron(III))
Ferric thiocyanate complexes, 100
  in Volhard's method, 77, 78, 197
Ferrocyanide, 47, 243
Ferroin, 83, 85, 199
Ferrous ion (*see* Iron(II))
Ferrous 1,10-phenanthroline sulfate (*see* Ferroin)
Filter crucible, 15
Filter paper, ashless, 15, 269, 281
Filtration, 15, 268
  suction, 268
Fischer, Karl, 122
Fluoride, 245
Formal extinction coefficient (*see* Extinction coefficient)
Formality, 17

Formation constant, 27
  Table, 92, 358
Formula weights, *Table* (*see inside back cover*)

Galvanic cells (*see* Cells, electrochemical)
Glass electrode, 245, 246
Glassware, 269
  cleaning, 263, 273
  volumetric, 269
Gold(III), 89, 242, 274
Gooch crucible, 268, 278
Gravimetric analysis, 12, 276
  by decomposition, 13, 277
  by gas absorption, 15
  by precipitation, 14, 278
Gravimetric factors, 21

Half-cell, reference, 152, 241, 247
Half-cell, sensing, 151, 241
  emf, equation for, 239
  redox-couple, 243
  specific-ion, 242
Hardness, of water, 99, 308
Hue, 192
  and spectral wavelength, 192
  complementary, 192
  saturation of, 193
Hydrazine, 88
Hydrochloric acid, 274 (*see also* Hydrogen chloride)
  concentrated, 289
  constant-boiling, 74, 80
  standard solution, 74, 289
Hydrogen, replaceable, 291
Hydrogen bond, 116
  acceptor, 116, 117
  donor, 116
Hydrogen bromide, 134, 139
Hydrogen chloride, 110, 134, 139, 140
  acid ionization, 131
Hydrogen electrode, 152, 243
Hydrogen peroxide, 81, 84, 89
Hydrogen sulfide, 88
Hydrolysis, 71, 159
  of iodine, 87
  of metal ions, 92
Hydrolysis constant, 71
Hydroxide ion:
  hydration, 119
  in Me₂SO, 119, 135
Hydroxylic solvents, 116, 212, 213
tris-Hydroxymethylamine, 74
Hypoiodite, 87
Hypoiodous acid, 88

Indeterminate error, 33, 34–36
Index of refraction, 217, 220, 221
Indicator (*see* Color indicator)
Indicator blank, 286
Indicator ratio, 201

Indium(III), 47
Innate titration error, 33, 36, 172, 226
  and algebraic order, 178–182
  and equilibrium constant, 37, 38, 165
  and ion product of solvent, 139
  and slope of titration curve, 173
  effect of color indicator, 197
  effect of dielectric constant, 114
  engineering formulas for, 175, 179–182
  estimation of, 37, 38
  in acid-base titration, 69
  in selective titration, 184, 185
  in Silver(I) titration, 76
Instrumental analysis, 216, 339
  by potentiometric titration, 251, 330
  calibration curves, 219
  control experiments, 217
  linear titration methods, 222
  references, 255
  specificity, 222
  teamwork in, 217
  temperature control, 219
  with electrochemical cells, 241, 330, 338
International atomic weights, *Table* (*see inside front cover*)
Interpolation, 219
Iodate, 81, 89, 90, 318
Iodide ion, 26, 75, 81, 242, 243, 245
  as reducing agent, 86, 88, 313, 318
  titration of, 78, 79
  potentiometric, 337
Iodimetry, 317
Iodine, 26, 81 (*see also* Triiodide)
  as oxidizing agent, 88, 317
  standard solution, 320
  titration with, 86, 313
Iodine chloride, 83, 186
Iodometry, 318
Ion:
  charged sphere model, 102, 112
  chemical model, 115
  hydrogen bonding, 117
Ionic association, 107, 108, 145
  effect on extinction coefficient, 212, 213
Ionic association constant, 107
  Tables, 108
Ionization constant, 131, 145
  (*see also* Acid *or* Base dissociation constant)
Ion pair, 106, 159
  charged sphere model, 103, 106
  chemical models, 115
  hydrogen bonding, 117, 119
  in chemical reactions, 109
  in gas phase, 106
  light absorption, 212, 213
  metathesis, 111
  solubility of, 110

Ion product, of solvent, 26, 27, 28, 128
  Table, 139
Ion-selective membranes, 244
  Table, 245
Iron(II), 81, 84, 92, 199, 243
Iron(III), 81, 92, 243
  hexahydrate, 121
Iron ore, 84
  analysis, 316
Iron oxide analysis, 316

Jones reductor, 84
Junction, liquid (*see* Liquid junction)

Karl Fischer titration, 122
Kinetics (*see* Chemical kinetics)

Laboratory notebook, 264
Lactate, 324
Lanthanum fluoride membrane, 245
Law (*see specific law*)
Lead, 81
Lead(II), 242, 245, 348
Lead-acid battery, 292
Lead dioxide, 81, 82, 89
Leveling effect on acidity, 133, 153
Lewis acid-base reactions, 18, 120
  and *p*H, 49
Ligand, 120
Light:
  absorption, 5, 192
  plane polarized, 222
  perception (*see* Color)
  scattering, 5
  transmission, 5, 193
  wavelength, 6
  white, 5, 192
Limestone, 13
Linear titration curves (*see* Titration curves)
Liquid junction, 239, 241, 248
Lithium bromide, in the gas phase, 106
Litmus, 200
Logarithms, *Table*, 362
Logarithmic titration curves (*see* Titration curves)
Lyate ion, 128, 135
  structural formulas, 129
Lyonium ion, 128, 135
  structural formulas, 129

McBride procedure, 315
Magnesium(II), standard solution, 304
Magnesium carbonate, 13
Magnesium-EDTA solution, 307
Magnesium-EDTA titration, 94, 199, 304
Magnesium perchlorate, 14
Magnetite, analysis, 22

Manganese, in steel, 342
Manganese dioxide, 89
Mass action expression, 25, 26, 178, 239 (*see also* Equilibrium constant)
  algebraic order, 178
Mass balance, 155
Maximum slope method (in potentiometry), 252
Mean, algebraic, 34
Mean deviation, 34
Membrane emf, 244
Membranes, ion-selective, 244
  Table, 245
Meniscus, 271, 272
Mercuric chloride, 85
Mercury, 84, 348, 349 (*see also* Dropping mercury electrode)
Mercury(I), 242
Mercury(II), 99
Mercury-calomel electrode, 243
Metal hydroxides, precipitation, 92, 312
Metathesis, 111
Methanol, 104, 128, 129, 137, 139, 144, 145, 295
  acid-base titrations in, 141, 226, 295, 296
Methanol-benzene, 142
  specific solvation effects, 142
Method of successive dumpings, 267
Methyl orange, 73, 204, 210, 213
  modified, 290
Methyl yellow, 204, 210, 213
o-Methyl red, 204, 213, 290
p-Methyl red, 213
Mixed indicator, 290
Mohr method, for halide analysis, 76, 298
Mohr pipet, 271
Molar extinction coefficient (*see* Extinction coefficient)
Molarity, 17, 155 (*see also* Concentration)
Molecular complexity, 44
Molecular equation, 45
Molecular species, 44
  and formal components, 155
  principal (*see* Principal species)
Molybdenum, 84

Nernst's equation, for emf, 151, 239
Neutral red, 213
Nickel, 46
Nickel(II), 242
Nitric acid, 274
  acid strength, 130, 134
Nitrite, 84, 89
Nitrobenzene, 77, 300
para-Nitrophenol, 229
  titration curves, 149, 150
Nitrous acid, 89
Normality, 19
Notebook, 264

Optical activity, 222
Optical density, 5
Oscillometric titration, 227
Osmium tetroxide, 83
Oven, drying, 15
Oxalate, 81, 84, 89, 315
Oxalic acid dihydrate, 74
Oxidation potential, standard, 27
   (*see also* Electrode potential, standard)
Oxidation-reduction, 80
  and complex ion formation, 48
  and *p*H, 49, 81
  balanced equations, 11, 12
  equilibrium constants, 30, 31
  equivalent in, 18, 81
  half-reactions, 11, 29
  in electrochemical cells, 236
  interference by oxygen, 84
Oxidation-reduction titrations,
  80, 313 (*see also name of*
  titrant)
  potentiometric, 336
Oxidimetric standards, 40, 89
Oxygen, 84, 88 (*see also* Air oxidation)

Palladium(II), 242
Parallax, 272
Perchloric acid, 134, 139, 140
Periodate, 89, 342
Permanganate, 81, 89
  decomposition in solution, 83,
  314
  oxidation of chloride, 83, 85
  spectrum, 6, 342
  standard solution, 83, 314
  titrations with, 82, 195, 313
Peroxide, 81, 84, 89
*p*H, 41, 133, 135
  at equivalence point, 185, 205,
  206
  glass electrode for, 246
  in acid-base titrations, 69, 70,
  72
  in EDTA titrations, 94, 95
  in iodine titrations, 88, 89
  in redox, 81
  measurement, 151, 247, 249,
  331
*p*H buffers (*see* Buffers)
*p*H indicators (*see* Acid-base
  indicators)
*p*H meter, 331
Phenol:
  in presence of acetic acid, 185
  in water, 53, 55, 59, 62
Phenolphthalein, 73, 204, 208,
  213
Phenol red, 141, 204, 209, 213
Phosphoric acid, in water, 57, 63
Phthalein dyes, 207
Physical methods, 4 (*see also spe-*
  cific method)
  critique, 8
Physical properties, 8
Picric acid, 226
Pipet, 270, 272

Pipet bulb, 272
Platinum, 13, 274
  for electrodes, 242, 244, 336,
  337
Polarization, electric, 102
Potassium acid phthalate, 74,
  140, 141, 288
Potassium bromide, in acetic
  acid, 110
Potassium dichromate, 85, 90,
  273 (*see also* Dichromate)
Potassium iodate, 90
  standard solution, 318
Potassium permanganate (*see*
  Permanganate)
Potentiometer, 152, 238, 239, 331
Potentiometric titration, 249, 331
  (*see also name of titration*)
  emf at equivalence point, 251
  maximum slope method, 252,
  253
  procedure, 332
  sensing electrodes in, 250
  sufficient conditions for, 249,
  331
Precipitation, 15
Precipitation titration:
  conductometric, 346
  with silver nitrate, 75, 297
Pressure, decomposition, 13
Preventive solution, 316, 317
Principal species, 46, 47
  change of, 47
  effect of *p*H, 49
  of electrolytes, 109
Principal species diagram, 49, 50
  and selective titration, 56, 57,
  58
  for EDTA, 93
  with conversion intervals, 55,
  57
Prism, 5
Properties:
  bulk solution (*see* Bulk solution properties)
  physical, 8
Protic solvents, 127
Proton acceptor (*see* Base)
Proton donor (*see* Acid)
Pseudo-equilibrium constant, 41,
  181
Pyridine, 104, 131, 137
Pyridinium perchlorate, 219, 220,
  221

Rate law, 324, 329
Rate constant, 329
Reaction capacity, 17
Reaction rate, 324, 327
Redox (*see* Oxidation-reduction)
Reduction potential, standard,
  30 (*see also* Electrode po-
  tential, standard)
  Table, 359, 360, 361
Reference electrode, 152, 241,
  247, 248, 337
Refractive index, 217, 220, 221
Reports, 264

Resistance, electrical, 218, 222,
  344
Rhodamine 6G, 79
Rule, effective titrant, 63

Salt bridge, 248, 335
Samples, analytical, 274
  preparation, 274
Saturation, of hue, 193
Saturated calomel electrode, 152,
  248, 337, 348
  reduction potential, 348
S.C.E. (*see* Saturated calomel
  electrode)
Schwarzenbach's magnesium-
  EDTA titration, 94, 199
Selective titration, 52, 54, 184,
  187
  at controlled potential, 348
  color indicator for, 207
  innate titration error, 184
  of dibasic acids, 59
  of polybasic acids, 59
  of weak acids, 55, 56
  redox, 85
  with EDTA, 92, 186
Self-indicator, 195
Self-ionization constant (*see* Ion
  product, of solvent)
Significant figures, 35
Silver(I), 92, 242, 245
  in acetonitrile, 120
Silver bromide, 25, 76, 187, 337
Silver chloride, 187
  in acetonitrile, 121
  photoreduction, 301
  precipitation, 278
  solubility, 48, 76
Silver chromate, 76
  solubility and *p*H, 49
Silver iodide, 76, 337
Silver nitrate, 25, 80
  conductometric, 346
  potentiometric, 337
  standard solution, 298
  titration with, 75, 187, 197
Silver-silver chloride (halide)
  electrode, 243, 247, 337
Silver sulfide membrane, 245
Silver thiocyanate, 76, 78, 197
Soap titration, 196
Soda ash, 294
Sodium bicarbonate, analysis,
  277
Sodium carbonate, 74, 277, 289,
  294
Sodium chloride, standard solution, 298
Sodium diphenylamine sulfonate,
  86
Sodium hydroxide (*see also* Hy-
  droxide ion):
  carbonate-free, 287
  concentrated, 287
  standard solution, 73, 288
Sodium oxalate, 89, 315 (*see also*
  Oxalate)

Sodium thiosulfate (*see* Thiosulfate)
Solubility, of electrolytes, 110, 111, 114, 119
Solubility product constant, 27, 29, 110, 111, 243, 245
Tables, 356, 357
Solution properties (*see* Bulk solution properties *or name of property*)
Solvents, 103, 116, 117, 127, 274
acid strength, 126, 135, 137
base strength, 126, 135, 137
Brønsted's classification, 126
dielectric constant, 104
for analytical samples, 274
ion product, 139
leveling effect, 133
mixed, 142
nonaqueous, 103
prediction of pK_A in various, 143
Tables, 104, 105, 137, 139
Species, molecular (*see* Molecular species)
Specificity, of analytical method, 4, 222
Spectrophotometer, 5, 8, 222, 340
Spectrophotometry, 5, 8, 339
blanks, 341
Spectrum, 5
absorption (*see* Absorption spectrum)
Stability constant, 27
Table, 92, 358
Standard electrode potential (*see* Electrode potential, standard)
Standard solutions (*see name of reagent*)
Standards, chemical (*see* Chemical standards)
Starch indicator, 87, 195, 319
Steel, manganese in, 342
Stepwise equilibrium constants, 56
Stepwise reactions, 196, 227
Stoichiometric equation, 45
Stoichiometry, 9, 226
level of complexity, 44, 51
Substrate, 9, 164
as self-indicator, 195
effective, 60
with several reactive sites, 56
Successive approximations (for solving equations), 161
Sulfate, gravimetric, 282
Sulfide, 245
Sulfite, 84
Sulfonphthalein dyes, 209
Sulfuric acid, 137, 139
anhydrous, 104, 128, 129, 135, 139

Sulfuric acid (*contd.*)
density of aqueous, 293
titration of, 153, 154, 293, 334

Tartaric acid, in water, 58
Teeth, 90
Temperature, control of, 219
in chemical kinetics, 324
Tetrabutylammonium hydroxide, in methanol, 335
Tetrabutylammonium methoxide, 143
Thallium(I), 242, 243
Thallium(III), 243
Thiocyanate, 77, 78, 197, 245
standard solution, 78, 300
Thiosulfate, 26, 81, 87, 88, 157, 183
as titrant, 89, 195, 318
bacterial action on, 87, 318
standard solution, 87, 318, 319
Thymolbenzein, 213
Thymol blue, 141, 142, 204, 213
Thymolphthalein, 204
Tin(II), 81, 84
Titanium, 84, 243
Titer, 99, 303
Titrant, 16, 164
as self-indicator, 195
effective, 61, 63
Titrant/substrate ratio, 164, 181
Titration, 16, 19, 148, 273 (*see also name of titrant*)
acid-base (*see* Acid-base titration)
amperometric, 339, 346–349
conductometric, 150, 225, 343, 345
design of, 182
effective equation for, 61
Karl Fischer, 122
nonselective, 52, 57, 184
of mixtures, 52, 56, 183
oscillometric, 227
photometric, 149, 223
potentiometric (*see* Potentiometric titration)
references, 96, 146
selective (*see* Selective titration)
stepwise, 52, 54, 57, 186
Titration curves, 148
acid + base, 69, 153
algebraic equations for, 156
and innate titration error, 173, 175, 226
at equivalence point, 70, 165, 173, 176, 177, 179–182, 205, 225
calculation of, 164, 167
conductometric, 150, 225
linear, 151, 165, 222

Titration curves (*contd.*)
logarithmic, 151, 168, 174
photometric, 149, 150, 223
potentiometric, 152
truncated, 188
Titration error, 33, 43 (*see also* Innate titration error)
p-Toluenesulfonic acid, 334
nonaqueous titration, 335
Transition range, 73
of color indicators, 73, 203, 213
Transmission, of light, 5
Triethylamine, 118, 226, 229
Triiodide, 26, 40, 86, 157, 183, 243
color, 87, 195
Trinitrobenzoic acid, 74
Triphenylmethane dyes, 210
TRIS, 74
Tropeolin 00, 204, 213
Turbidity, 5

Uranium(IV), 84, 223, 243

Vanadium, 84
Vanadyl, 84, 243
Vinegar, 292
Viscosity, 222
Volhard method, 77, 78, 197, 299
Voltaic cell (*see* Cells, electrochemical)
Volumetric analysis (*see* Titration)
Volumetric flask, 269

Wash bottle, 15
Washing, of precipitates, 15
Water, 104, 137
hardness, 99, 196, 308
ionic association in, 108, 111
Karl Fischer titration of, 122
refractive index, 221
test for purity, 326
Wavelength, 6
Weighing, 265
by difference, 267
Weighing bottle, 267
Weighing paper, 268
Wheatstone bridge, 344

Δz^2, for ionic reactions, 114
and effect of dielectric constant, 115, 141, 145, 183
Δz^2 test, 115
Zimmermann-Reinhardt method, 85, 316, 317
Zinc, 81, 84
Zinc(II), 90, 92, 242
titration, 304
Zwitterion, 209, 210, 212